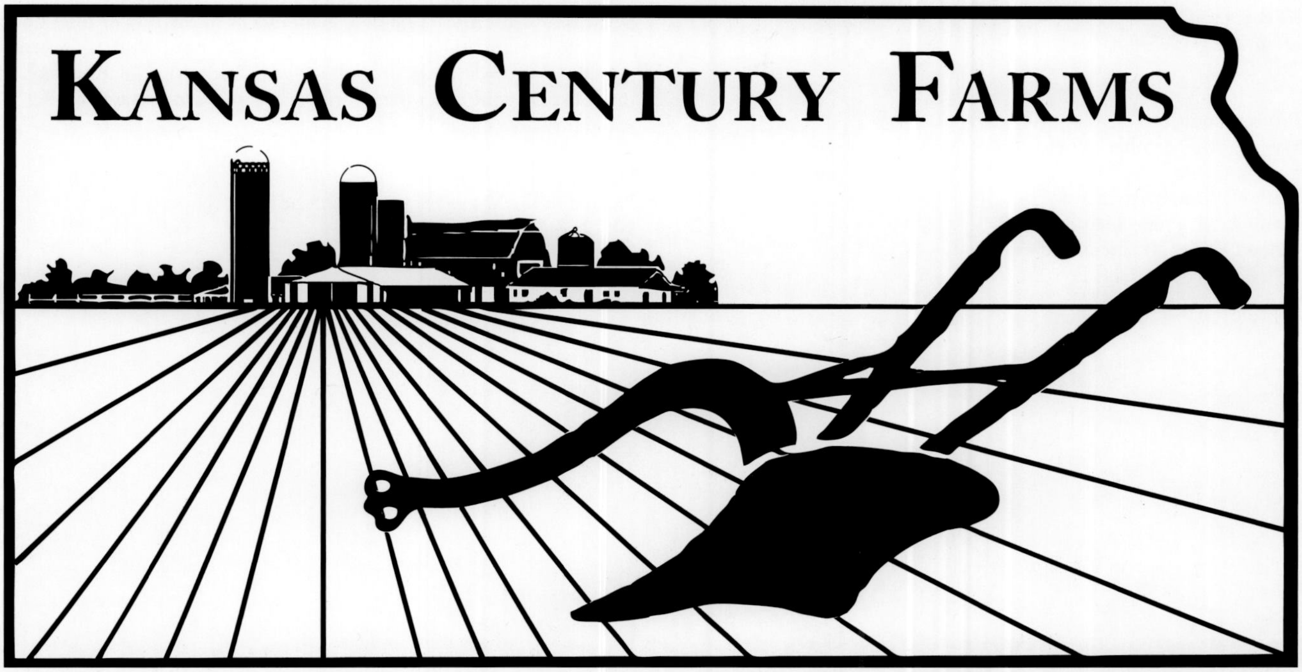

PRESERVING OUR AGRICULTURAL HERITAGE OF FAMILY OWNED FARMS

Acclaim Press
MORLEY, MISSOURI

Acclaim Press
— Your Next Great Book —

P.O. Box 238
Morley, MO 63767
(573) 472-9800
www.acclaimpress.com

For more information on the Kansas Century Farm Awards program, contact:

Kansas Farm Bureau
2627 KFB Plaza
Manhattan, KS 66503-8508
(785) 587-6118

Book Project Coordinator:
Debbie Hargrave, Member Benefits Director, KFB

Book Design: Rodney Atchley
Cover Design: M. Frene Melton
Cover Logo Design: M. Frene Melton

ISBN: 978-1-942613-96-1 | 1-942613-96-2
Library of Congress Control Number: 2018903390

First Printing: 2018
Printed in the United States of America
10 9 8 7 6 5 4 3 2 1

The Kaad Farmstead, early 1900s. Courtesy of Robert Hart

Contents

Foreword

By Rich Felts, Kansas Farm Bureau President

In 2019, Kansas Farm Bureau celebrates 100 years or the centenary for this state's largest general farm organization. This year's celebration will commemorate the vision and voluntary spirit displayed by our founders and continued throughout the organization's history.

Our organization began on Feb. 10, 1912, in Leavenworth County when J.M. Gillam and Albert Wulfekuhler were elected to lead the small group of farmers who organized to acquire the latest production-agriculture information from Kansas State Agricultural College.

This original group was called *The Progressive Club.* Fortunately, the organization kept this positive outlook focused in the consecutive decades following the organization of Leavenworth County Farm Bureau's association.

As the movement caught fire in other counties and states, Farm Bureau spurred the interests of elected leaders at the local, state and federal levels of government. In 1919, a group of like-minded individuals established Kansas Farm Bureau to enhance communications and political strength.

Farm Bureau's early efforts focused on achieving fair rail rates and strong wool prices. The first economic service to members included a program that allowed Farm Bureau members a chance to buy Army surplus horse harnesses.

Later, the farm and ranch organization played a key role in the passage of the Agricultural Adjustment Act and other economic measures during the Great Depression. As Farm Bureau grew in Kansas, the organization influenced state and federal legislation, including farm bills and export food assistance acts.

As farmers' needs changed, so did the organization. As horses gave way to automobiles and petroleum-powered farm implements, economic services shifted to provide insurance for Farm Bureau members.

About this same time federal farm programs also increased the regulations of crop and livestock production and Farm Bureau members moved toward a market oriented agriculture. During the late '70s and early '80s, abundant crop production created huge surpluses and Kansas Farm Bureau pushed for increased exports and reduced carryover stocks.

Tradition and heritage remain an integral part of what makes agriculture such a noble profession. Farmers and ranchers provide us sustenance including food, fuel and fiber.

Almost 100 years later, Kansas Farm Bureau still strives to improve the lives of all Kansans. Our mission is to strengthen agriculture and the lives of Kansans through advocacy, education and service.

Rich Felts at home in a field of soybeans.

Kansas Farm Bureau is and will remain the *Voice of Agriculture* in the legislature, Congress and the public. We will inform when needed, protect when challenged and fight for our members to strengthen the lives of rural Americans and to build strong, prosperous agricultural communities.

Thank you all for being committed, participatory Farm Bureau members in Kansas. And remember there is meaning in this quote from Peter Ustinov, "The point of living and being an optimist, is to be foolish enough to believe the best is yet to come."

Let's make the next 100 years even better.

Rich Felts
Kansas Farm Bureau

PAT ROBERTS
KANSAS

109 HART SENATE OFFICE BUILDING
WASHINGTON, DC 20510–1605
202–224–4774

http://roberts.senate.gov

United States Senate

WASHINGTON, DC 20510–1605

COMMITTEES:
AGRICULTURE

FINANCE

HEALTH, EDUCATION, LABOR, AND PENSIONS

ETHICS

RULES

March 14, 2018

President Richard Felts
Kansas Farm Bureau
2627 KFB Plaza
Manhattan, KS 66503-8116

Dear President Felts,

When the calendar marks 100 years in Kansas agriculture for a farmer's family, the Farm Bureau bestows upon them the venerable Century Farm designation. Given the challenging nature of farming, especially in Kansas, nurturing an operation through generations of family requires hardy stuff.

We spend a lot of time recognizing farmers for this toughness. After all, farmers in Kansas must tackle drought, wind, freezes, floods, fire and violent storms and those are just some of the challenges from mother-nature. All Century Farms have had to survive economic hardship well. There is no question that Kansas farmers and ranchers are tough and resilient.

I want to recognize something deeper, the courage of the farmer and their family down through the generations. It is courage that allowed our ancestors to load up a wagon and head into unknown and uncharted lands to make a better future for their families.
It is courage that allows the farmer to assume the risk of tending crops and animals. Producers must have courage to try new ideas to prosper. We have seen revolutionary changes in farming in our lifetime whether these changes are in the fields, in technology, in the farm economy or in the laboratory. It takes courage to embrace these changes and to take these risks, over and over again.

When reflecting upon the hallmark of a century of agriculture, this is a natural time to look ahead to our future Century Farms. What will they need to address in order to remain viable? The challenges and the advancements are impossible to predict. But I know Kansans will be up to the task. We have hardy stuff… and courage.

With every best wish,

Pat Roberts, Chairman
U.S. Senate Committee on Agriculture, Nutrition and Forestry

1320 Research Park Drive
Manhattan, Kansas 66502
(785) 564-6700

Kansas
Department of Agriculture
agriculture.ks.gov

900 SW Jackson, Room 456
Topeka, Kansas 66612
(785) 296-3556

Jackie McClaskey, Secretary

Governor Sam Brownback

December 26, 2017

From its earliest days, Kansas always has been a state rooted in agriculture, and farming and ranching remain a way of life for thousands of Kansans. Today, agriculture is the foundation of the state, contributing 45 percent of the state's economy, nearly $68 billion annually. Farmers and ranchers in every county of the state are committed to continuing the tradition that keeps agriculture alive and well in Kansas.

The families who own and operate the Century Farms in Kansas represent that tradition and heritage which has built the foundation of Kansas agriculture. The people who grow the crops and livestock in our state are hard at work at every hour in order to provide the food, fiber and resources we require on a daily basis. These individuals are determined, resilient and passionate about the work they do.

My family's farm is a Century Farm in Crawford County near Girard, Kansas. Five generations of McClaskeys have raised livestock and crops on our farm since 1893, and I am blessed to be part of the farming legacy that is so vital to Kansas. I am personally very thankful to the Century Farm program for honoring the heritage of this state and of agriculture which has been so meaningful to me and to my family.

During the 100 years these Century Farms have been in operation, they have witnessed many changes in agriculture, from technological advances to issues affecting farm policy to the development of livestock generics and new crop varieties. But through it all there has been one constant: the people who call farms their home. These people share common values of hard work, faith, family, persistence, loyalty and humility. I learned these values from my parents, Jack and Jean, who learned them from their parents, who learned them from theirs, just like the many generations of Kansas families represented by these Century Farms.

As we embrace the tradition of Kansas agriculture, we look to the future and the growth that lies ahead for the farmers and ranchers of Kansas. The resilience of these Century Farm families and the values passed down from generation to generation have stood the test of time in Kansas.

Jackie McClaskey
Kansas Secretary of Agriculture

Topeka • Manhattan • Garden City • Parsons • Stafford

KANSAS STATE UNIVERSITY | College of Agriculture
Dean of the College of Agriculture
Director of K-State Research and Extension

August 16, 2017

Kansas agriculture contributes significantly to the state's economic interests, a tradition that precedes its statehood. Farmers and ranchers across the state are responding to demands from consumers to raise wholesome and nutritious food, and are continuously striving to do better, to raise more food using fewer resources. As the world population grows, Kansas farmers and ranchers will continue to play a vital role in feeding Kansas, the nation, and people all around the world.

Maintaining a family farm for 100 years or more in Kansas is no small accomplishment. This book celebrates the successes and recognizes the difficulties Kansas family farms and ranches have had to contend with. It also celebrates the extraordinary contributions such family farms have made to Kansas agriculture.

On behalf of the College of Agriculture, K-State Research and Extension, and Kansas State University, I thank you for your continued efforts to develop a better agricultural and food system, a system that feeds many more people today than ever before, a system that produces better food, while using less water and other inputs, or having a smaller impact on the environment. I also want to thank you for your dedication to a way of life that's steeped in tradition, hard work, perseverance and love for the land, a way of life that many admire, but only the fortunate few get to experience.

With kind regards and best wishes for another successful century ahead,

John D. Floros

John Floros, Ph.D.
Dean and Director
College of Agriculture and
K-State Research and Extension

K·STATE
Research and Extension

113 Waters Hall, 1603 Old Claflin Place, Manhattan, KS 66506-4008 | 785-532-7137 | fax: 785-5

Snapshots in Time

An Introduction to the Century Farm

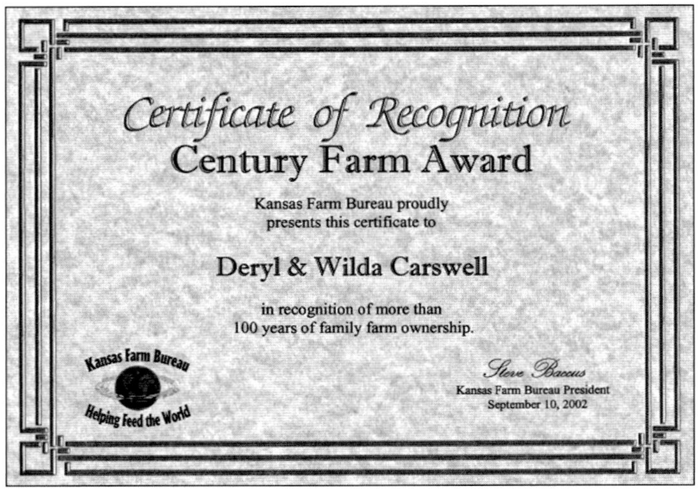
Certificate of Recognition

Century Farm Award

Kansas Farm Bureau proudly presents this certificate to

Deryl & Wilda Carswell

in recognition of more than 100 years of family farm ownership.

Kansas Farm Bureau — Helping Feed the World

Steve Baccus
Kansas Farm Bureau President
September 10, 2002

When most people think of a family farm, their thoughts turn to a plot of land farmed by a single family in the present. There is a warm, fond connotation associated with an honest perception. However, many farms across the state of Kansas go beyond that perception. The family connection to the farm goes back a hundred years or more. The land has been worked by several generations, and it has more than a few stories to tell. These farms are special. They are "Century Farms."

If the Century Farm could speak, you might hear stories of the first time the soil was tilled with a plow, the care the farm family took in preserving the land, or perhaps it might regale you with stories of humorous cow-milking lessons. It's these vignettes that tell the story of a place in a way that broad-stroke history accounts cannot.

The story of agriculture in Kansas began thousands of years ago, with people and crops that are gone but not forgotten; their stories still linger. They have been written down in journals, told on cold winter evenings and passed down from generation to generation. And there is a unique connection that is made when stories of traditions, ingenuity, progress, and ultimately evolution of rural life — of the family farm — are told, when you hear it from someone who lived it.

The stories included here span the state, from small farms to big farms, and families to organizations. Each tells their own story of agriculture in rural Kansas. A timeline provides a touchstone to easily find important agricultural dates, followed by histories of the Kansas State Fair, the annual cornhusking competion and special profiles.

Many people have fond memories of sitting on a relative's knee, huddling around a campfire or just talking to make a task go faster, when they listened closely to stories of grand adventures, local history or youthful escapades. As you move beyond organizational histories, stories such as these begin to highlight snapshots in time. They are interest vignettes about a person or place that gives us an understanding of a time and place we could never personally know. These personal recollections paint beautiful pictures of memories growing up in rural Kansas and living on land that has passed down through generations.

To the many who have contributed to this book to tell the story of the Century Farms and Kansas agriculture, we have the deepest gratitude. Your assistance and contributions ensure Kansas's Century Farms will be remembered. In particular, Kansas Farm Bureau was instrumental in preparing this book for publication.

As sponsor of the Century Farm Program, Kansas Farm Bureau is proud to represent many of the farm families who call a Century Farm home. This organization has had the honor of serving its members for 100 years, and in that time it has witnessed the ever-growing story of agriculture. At times, it has been a part of the story, but more importantly, its service has played a small role in perpetuating the tale.

Share the stories.

Courtesy of Riley Winkler

Farm Bureau
Century Farm
100 Years of Helping Feed the World

MEMBER
KANSAS
AFBF
KANSAS
FARM
BUREAU

Farm Bureau
Century Farm
100 Years of Helping Feed the World

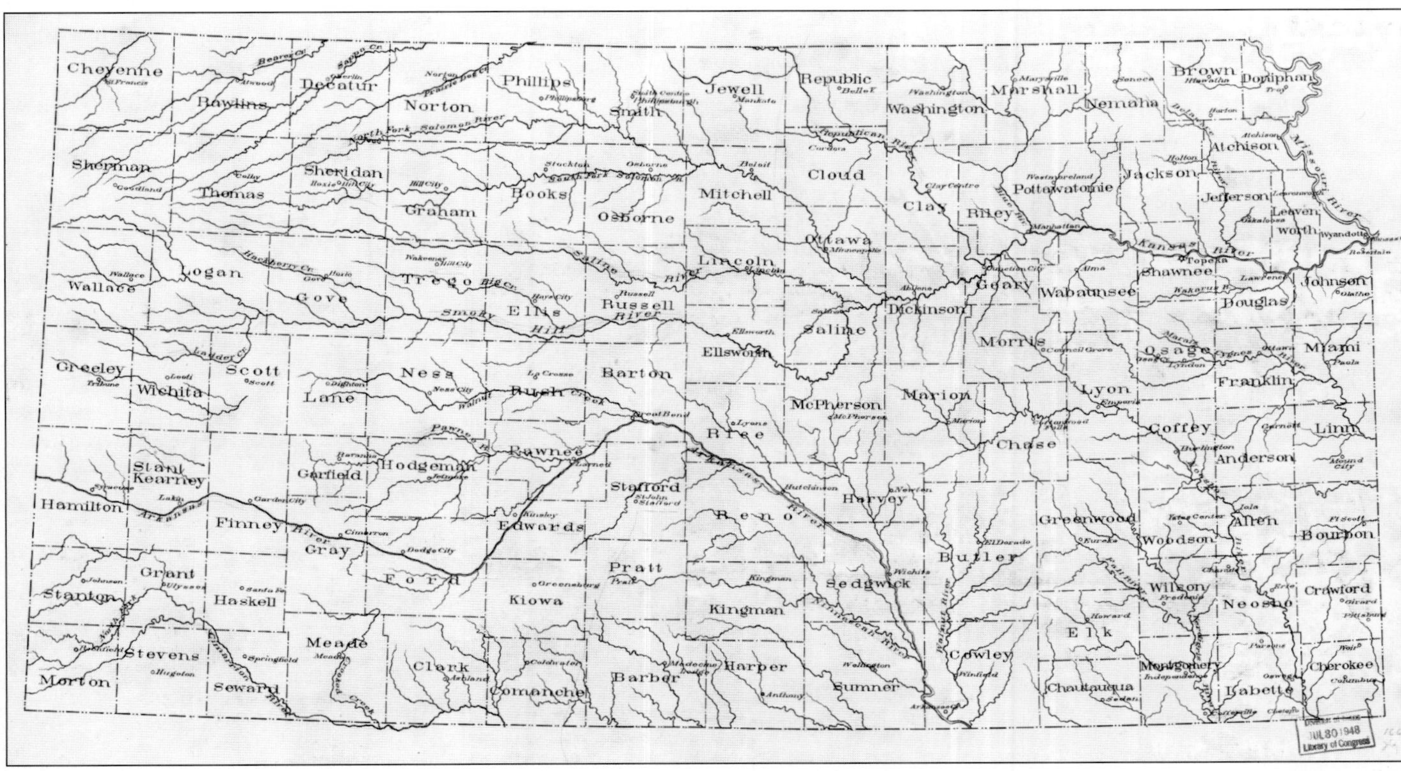
Cheyenne
Rawlins
Decatur
Norton
Phillips
Smith
Jewell
Republic
Washington
Marshall
Nemaha
Brown
Doniphan
Sherman
Thomas
Sheridan
Graham
Osborne
Mitchell
Cloud
Clay
Riley
Pottawatomie
Jackson
Atchison
Jefferson
Leavenworth
Wallace
Logan
Gove
Trego
Ellis
Russell
Lincoln
Ottawa
Saline
Dickinson
Geary
Wabaunsee
Shawnee
Douglas
Johnson
Greeley
Wichita
Scott
Lane
Ness
Rush
Barton
Ellsworth
McPherson
Marion
Morris
Lyon
Osage
Franklin
Miami
Hamilton
Kearney
Finney
Garfield
Hodgeman
Pawnee
Stafford
Rice
Reno
Harvey
Chase
Coffey
Anderson
Linn
Grant
Haskell
Gray
Ford
Edwards
Kiowa
Pratt
Kingman
Sedgwick
Butler
Greenwood
Woodson
Allen
Bourbon
Stanton
Stevens
Morton
Seward
Meade
Clark
Comanche
Barber
Harper
Sumner
Cowley
Elk
Chautauqua
Montgomery
Labette
Cherokee
Crawford
Neosho
Wilson

Kansas
Breadbasket of the World

Origins

The word "Kansas" comes from the native American Kaws or Kansa (Kanza) people (a Sioux tribe), whose name is derived from the Siouan aca, "Southwind". By the mid-18th century, the "People of the Southwind" were the predominant tribe in what became the state that was named in their honor.

Nicknames

The Sunflower State
The Jayhawker State
The Wheat State
Breadbasket of the World

Motto

Ad Astra Per Aspera (To the Stars through Difficulties)

Capital

Topeka

Largest City

Wichita

Total Area

82,278 square miles • 15th largest state by area

Population

2,907,289 • 35th most populous state (2016 est.)

Population Density

35.1 persons per square mile • 40th state ranked by density

Highest Elevation

Mount Sunflower • 4,041 feet

Lowest Elevation

Verdigris River at the Oklahoma border • 679 feet

Admitted to the Union

January 29, 1861 • 34th state

Boundaries

Nebraska on the north, Missouri on the east, Oklahoma on the south, and Colorado on the west

Flint Hills, Kansas, by Edwin Olson, U.S. Fish and Wildlife Service

Counties
105

Regions

Arkansas River Lowlands: generally encompasses the land flattened out by the Arkansas River, cutting through western and central Kansas; it is noted for sand and sediment carried in from the Rockies.

Chautauqua Hills: west of the Osage Cuestas, known for their thick layers of sandstone with densely vegetated oak and other timber.

Cherokee Lowlands: northwest of the Ozark Plateau, traditionally known for its coal, zinc, clay and limestone deposits.

Flint Hills: contains one of the few large areas of native prairie grassland left in the United States, over a bed of flint rock; used for grazing cattle.

Glaciated Region: this area in the northeast of the state is covered by large rocks and boulders from when glaciers covered the land; the area also includes a layer of sediment known as loess that is good for farming.

High Plains: an area of open expanses of flatlands and gently rolling hills that rises gradually in elevation east to west; once covered by a short-grass prairie, much of the land has been farmed and only small areas of prairie remain.

Osage Cuestas: north of the Cherokee Lowlands, the land includes rocks formed from sediment deposited by seas that covered the area long ago; known as cuestas, these hills have a steep slope on one side (a cliff) and gentler slopes on the other.

Ozark Plateau: the southeast corner of the state and part of the Ozarks, a hilly and densely forested region that include limestone and flint.

Red Hills: located in south-central Kansas, this area includes thick shales and soil containing iron oxide, giving the hills a rusty color; it is also known for many sinkholes throughout the region.

Smoky Hills: west of the Flint Hills, this area contains rocks formed from sediment deposits when it was a shallow sea; the eastern part of the region is covered with sandstone, the middle by limestone, and the western by chalk deposits; because wood was so scarce in this area, farmers used to quarry limestone to use as fence posts.

Wellington-McPherson Lowlands: this area is known for its thick salt deposits, with salt mining being a major industry in the region.

Average annual precipitation - rainfall
36.49 inches

Average annual snowfall
18 inches

Highest recorded temperature
121°F • July 24, 1936

Lowest recorded temperature
-40°F • February 13, 1905

National Agriculture Rankings*
Rank and Percent of U.S. Total in 2016

1st All Wheat Produced
Sorghum for Grain
Sorghum for Silage

2nd Cropland, Total Acres
Principal Crops Harvested

3rd All Cattle and Calves (Jan. 1, 2017)
All Hay Produced
Cattles and Calves on Feed (Jan. 1, 2017)
Land in Farms
Off-Farm Grain Storage Capacity (Dec. 1, 2016)
Principal Crops Planted
Red Meat Production, Commercial
Sunflower (Non-oil Type) Produced

5th All Sunflower Produced
Summer Potatoes Produced

6th All Beef Cows (Jan. 1, 2017)
Canola Produced
Irrigated Cropland
Other Hay (excluding Alfalfa) Produced
Sunflower (Oil Type) Produced

7th Calf Crop
Cash Receipts from Farm Marketings, 2015
Corn Grain Produced

8th Alfalfa Hay Produced
Bison

9th Goats, Meat and Other (Jan. 1, 2017)

10th All Hogs and Pigs (Dec. 1, 2016)
On Farm Grain Storage Capacity (Dec. 1, 2016)
Soybeans Produced

**USDA National Agricultural Statistics Service*

Kansas Farmers and a Timeline of American Agriculture

Note: Items with the state outline are specific to Kansas agriculture

1541

Coronado explores Kansas

Pre-1700s

The first Kansas farmers were Indian squaws who raised small crops of corn and beans to supplement the diet of game. They planted seeds in holes punched in the ground with sharpened sticks, and cultivated the crop with implements fashioned from buffalo bones.

The first American explorers who passed through the Territory reported that the region was totally unfit for human habitation.

Late-1700s

The first white farmers were Frenchmen who settled in the Wolf River country, now Doniphan County, during the latter part of the eighteenth century, and planted fields of corn in the rich glacial soil of this northeastern corner of the State.

1795

George Washington begins breeding mules after the King of Spain gives him two jack mules.

1803

Louisiana Purchase doubles the size of the United States and includes the region that would become Kansas.

1815

Frontier farmers ship wheat and meat East.

1819

Jethro Wood patents iron plow with interchangeable parts.

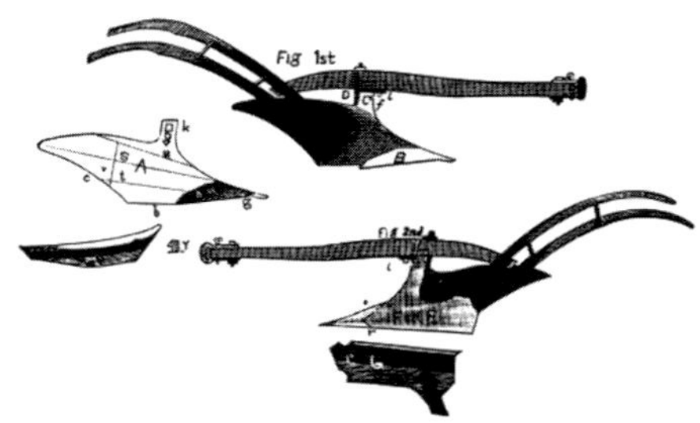

Drawing of the cast iron moldboard plow invented by Jethro Wood in 1819.

1819-25

U.S. food canning industry established.

1820s

New breeds of swine become available (Poland-China and Duroc-Hersey swine developed and Berkshire swine imported.)

1827

The U.S. Government conducted agricultural experiments in the Territory ac-

Sketch featured in "Harper's Weekly" of settlers curing buffalo hides and bones, April 4, 1874, by Frenzeny & Tavernier. Library of Congress image

Advertising card produced by Warder, Mitchell and Company, manufacturers of the Champion line of agricultural machinery, depicting a farmer operating a Champion self-rake reaper with a traveler looking on, ca. 1875. Printed by Major & Knapp, New York.

quired through the Louisiana Purchase and sent Daniel Morgan Boone to teach farming to the Native Americans in Kansas. The son of Daniel Boone, he took a farm of one hundred acres on Stone House Creek in the present day Jefferson County, less than 50 miles from the land broken by the Wolf River Frenchmen of the previous century.

Pioneer farmers supplemented their meager crops by killing buffalo and selling their hides, or gathering buffalo bones and selling them to fertilizer manufacturers.

1830

About 250-300 labor hours are required to produce 100 bushels (five acres) of wheat with walking plow, brush harrow, and hand broadcast of seed, sickle and flail.

1834

McCormick Reaper patented for wheat.

John Lane manufactures plows faced with steel saw blades.

1837

John Deere begins making steel plows.

1840s

Factory-made agricultural machinery increases farmers' need for cash and encourages commercial farming.

New breeds of cattle become available (Hereford, Ayshire, Galloway, Jersey, and Holstein imported and bred).

1841

Practical grain drill patented.

1843

First grain elevator, Buffalo, New York.

Sir John Lawes founded the commercial fertilizer industry by developing a process for making superphosphate.

1849

Boston hosts the first American poultry exhibition.

1850

About 75-90 labor hours are required to produce 100 bushels (2.5 acres) of corn with walking plow, harrow and hand planting.

1850s

Pioneer farmers broke the sod for planting with ox

The McCormick Reaper was designed by Robert McCormick in Walnut Grove, Virginia and patented by his son, Cyrus McCormick, in 1837 as a horse-drawn implement to cut small grain crops.

Farmer's Cooperative grain elevator, Richmond, Kansas, May 1936. Arthur Rothstein photo courtesy of the Library of Congress.

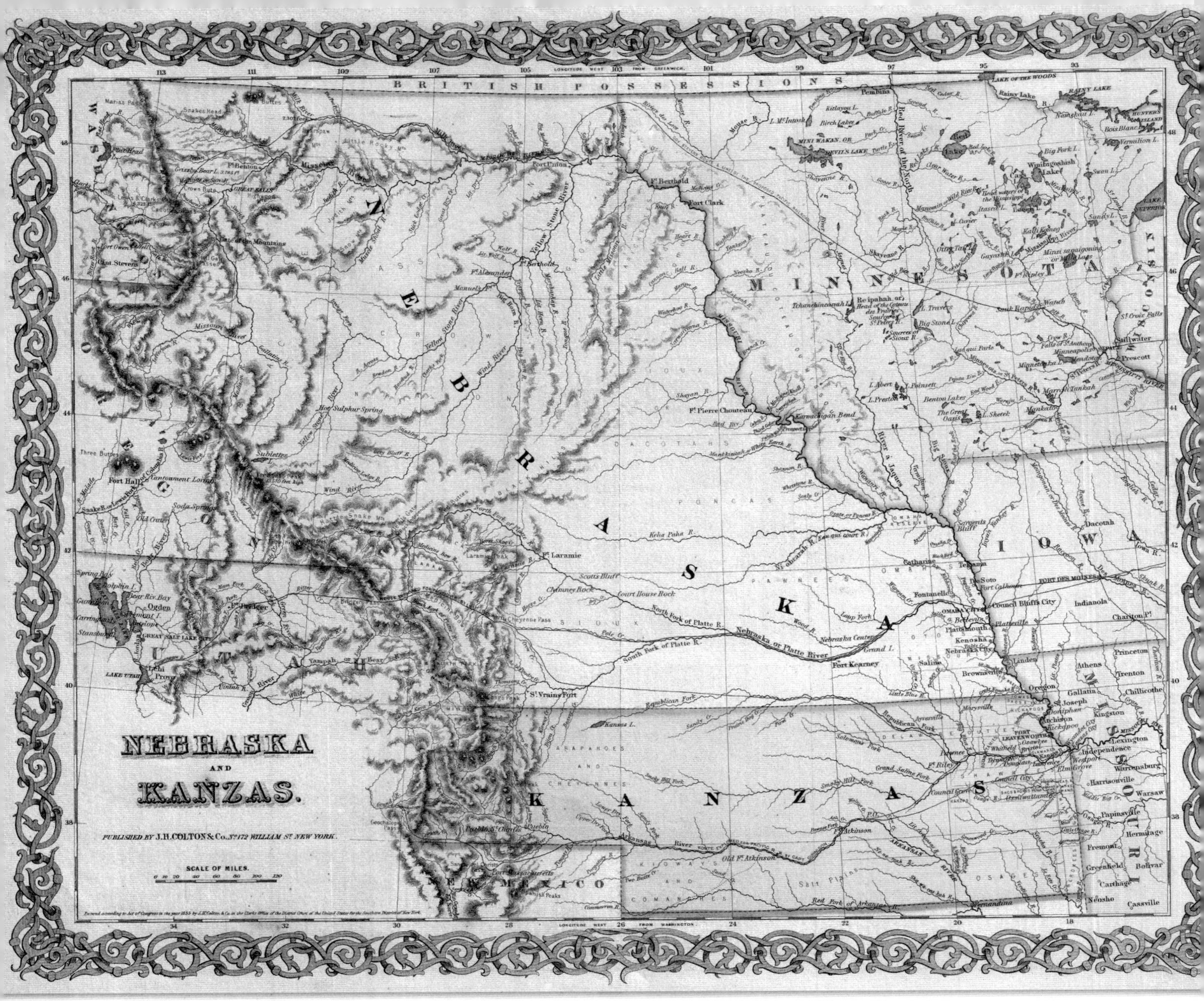

Map of the Nebraska and Kansas Territories by J.H. Colton, 1855.

teams hitched to crude plows, planting corn by slitting the sod with an axe, dropping the kernels into the slits, and closing them by stamping. Some of these corn planters averaged a crop of nearly one hundred bushels to the acre.

The folktale of "Sod Corn" Jones is circulated, which told the story of a farmer who said the Kansas sod was fertile enough to raise corn even when others said it couldn't be done.

Portable steam engine with drive wheel is adopted for threshers, saws and dairies.

Threshing wheat with a sickel and manpower.

1854

The Territory was opened to settlement under the Kansas-Nebraska Act.

1855

Beginnings of the first Department of Agriculture, when an agriculture committee was formed and laws passed to protect and aid farmers in Kansas Territory.

1856

Two-horse straddle-row cultivator is patented.

1857

The Chicago stockyard ships iced pork.

1858

Mason home canning jar invented.

1859

Kansas constitution is adopted by convention.

1860

Farmers are 58 percent of the labor force.

The change from hand power to horses characterizes the first American agriculture revolution.

1861

Kansas enters the Union as the 34th state on January 29.

1862

The Kansas State Agricultural Society was formed on March 5, 1862.

1866

At the close of the Civil War, the U.S. Government offered homesteads in Kansas to Union Army veterans, and more than 100,000 moved to the Territory. These veterans became the first real pioneer farmers of Kansas.

A team of oxen with one man to drive and one to hold the plow could not break more than one acre per day.

Horse-drawn planter

Horse-drawn planter

1870s

The resident population grows from just under 365,000 to nearly a million people during this decade.

1872

The State Board of Agriculture is created based on the structure of the Agriculture Society.

1873

The Reno County Agricultural Society creates the first fair association on January 18, and later hosted the first fair in Hutchison on September 23-24 that same year.

1874

The grasshopper plague of 1874 and subsequent droughts lead to the decline of corn in Kansas. That same year, Russian Mennonite settlers introduced Turkey Red hard wheat to the central plains, which grew better in Kansas because it was more drought resistant and hardy. This was the beginning of the state's greatest industry.

Prior to this year, Kansas had never produced as much as 5,000,000 bushels of wheat per year.

Glidden barbed wire patented; fencing of rangeland ends era of unrestricted, open-range grazing.

1876

Kansas surpasses other states in the production of winter wheat.

1881

Hybridized corn produced.

1881

Kansas becomes the first state to constitutionally prohibit alcohol.

1888

The *Topeka Daily Capital* claims, "In wheat, Kansas can beat the world."

First long-haul shipment of a refrigerated freight car was made from California to New York.

1889

Corn is still the leading crop in the state, reaching its peak this year with a yield of 273,000,000 bushels.

Carrier of cattle tick fever is identified.

1890

Hundreds of Kansas farmers were facing foreclosure, hampered by heavy mortgages, periodic drought conditions, and a nationwide depression that lowered the price of corn to 10 cents a bushel. Farm tenancy begins as many acres revert back to land speculators.

A woman (front, center left) is dwarfed by stalks of corn in Eastern Kansas, 1897. Strohmeyer & Wyman photo courtesy of the Library of Congress.

Superintendent of the U.S. Census declares the American frontier closed. Farmers are 43 percent of the U.S. population. Farming produces 71 percent of total U.S. exports.

Forty to 50 labor hours are required to produce 100 bushels (5 acres) of wheat with gang plow, seeder, harrow, binder, thresher, wagons and horses; 35-40 labor hours required to produce 100 bushels (2.5 acres) of corn with 2-bottom gang plow, disk, and peg-tooth harrow and 2-row planter.

1890-99

Average annual consumption of commercial fertilizer: 1,845,900 tons.

1891-92

First gasoline tractor is built by John Froelich.

1896

Rural free delivery of the mail.

1900

Beginning of The Farmers' Alliance, which later became

The original Froelich tractor, 1892.

Farm laborite in demonstration at Columbus, Kansas, May 1936. Arthur Rothstein photo courtesy of the Library of Congress.

John Deere tractor binder

the Populist party, to protect farmers from capitalistic and industrial powers.

1901

The Central Kansas Fair Association is created and a fair is held on 50 acres of land in Hutchison; two years later, in 1903, the state legislature gave the fair association license to legitimately call their event "The Kansas State Fair."

1902

The Reclamation Act of 1902 earmarks federal aid for irrigation projects.

1905

The first business devoted exclusively to making tractors is established.

1906

Food & Drug Act.

1900s

Kansas adopts the new nickname, the "Wheat State", and becomes known as the "breadbasket of the world."

1910

There are still 1,150,000 horses and mules being used in Kansas farming, providing a home market for $50,000,000 worth of Kansas' corn and other feed.

Commercial fertilizer use is 6,116,700 tons per year.

Farmers are 31% of the labor force. One farmer feeds seven people.

1914

Kansas farmers planted 9,000,000 acres and harvested 181,000,000 bushels of wheat, which they sold for $151,500,000.

Wheat acreage surpasses corn for the first time.

1915

The number of tractors and motorized harvesters is on the rise.

1916

Stock Grazing Homestead Act.

1919

Kansas Farm Bureau is established.

1920s

Charles Angell Sr. develops the one-way disc plow, the first minimum tillage implement to successfully practice the dryland farming technique of stubble mulching.

1929-39

The Great Depression.

Commercial fertilizer use is 6,599,913 tons per year.

1930s

Frozen food is readily available.

A farmer prepares his sumer fallow with a one-way disk plow, April 1942. Irving Rusinow photo courtesy of the Library of Congress.

Old-fashioned pulley driven hay bayler.

Miss Vada Watson, the Kansas Wheat Girl, with President Calvin Coolidge, January 1925. Miss Watson presented the president with a sack of wheat in observance of the 64th anniversary of Kansas statehood. She was the first "wheat girl" elected in a contest sponsored by Kansas State University, and toured the country to promote the Kansas wheat industry. Library of Congress photo.

All-purpose, rubber-tired tractor with complementary machinery popularized.

1930

Fifteen to 20 labor hours required to produce 100 bushels (2.5 acres) of corn with 2-bottom gang plow, 7-foot tandem disc, 4-section harrow, 2-row planters, cultivators and pickers; 15-20 labor hours required to produce 100 bushels (5 acres) of wheat with 3-bottom gang plow, tractor, 10-foot tandem disc, harrow, 12-foot combine and trucks.

1931

Kansas farmers planted 12,000,000 acres and raised 240,000,000 bushels of wheat, which they sold for $81,500,000. With machines the farmers raised more wheat, but received less money. The price per bushel is 30 cents, compared to 90 cents in 1914.

1933

There are 19,353 members in 4-H Club in 100 countries; this year there are 4,321 entries at the Topeka and Hutchinson State Fairs with $4,325 in prize winnings.

1936

Motorized farming was at its height with 63,000 farm tractors and 24,000 combines, and only 545,000 draft animals still in use.

This year there were 174,580 farms in cultivation throughout the state, averaging 275 acres in area. Of these 96,896 were wholly or partially owned by their occupants, while 76,771 were occupied by tenants.

1937

The state's total wheat acreage is 13,549,000, with a value of $170,000,000. That same year, it is estimated the state had nearly 2,500,000 beef cattle and 500,000 dairy animals, with approximately 2,000,000 hogs and 300,000 sheep.

More than 90,000 combines working wheat compared with 4,000 in the 1920s.

Soil blown by "dust bowl" winds is piled up in large drifts near Liberal, Kansas, May 1936. Arthur Rothstein photo courtesy of the Library of Congress.

Planting corn in Jefferson County, May 1936. Arthur Rothstein photo courtesy of the Library of Congress.

Farm for sale in Jefferson County, October 1938. John Vachon photo courtesy of the Library of Congress.

Harvest time in Sheridan County, August 1939. Lee Russell photo courtesy of the Library of Congress.

Work horses near Junction City, 1942-43. John Vachon photo courtesy of the Library of Congress.

Harvester combine in action in a Kansas wheat field, ca. 1900. Note the man at left packing grain in cart from the pouring chute. Library of Congress photo.

1938

Preventive measures against soil blowing adopted by the federal government and the state Department of Agriculture lead to the decline of the Kansas dust bowl, with soil drifting confined to three or four counties in the southwestern corner of the state.

1930s

Development of suitcase farmers, where the farmer worked the land and harvested the crops for bankers and businessmen who owned the land but lived elsewhere.

1940

One farmer feeds 19 people.

Commercial fertilizer use is 13,590,466 tons per year.

1945-70

Change from horses to tractors and increasing technological practices characterize the second American agriculture revolution.

1948

Farmers shipping by truck.

1950

One acre produces 38 bushels of corn.

Commercial fertilizer use is 22,340,666 tons per year.

One farmer supplies 15.5 persons.

1951

The Great Flood of 1951 occurs in July in the Kansas, Neosho, Marais Des Cygnes and Verdigris river basins.

1954

More tractors than draft animals on the farm.

1955

Six and a half labor hours required to produce 100 bushels (4 acres) of wheat with tractor, 10-foot plow, 12-foot row weeder, harrow, 14-foot drill, self-propelled combine and trucks.

A farmer on his tractor in the 1940s.

1950s-style open-cab combine.

A farmer preparing to plant seeds in the 1950s.

1960

Farmers are 8.3 percent of the labor force. One farmer feeds 46 people. Farming produces 23% of total U.S. exports.

Hybrid corn planted in 96 percent of corn acreage.

1965

Five labor hours required to produce 100 bushels (3 acres) of wheat with tractor, 12-foot plow, 14-foot drill, 14-foot self-propelled combine and trucks; 99 percent of sugar beets harvested mechanically.

1966

Fair Labor Standards Act extended to farm workers.

1970s

Center pivot sprinkler systems introduce a fundamental change in the history of agriculture, making it possible to cultivate land that previously could not be irrigated.

Commercial fertilizer use is 43,643,700 tons per year.

No-tillage agriculture popularized.

1975

Three and three-quarter labor hours required to produce 100 bushels (3 acres) of wheat with tractor, 30-foot sweek disk, 27-foot drill, 22-foot self-propelled combine and trucks; 3.33 labor hours required to produce 100 bushels of corn with tractor, 5-bottom plow, 20-foot tandem disk, planter, 20-foot herbicide applicator, 12-foot self-propelled combine and trucks.

1976

Kansas Farm Bureau first recognized farms across the state that had been in the same family's ownership for one century or longer; the program would stay dormant for the next 24 years before continuing annually in 2000.

1980s

Commercial fertilizer use is 47,411,166 tons per year.

More farmers use no-till or low-till methods to curb erosion.

1980

One farmer feeds 75 people.

1981

USDA founds Agriculture in the Classroom for school children.

1983

American eat 25 percent more red meat than in 1945.

1989

After several slow years, the sale of farm equipment rebounds; more farmers begin to use low-input sustainable agriculture (LISA) techniques to reduce chemical applications.

1990

Farmers are 2.6 percent of the labor force.

One farmer feeds 100 people.

1994

Farmers begin using satellite data to manage crops.

One farmer feeds 129 people.

Farming produces almost 19 percent of total U.S. exports.

Farm Bureau celebrates its 75th anniversary.

1997

First weed and insect resistant crops appear (soybeans and cotton).

2000

One acre produces 137 bushels of corn.

2000s

One farmer feeds 155 people.

2014

Kansas is home to 143,000 dairy cows that produce more than 2.9 billion pounds of milk.

2015

Kansas is one of the most productive agricultural states, producing high yields of wheat, corn, sorghum and soybeans.

2016

Kansas exported nearly $3.4 billion in agricultural products. The top five exports include wheat, beef and veal, soybeans, feeds and forage, and corn.

There are 61,773 farms in Kansas, which generated more than $18.5 billion in agricultural output.

The average age of the Kansas farmer is 58.2. The average size of a Kansas farm is 747 acres.

Agriculture accounts for 44.5 percent of the state's total economy.

There are 46,137,295 acres of farmland, which accounts for 88% of all Kansas land. More than 21 million acres is harvested for crops and more than 16 million is pastureland for grazing animals.

The agriculture sector employs more than 246,877 people through direct, indirect and induced effect careers, or 13 percent of the entire workforce in the state.

2017

Kansas is ranked first in grain sorghum production, growing more than 46 percent of the nation's crop.

Kansas is ranked second in sorghum for silage production.

Kansas is ranked second in wheat production.

Kansas is ranked third in cattle production and beef processing.

Kansas is ranked fourth in sunflower production.

Harvesting wheat in Kansas, ca. 1920s-30s. Library of Congress image.

Century Farm Recognition Begins in 1976

By John Schlageck, Kansas Farm Bureau

Tradition and heritage remain an integral part of what makes agriculture such a noble profession. Farmers and ranchers provide us sustenance including food, fuel and fiber.

The history of these farms and ranches is rich, with many stories to tell. In that spirit, Kansas Farm Bureau recognized farms across the state that had been in the same family's ownership for one century or longer beginning in the summer of 1976.

At that time, more than 40 years ago, the oldest *Century Farm* in Kansas still held by descendants of the same family, was that of Elmer and Helen Roach in Leavenworth County. The Roach family lived in Kickapoo township.

The original purchase from the U.S. government was made May 6, 1854 and consisted of 160 acres. The first owner of the farm was Elmer Roach's great-grandfather, Henry T. Roach, a native Kentuckian. The farm later passed to Elmer's grandfather, John H. and to his father, Albert N.

In 1976, the farm's primary crops included corn, milo, wheat, soybeans and hay-pasture. A barn from the 1860s was still standing.

Running a close second to the Roach farm was the Douglas County farm of William Meairs. Only four days separated the two oldest century farms. The Meairs farm was purchased on May 10, 1854.

In addition to the Roach and Meairs farms, eight additional district winners were named across Kansas.

After a lapse of several years, the Kansas Farm Bureau *Century Farms* started again in 2000. Only the 1976 program recognized a state and district winners.

During its centennial celebration, Kansas Farm Bureau, the voice of agriculture, celebrates its 19th year of recognizing *Century Farms* – families whose current owner/operator is related to the owner/operator of the farm in 1918 or before.

As farmers and ranchers, we are proud of our legacy and heritage. The Century Farm program will help others better appreciate the strong family ties and tradition that we hold so dear.

Kansas Century Farm
State & District Recognition in 1976

1st District
Elmer M. Roach, Easton, Leavenworth County, May 6, 1854

2nd District
William R. Meairs, Lawrence, Douglas County, May 10, 1854

3rd District
Arch L. Ramsey, Uniontown, Bourbon County, 1855

4th District
William Norton, Madison, Greenwood County, August 1857

5th District
Wesley S. Staatz, Junction City, Dickinson County, May 1857

6th District
Clifford Houghton, Tipton, Mitchell County, August 8, 1867

7th District
Arthur Joyner, Freeport, Harper County, February 22, 1864

8th District
Mrs. Lon E. Wells, Alexander, Rush County, 1873

9th District
W. T. Kells, Satanta, Haskell County, 1886

10th District
John M. Bieber, Almena, Norton County, 1872

State Winner
Elmer M. Roach, Easton, Leavenworth County, May 6, 1854

Home on the High Plains

by John Schlageck, Kansas Farm Bureau

Settlers first ventured to the wide-open, treeless High Plains in the late 1870s. Most of the new arrivals came from Europe, the Midwest and the eastern United States.

Drawn by the promise of, "owning their own land," offered through the Homestead Act of 1862, these determined Americans and immigrants headed west. They were determined to stake their claims in soil with deep sandy loam and few rocks and grass as far as the eye could see.

The High Plains challenged these early settlers to their core. At 3,000 feet above sea level, this region of Kansas experiences day-night temperature variations of 30 degrees or more. It lies within the rain shadow of the Rocky Mountains, creating a semi-arid environment.

Average precipitation amounts to 20 inches or less and a wide variation from year to year – not to mention strong winds that blow year-round.

While many did not know where they would stop, most knew they would wind up out west. Along the way they passed places with more reliable rainfall, trees and abundant surface water but they liked what they saw in western Kansas and there seemed no reason to farm differently than they had back home.

These early settlers lived mostly on hope, with many disappointments and heartaches. They welcomed the generous years of good rains and dreaded the bad years that brought bitter winter blizzards that froze livestock and sometimes their fellow man.

Too often they endured summers of searing winds that sucked up what little rain had fallen, and nothing grew.

Otterbourne School, an early sod structure.

"We would plant and see things wither and die from the scorching rays of the sun and sometimes hot winds would blast our hopes and quicker yet, the bad hailstorms would cut everything to the ground," Mrs. C.E. Moore, Thomas County settler wrote.

For the early pioneers who settled the High Plains, ups and downs became a way of life. Populations ebbed and flowed with the drought cycles. Those who ran cattle, kept a few chickens as well. They grew their own gardens and managed to ride out the dry years.

On the other hand, those who gambled on a single cash crop were often forced to leave when drought turned their hopes to dust.

"Farming in western Kansas is a gamble," Tom McNeal, *Colby Free Press* wrote on July 19, 1917. "A lottery in which nature turns the wheel. Sometimes the players draw prizes of astonishing value and sometimes they draw blanks."

Adaptation to the physical characteristics of the sprawling grassland became the greatest single hurdle facing the settlers. This led to adaptation of new tools for this new land.

Barbed wire, the grasshopper plow, the sod cutter and the windmill ranked at the top of the new innovations. Introduced in the 1870s, barbed wire became an effective and comparatively inexpensive way to fence off large areas of land. Based on a relatively simple design of two wires twisted together with metal barbs attached at regular intervals, fences sprung up across the High Plains.

The advent of the windmill provided a steady water supply in this land of scarce and precious moisture. Hauling water from a distance or digging a well by hand was back-breaking work. Pulling water from a well by hand took time and effort to provide enough water for a family and livestock.

Invented in 1854 by Daniel Halladay, the windmill became a God-send for thirsty High Plains' settlers. The spinning silver blades towered above the prairie and dominated the landscapes from the 1880s well into the 20th Century.

During the decade of the 1880s, the grasshopper plow also came into its own. Pulled by horses, this implement used a set of adjustable rods instead of the iron moldboard found on a regular turning plow. These rods allowed the plow to cut sod into long continuous strips to be used to build sod houses.

The sod cutter sled followed the plow and enabled settlers to cut a long 1-foot wide strip of sod that could then be cut into 2-foot lengths to make the sod "bricks" for their homes. A man would often stand on the sled to weigh it down as a team of horses pulled it through the sod.

The sod house came to epitomize that flinty determination by which these hearty souls won out on the wind-swept High Plains.

"The fellows out here are the most outrageously optimistic folks I ever saw," Vance Johnson wrote in *Heaven's Tableland.* "They will stand in the wind, eating the dirt that blows and telling you what good soil it is."

And with prairie as far as the eye could see, they built their homes out of the native grass sod. The sod bricks were cut wet and laid wet immediately, crosswise on the walls, jammed snugly together laterally and each layer shaved smooth with a sharp spade or push hoe, according to V.A. Kear, Thomas County noted historian on sod houses.

Most sod buildings were rectangular and many sported walls two feet in thickness. Once the walls were up and the door and windows in place, roof boards were added usually 12 to 18 inches wide and 10 feet long for a home 14-feet wide inside.

Next came the tar paper and more sod laid above that. Kear and other sod house historians claim such a carefully laid roof did not leak.

The tar paper kept the roof boards from rotting for approximately five years. Then the old tar paper was removed and replaced with new.

Firmly packed and dampened dirt made up the floors in most soddies. As soon as the home owners could afford to buy materials, a wooden floor was added. A few sod houses contained wall-to-wall homemade carpet.

Six months after a sod house cured, the occupants plastered the walls with sticky clay mixed with sand. When the plaster dried, it was whitewashed with lime mixed with water.

In some early sod homes, walls were covered with pages from newspapers and magazines. Children and some adults enjoyed reading their walls.

Most pioneers started with a sod house of one room with a window in each end and one door generally facing east because the wind rarely blew from that direction. Additional rooms were added as needed by building on at either end by cutting out the wall and window to make a doorway into the new room. Some soddies sported five rooms.

First-hand, written accounts of life in a sod house contend homes were warm in winter and cool in summer. They withstood strong winds, including tornadoes, with little or no damage.

Prairie fires and fire arrows could not burn a sod building. With outside temperatures dropping to 10 or 20 de-

Blizzards and heavy snow storms often threatened these prairie structures. They left soddies surrounded by high snow drifts.

Once the storms finished, inhabitants shoveled drainage channels through the packed drifts to free pools of water trapped near the sod walls. Failure to do so could cause the walls to soak through with serious damage to the building.

During the early days of homesteading and pioneer settlements, more than a million sod and dugout buildings were built throughout the prairies of North America. An estimated 25 states built sod buildings of one kind or another.

grees below zero, inside water in pans did not freeze in these well-insulated structures.

Sod homes in northwest Kansas lasted from 25 to 50 years depending on how well the structure was originally built and maintained. Rodents, snakes and sparrows bored holes into the sides of soddies. These needed to be tended to and plugged immediately. Fences were often erected around the home to keep cattle and other animals from rubbing or digging holes and notches in the sod walls with their horns.

Sod houses enjoyed a long and useful life span during the homesteading of the west. The art of building with sod by the immigrants and other American pioneers led to shelter for them and their families. Their ingenuity and determination made it possible to conquer the vast prairies and create a better life for themselves and future generations.

The Kansas State Fair

In 1871, Indian Agent Clinton "C.C." Hutchinson contracted with the Santa Fe Railway to build a town where the railroad crossed over the Arkansas River. Named in his honor, the town was incorporated in August 1872, and months later, in January 1873, town business leaders organized the Reno County Agricultural Society.

One of the society's first activities was to plan a fair, which was held later that year on September 23-24. Held in a small wooden livery stable behind the town's only bank on the northwest corner of Sherman and Main, the event was a success and plans were made for an even bigger fair the next year; in fact, a proposal was made to issue a tax levy in support of the event.

However, the 1874 fair was not to be, either because of public disapproval or the devastating "grasshopper plague" that same year that impacted many states throughout the Midwest.

State Fair, ca. 1900 - 1919.

The society pressed ahead, buying acreage southeast of where the state reformatory would be located, and hosted the First Annual Reno County Fair on September 28-30, 1875. The event featured 20 classes for entries, with most awards presented as certificates, with a few $5 cash prizes.

This fair, one of only a few throughout the state at that time, was the forerunner of what would become the Kansas State Fair.

In 1878, the society purchased new grounds just north of Eastside Cemetery, and fairs were held there through the early 1880s. By 1885, the group would be reorganized and renamed the Arkansas Valley Fair Association, with the event moving back to its 1875 location. Over the next several years the fair would continue to grow, with new buildings added every year, a security fence built around the fairgrounds, and a half-mile racetrack added. Special streetcar tracks were also extended from the city proper to the fairgrounds for the ease of visitors.

In 1900, the association was reorganized as the Central Kansas Fair Association, and a large, underdeveloped park was purchased that stretched along the east side of Main Street to Poplar, from 11th Avenue north to 17th Avenue.

In 1903, the Central Kansas Fair was formally recognized by the state legislature, a move that gave the fair association the license to legitimately brand their event "The Kansas State Fair."

In 1911, the state of Kansas celebrated its semi-centennial (50th) anniversary of statehood, and the fair became a focus of this celebration with President William Taft on hand on September 26. This year the fair, which lasted nine days, became the largest 50th Anniversary celebration held in the state as well as the largest fair ever held in the city of Hutchinson, with paid admissions topping $183,000, and a profit of $11,689.49.

The fair continued to expand, with 112 acres purchased north of 17th Avenue and east of Main Street in 1912. The Sante Fe Railway also laid a switch track from the southeast to bring rides and shows directly into the grounds. To help with these expansion efforts, in April 1913 the public voted in favor of the $50,000 in bonds needed to cover the costs. That same year, the state legislature passed a bill to lend monetary support for the fair. In return, the city of Hutchinson gave the fairgrounds back to the state.

The first "Official" Kansas State Fair was held September 13-20, 1913. Also around this time, the House of Capper was opened, which offered a shaded veranda with rocking chairs, cool water to drink and rest rooms. This building was restored in 2007 to its original purpose, after serving as the Professional Arts Building for many years in between.

The 1915 state fair saw the introduction of Ye Old Mill, which fed 1,000 feet of water-filled channels to carry boats and passengers through an elaborate tunnel system. Initially, the tunnel was built of wood with a canvas cover, and in 1923 the wooden walls were replaced with rusticated cast concrete and a wood-frame roof. Modern steel boats were introduced in 1935, and by 1937 the State Fair Bulletin reported that nearly 25,000 visitors had enjoyed the waterways. Ye Old Mill has been in continuous operation each year at the fair since it originally opened.

The 1920s were noted for the beginning of several expansion projects, including the Grandstand, Sheep and Swine Barn, Beef Tie Barn and Domestic Arts Building, to name a few. The beef barn was converted to a rabbit and poultry building and a new building established, the Prairie Pavilion. The Domestic Arts Building was renovated in

Aerial view of the State Fair, 2014.

2003 and includes a full-service kitchen and is available for fundraisers, weddings, banquets and family events.

The Grandstand was completed in two stages, with a 256-foot section completed in 1928, that included a seating capacity of 6,600 visitors. The finished grandstand was 448 feet long and 100 feet long and could seat up to 10,000 customers. The 1929-30 State Fair Biennial Report noted that a portion of Grandstand funding was used to build indoor restroom facilities, rather than outside toilets, showing that public comfort was important to members of the fair board.

Through the years, the Grandstand has featured automobile and other racing venues to include horse, harness and dog racing, concerts, tractor pulls and other events.

Judges inspect horses at the 1906 State Fair.

Lake Talbott, once a neglected sandpit from which laborers had taken material to make brick, concrete and asphalt for the fairgrounds, was developed and landscaped into a garden spot at the fairgrounds in 1931. Named in honor of Joe Talbott, a Hutchinson businessman and supporter of the fair for 50 years, Lake Talbott is available to rent for many occasions.

On September 17, 1935, Governor Alf M. Landon was on hand to dedicate the new $100,000 4-H Encampment Building, with delegates from across the state in attendance. Heralded as the finest of its kind in the United States at the time, the building was used to house National Guardsmen, Navy personnel, and German POWs who worked for local farmers during World War II. On June 23, 1957, the first mass for Holy Cross Catholic Church was held there, and it is still in use today for a variety of non-fair activities including meetings, wedding receptions and similar events.

Throughout the World War II years, the fair openly supported the troops with special booths set up to sell war bonds and stamps, and a "Scrap Day" in 1942, where adults could earn free admission to the fair for donating 100 pounds of metal for the war effort, and a free child's admission for 50 pounds donated. Because of this effort, more than 32 tons of metal help support U.S. efforts in the war.

Today, the Kansas State Fair still strives "to promote and showcase Kansas agriculture, industry and culture, to create opportunity for commercial activity, and to provide an educational and entertaining experience that is the pride of all Kansans." The event takes place each year the first Friday after Labor Day and lasts 10 days. Approximately 350,000 people attend the fair on average each year.

The Dairy Bar is located within the Grandstand Building and was opened in 1930 by the state Board of Agriculture. Today, it remains a longstanding hot spot on the fairgrounds. Today it only serves ice cream items. It has recently been remodeled to accommodate more fairgoers and is one of the most popular stops while at the fair.

The Governor's Cookie Jar is being presented by Fred Hall to the winner in 1955. The contest has been a tradition for many years at the fair and it has a new theme each year.

The House of Capper is one of the oldest buildings on the fairgrounds. It was donated by Kansan, Arthur Capper, journalist, Governor and Senator. It was built on the new fairgrounds in 1913 and was a place to rest and enjoy a cool drink of water. It was used for many years as a bandstand. Later the building was enclosed in the 70's and used for the Professional Arts Building until 2005. It has been recently renovated and taken back to the original look. Today, it again serves as a place to rest and enjoy watching the fair crowd.

The 4-H Encampment Building was completed in 1935. The two-story structure could house 800 boys and girls. The main floor had a dining hall, kitchen, auditorium, a recreation hall, and executive offices. On the second floor were the dormitories. It still gets a great deal of use today.

Harness Racing and horse racing was a big event at the KSF. This picture was taken in 1939. It received a great deal of popular support.

The Ye old Mill ride was 100 years old in 2015 and still is one of the most popular rides on the KSF midway. This picture was from the 1960 fair. There are only two of these rides left in the US.

The Agricultural Building was built in the 1920's. It housed various agricultural exhibits, county and individual. It has been replaced by the Pride of Kansas Building.

A grandstand crowd waiting in the 1930's waiting to enter the Grandstand seating area to see one of the live performances.

The Kansas State Cornhusking Contest

The month of October is widely celebrated as the end of the growing season, a time of harvest, and for the western Kansas town of Oakley, the annual Corn Festival, which has celebrated the unusual sport of competitive cornhusking since 1972.

Cornhusking competitions actually began in the 1920s as a way to preserve the traditional way of harvesting corn before mechanized methods became popular. In those times, farmers would invite friends and neighbors to help husk the fields, with food provided, and it was only natural for good-natured competition to take shape to see who was the fastest shucker.

The tools for shucking corn were pretty simple: a pair of sturdy gloves, a hook or peg to remove the husk, and a horse-drawn wagon to throw the ears of corn.

These contests later became annual events that were well attended. In fact, in 1930, the town of Norton held the first national contest with an estimated 25,000 people in attendance.

In 1937, the city of Lawrence, began making plans for such an event, and made a pitch to host the national event for 1939, which they were awarded. The event was held on November 3 on land owned by Kansas University and operated by Frank H. and Lawrence Leonhard. Students from Haskell Institute, now Haskell Indian National University, helped plant the competition field earlier that year. To invite participants, the *Lawrence Daily Journal-World* printed a special "National Cornhusking Edition" and mailed copies to every state and seven foreign countries.

On the big day, a special 65-acre-tent was erected and dubbed "Cornville on the Kaw," with special concession stands, vendors and agricultural displays. It was noted that the Cornville tent and grounds included approximately 300,000 square feet of canvas. A Jayhawk perched upon an ear of corn became the symbol of the event. Officials estimated more than 100,000 people attended the event.

The contest was for who could husk the most corn in 80 minutes, and the shuckers were penalized three pounds of corn for every pound of corn a husker missed on the stalk or left on the ground. The winner was Lawrence "Slim" Pitzer of Indiana, who

The 1939 Corn Husking Contest, 1st row (l-r): Clarence Schaake, Edwin Schowengerdt, Tom Cawley, Walter Heck, Elmer Wise, Ben Schaake, Leo Ousdahl, Albert Schaake, and Gus Brune. 2nd row: Elmer Ousdahl, Raymond Pine, Frank Walters, Homer Weeks, Will Hayden, Paul Brune, Eugene North, Charles Wise, Alfred Heck, and Herb Nunemaker. 3rd row: Paul Smart, unknown, Ralph Pine Sr., unknown, John Vogel, Arthur Heck, Roscoe Pine, Myron Dicker, Bill Rice Sr., John Morgan, and Fred Heck. Courtesy of Nancy Vogel.

picked 28.39 bushels of corn, which was notably a low number because the corn was dry and weighed less than normal.

Since 1972, the town of Oakley is the only place the Kansas State Cornhusking contest is held, hosted at the Buffalo Bill Cultural Center. The contest is simple: competitors try to husk the largest amount of corn as possible during a specific amount of time, tossing each in a wagon as they go along.

There are individual contests, three-person team events, and even divisions for novice pickers and children. Following the events, the shucked corn is put into a combine that fills a bin with corn kernels, then coins are mixed into the kernels and children get three minutes to search for the coins before the corn is taken away to the elevator.

Shucking corn, October 1938. John Vachon photo courtesy of the Library of Congress.

The event has even expanded in recent years to include vendors, live demonstrations of pioneer skills like rope making, wheat weaving, and crocheting, pumpkin carving contests, corn ear throwing, and of course, plenty of things to eat.

Competitive shuckers may also participate in the National Cornhusking Contest sponsored by the National Cornhusking Association that includes members from nine states: Iowa, Indiana, Ohio, Minnesota, Kansas, Missouri, Illinois, South Dakota and Nebraska.

Today the combine harvester is used in place of manual labor at harvest time, but thankfully, cornhusking events like these will keep the old ways alive, teach future generations how things used to be and build that sense of community through rural Kansas and other agriculture states.

Kansas corn, 1938. John Vachon photo courtesy of the Library of Congress.

Couple Builds Monument to Farming Heritage & Tradition

Photos and story by John Schlageck

FOUR-MILE CREEK — If ever there was concrete testimony of pride in farming and Farm Bureau, Jim and Haroldene Everley's stone-wall memorial surely exemplifies it.

In November 2000, the Harvey County Farm Bureau couple was recognized as one of 385 "Century" Farm Bureau family farms in Kansas. Shortly after receiving the award, Mr. Everley planned to display his Farm Bureau Century Farm sign, and decided to showcase it in a stone wall. With his 400 acres of beans and 400 acres of milo harvested, Mr. Everley had some free time on his hands.

While extremely important, determining the exact site of the memorial was easy. The Everleys decided to build the wall on the bank of Fourmile Creek, approximately 150 feet from where C. D. (Charles) and Cynthia Wilson built their first homestead in 1891. Mrs. Everley is the great granddaughter of the Wilsons.

"It's all gone now but some of the stones," Mr. Everley says pointing to the original site of the homestead. "This wall is built with some of the stones of the fence that used to run a full mile north from the creek."

The Everleys decided to build the wall on the bank of Four-mile Creek, approximately 150 feet from where C.D. (Charles) and Cynthia Wilson (pictured) built their first homestead in 1891. Mrs. Everley is the great granddaughter of the Wilsons.

The 24-foot long stone-wall memorial also contains stone from Mr. Everley's side of the family. A heavy stone approximately two-feet square serves as the centerpiece of the memorial and anchors the 18 by 30 inch Century Farm sign. This stone came from Mr. Everley's great grandfather's homestead in Cloud County from a stone house that was built shortly after 1852. Mr. Everley collected 22 pallets of stone from his family's homestead site and transported them to his farm. He plans to use the rest of the stone on projects around his farm. Rock posts from a farm near Lucas frame the sign and family names are in the center of the memorial.

While Mr. Everley downplays the significance of his efforts in building the stone wall, this project took nearly three full days of hard work and the skill of a mason.

On the first day of construction, Mr. Everley ran the footing for the wall. He mixed the mud (cement) by hand in a wheelbarrow — about two and a half yards.

"I couldn't see any reason to bring over a machine," Mr. Everley says.

He's used to doing things by himself. The truth be known, he kind of prefers it that way. The 65-year-old farmer has spent most of his life farming fulltime. He is the sole owner and operator.

The stones Mr. Everley used in the wall are approximately eight inches wide by 16 inches long. When he finished building the mortared structure it stood nearly waist high.

To erect the five-foot long, 400-pound post rock, Mr. Everley called on his friend, John.

"John is my John Deere tractor, equipped with a front-end loader and a log chain," Mr. Everley chuckles.

After the mortar dried, Mr. Everley bolted the Century Farm sign onto the wall. Next came the quarter-inch, sheet metal with the three family names. The names Wilson, Lang and Everley are cut into the sheet and it is bolted below the Century Farm sign. There is plenty of room on the sign to add the names of future generations.

"We hope to leave this (the farmland) to our daughter, her husband and our grandson," Mrs. Everley says. The Everley's only child lives in Minneapolis, Minn.

Jim and Haroldene Everley's stone-wall memorial serves as the foundation for their Century Farm sign. It also exemplifies their dedication and pride in family, farming and Farm Bureau.

The Everleys are proud of their farming heritage. They appreciate the family farming tradition they are part of. That's the main reason they chose to build the memorial next to the original farmstead site. Their land is special to them.

While there are many tales about this rich tradition and heritage, there are just as many accounts of the strife and challenges these early pioneers faced while carving out their existence in the fierce frontier. One particular anecdote that remains forever etched in Mrs. Everley's mind is the story of how her family came to Kansas.

"It's been passed on to me and it involves my Grandmother Jimmie," Mrs. Everley says. "When my great grandmother was six months pregnant with my Grandmother Jimmie, her husband died of tuberculosis. My great grandmother had two children at the time, one boy and one girl. She loaded the two children and everything they owned in a covered wagon and came to Kansas. Grandmother Jimmie was born after they arrived."

That was back in the late 1800s and this same fertile bottomland has been home to Mrs. Everley's family every since. The land near Four-mile Creek has provided the Everleys with some outstanding crops during their 40-year history on the farm.

Mr. Everley hopes to harvest 100-bushel milo in his mile long field next to the Century Farm sign this fall. The way the beans have podded out, he believes this crop could produce 50-bushels-per-acre.

"Harvesting crops is a real joy when they look like this," he says with a smile.

The bottom ground holds the moisture well, according to the veteran Harvey County Farm Bureau member.

"I always remember what Haroldene's mother used to say, 'You know this is some of the best pieces of dirt that a crow ever flew over,'" Mr. Everley says, a twinkle in his eye. "She was quite a lady and she and her husband loved this land. She also baked the best beans in the country."

The Everleys are proud of their roots. They love their family and its more than 100 years in the farming profession. Like their ancestors who farmed the land before them, the Everleys understand the soil will provide a bountiful crop if they care for it properly. And like the stone-wall fence that serves as the foundation for their Century Farm sign, the Everleys dedication to family and farming is rock solid.

Profile of a Family Farm

The Vogel Family

The Vogel Farm consists of more than a section of land in Douglas and Leavenworth counties. The Kansa and Delaware were here long before William (Wilhelm) Schaake (1852-1925) of Vöhl, Germany, married Katherine (Catharine) Dude (1856-1907) in Illinois in 1877. This son of John Henry and Margaretta Schaake of the Kassel area bought land in the Kansas (Kaw) Valley east of Lawrence, first on the south side of the river and then on the north. A farmer-stockman who also sold mules to the military, he acquired the beginning of the Vogel farm when he purchased acres in Grant Township from the estate of Charles Robinson, first governor of Kansas, in 1913 (28-12-20, 33-12-20). He had already bought acreage from renowned fossil hunter Charles H. Sternberg and Anna Sternberg in 1909 (4-13-20).

William and Catharine Dude Schaake, 1877.

Edward C. and Magdalena Altenbernd Schaake, 1911.

The Vogel Farmstead, before the flood of 1951.

William's headquarters were at the Kaw Valley Stock Farm in Wakarusa Township. In those days, the Kaw River could freeze enough to allow livestock to be driven across from one side of the river to the other. On the day wolf season opened in Grant Township on January 13, 1913, the old white mare broke out of a fence on the north side and headed home. The south slope stopped her, but livestock kept following. The ice broke. Most were lost ("Old Home Town," *Lawrence Journal-World*, January 14, 2013). To survive that disaster showed the mettle of William, who had already lost his wife in 1907, plus a barn to fire and 125 hogs to cholera in 1912. He was a survivor and a thriver, an inspiration.

Today satellite photography reveals meander scrolls in the farm's class 1 and class 2 soils; the skyline of the University of Kansas (Fraser Hall) is visible on the western horizon. Edward C. Schaake, William's second son, married Magdalena S. Altenbernd in 1911, and they, along with Frieda (1913) and Irene (1915), moved from Leavenworth County into "Wilder Wild," the new home in Grant Township ("Women Get Help, Too," *The Weekly Kansas City Star*, August 16, 1922). Built by Constant, the house has oak woodwork and two stairways, the front one with a landing and a back one. A print of Millet's *The Angelus* always hung in the country kitchen. There was said to be a buffalo wallow in a pasture west of the house. Magdalena's brother Will would row across the Kaw to see his sister, and Frieda and Irene would ride the interurban railroad from Bismarck Station into Lawrence for their piano lessons. Magdalena allowed herself one luxury a year, namely the latest iris, ultimately growing 150 irises along the south border of the lawn, some of which still survive in family gardens. On the Schaake family crest is the fleur de lis. Magadalena's parents, William and Katharine Hill Altenbernd, had hidden in a cornfield in 1863 lest William Quantrill ride out of town the same way he had come in.

Years later, the iconic ABC-TV movie about nuclear war was filmed on the farm in 1982 when it temporarily became the home of the Dahlberg family. Schaake relatives in Germany, who once had a *Blumenhaus* (flower shop), reported long lines to see *Der Tag danach*. According to CNN cover-

Irene at the Watson Library, KU, 1930s.

John H. and Irene Schaake Vogel, 1939.

age on June 7, 2004, one of the reasons reported for President Ronald Reagan's mission to end the Cold War was said to be his viewing of *The Day After*.

Meanwhile, the infamous flood of 1951 had taken place, and *Collier's* reported on the challenge of recovery in "The Flood Lands Fight Back" (December 22, 1951). John Vogel is featured, and George Butell and Emil Heck are quoted. Gone was the grape arbor with sweet little pink grapes, yet some of the iris survived, as did the cherry orchard. Two caterpillars pulled a deep plow to turn over some four feet of soil covered with deposits of sand, returning the soil, sometimes said to rival that of the Nile, to rich valley potential, enhanced by irrigation. The Kaw Valley bottom land has grown barley, wheat, potatoes (seed potatoes from the Red River Valley); much later, potatoes were marketed at Frito Lay. Also grown were watermelons (Black Diamond) and cantaloupe, alfalfa (W.J. Small), peas (Stokely), corn, seed corn (Pioneer), soybeans, pasture hay, swine (Hampshire, SPF), cattle (largely Hereford). For many years, a Jersey cow was kept for the rich cream (used in coffee and in Irene's icing for her burnt sugar cake: l cup each of brown sugar, white sugar and cream).

There was a rope-operated elevator to the potato cellar in the north barn, and the south barn held hay and straw for the livestock below. A machine shed, farrowing house, brooder and chicken houses and in-ground silos were part of the operations.

John Henry Vogel, Jr., of Phillips County, married Irene Dorothy Schaake and took over operation of the farm after he had graduated from KU in 1939. Along with many neighbors, he took part as an official in the National Corn Husking Contest held near Lawrence. He is the one who replaced mules and horsepower, beginning with a John Deere B tractor, one of the first in the valley ($950, 1939). In the 1940s the young family would go on Sunday afternoon drives to the country past the Chamney Farm. For 14 years John chaired the House Agriculture and Livestock Committee in the Kansas legislature.

After John retired from active farming, he and Irene built on El Dorado in Woodfield at Alvamar in 1980. Pine, Congrove, and Heck; Pine Family Farms; and Nunemaker-Ross, transitioning into Bismarck Farms, have cared for the land and crops. Some farm artifacts are in the Kansas Museum of History in Topeka and the Territorial Capital Museum in Lecompton.

The Vogel offspring, Nancy and Jerry, who had once hunted for Easter eggs in the walnut grove, incorporated their quarter section as Vogelland LLC, south and east across the road from where Bismarck School stood not far from Bismarck Station. Jerry operated WAØOWH from the farm. Conrad and John McGrew ran hunting dogs at one time and also hunted ducks in the bottom. Art Weaver hunted from a concrete duck blind at the pond, and Skipper Williams hunted along the Kaw for mushrooms before the levee was built.

The nationwide CBS broadcast on "The Story of Wheat" began at the Schaake farm on July 15, 1935; Governor Alf Landon greeted the listeners: "Here in the nation's breadbasket the song of the harvest is the hum of the binder and combine, and the chugging of giant threshing machines. But a few years ago harvest time meant an influx of thousands of harvest hands." Carl B. Warkentin, whose family in Prussia had taken hard winter wheat to Russia and then on to Kansas, also spoke. Ed Schaake mentioned Kansas State University and wheat improvements (Kanred and Tenmarq). Less than a decade later, the government was calling for increased farm production of protein (meat) for troops; friendly German POWs worked on the farm during World War II. Back in Phillips County, John's family, J. H. and Margaret Veeh Vogel, received a Certificate of Farm War Service in 1943.

Irene's realm was the hearth and home, which she enjoyed, especially on "ordinary days." No longer did she need to ride Rex the pony bareback to deliver water to the hands in the potato fields. She firmly believed that Kaw Valley potatoes

John cultivating potatoes.

Frieda Schaake Heck, Magdalena Altenbernd Schaake, Nancy S. Vogel, and Irene Schaake Vogel, at Hecks (later Bismarck Gardens).

Jerry and Nancy hunting Easter eggs.

Jerry and Nancy holding chickens, with their dog Skipper watching.

Left: John, Irene, Jerry, and Nancy in California. Right: Skipper, Jerry, and Nancy — with red straw hats.

Jerry in the snow, with grape vines, the machine shed, north barn and garage.

make the best potato salad (with mustard too). Her daughter's chore as a girl was to gather eggs [how fast those hens could peck], and later she drove hot meals to the harvest fields. Once, the four Vogels spent breathless moments in the basement in the quiet eye of a tornado.

Jerry M. Vogel taught computers at Independence, Inc., where he managed the extensive resource page for the website. A great sports fan, he followed the KU Jayhawks and enjoyed going to KU football and basketball games. He recruited and scheduled lectors for church, and he helped engineer the 50th anniversary reunion of his class at Lawrence High School.

As Nancy S. Vogel grew up, she cherished her grandmother who would read her an iris book on her lap (and re-read it); later she saw iris bloom at Schreiner's and Giverny (and found *The Angelus* in Paris). At Bismarck she won a blue ribbon and $5 in the county spelling contest for third grade, which set her course in life to becoming the first woman both to publish a book with Phi Delta Kappa and to be named the President's Distinguished Scholar at Fort Hays State University. Her rural roots had led her to an early thesis on the affinity between Robert Frost and Andrew Wyeth.

Lots of challenges (heart attack, polio, cancer plus floods, tornado, microburst, blight) came to those on the farm. Mindful of blessings, they continued in their heritage of faith.

- John: Kansas Legislature (House), 18 years; University of Kansas Ellsworth Medal; KU School of Business Distinguished Alumni Award; Kansas State University Distinguished Service Award in the field of agriculture; leadership, 4-H Fairgrounds and Audio-Reader and Citizen of the Years, Lawrence (1988); original director and interim president of Douglas County Bank (bought by Central Bank of the Midwest).
- Irene: Homemaker, Progressive Circle, Oak Ridge Birthday Club, *The Kansas Cookbook* (1989), crewel and needlepoint; her mother, Magdalena, had done crochet.
- Jerry: Lawrence Lions "L" (football manager under Coach Al Woolard), FFA, Champion Pen of 10 at American Royal; Vogel Lobby at Independence, Inc., FHSU and KU supporter
- Nancy: *Robert Frost, Teacher* (1974), thesis on Frost and Wyeth, essay on "The Death of the Hired Man" in *Frost: Centennial Essays* (UP Mississippi)

Nancy, John, Irene, and Jerry, 1940s.

and entries in *The Robert Frost Encyclopedia* (Greenwood), publications on Maureen Daly of *Seventeenth Summer*, instigator of Plymouth School project and Vogel Pergola at FHSU; champion of Bismarck School in Dennis Schiel's mural of Kansas in the State House; editor, *P.E.O. in Kansas: Our Centennial History 1903-2003*; stories in *Lawrence Magazine, Topeka Magazine,* and *Kansas!*; recipe in *Chefs at the Library* cookbook (Palm Desert); supporter of KSU Gardens, FHSU, and KU.

- The Farm Flowers: iris (Stepping Out, Elmohr, Gudrun, Wabash), daylilies (Hyperion, Mary's Gold, Indian Chief, Chicago Knobby), and peonies (an old white one rimmed with pink petals and red streaks at the center).
- Farm Trees: cedar, elm, maple, oak, blue spruce; cherry tree(s).
- Farm Shrubs: forsythia, spirea, pink spirea, mock orange.
- Farm Color: White—white barns, white house, white livestock and pasture fences, white picket fences.
- Pets: Skipper, Rex, Senator R.C.J.H., Sparky, Aberdeen, Kilty, Monarch Mickey Mac: German shepherds, black and tan, white; Scotties, mixes, assorted puppies, rabbits, barn cats (Tux black and white), kittens; Betty and Violet (registered Angus).
- Equipment: Caterpillar (D2), John Deere, and Ford tractors, Massey Harris combine, Studebaker truck (with dump bed), plow, planter, harrows, drill, mower, rake, hay wagons, one-way, pivots, etc. John welded the mailbox to a sandpoint, and he constructed a matching flag pole. The flag pole made the move to town. The mailbox is in the Topeka museum.

Nancy drove her mother to the farm so that they could harvest some wheat stalks for bouquets for John and Irene's 50th wedding anniversary. That wheat has been used at subsequent family funerals.

Nancy and John Vogel with Magdalena's irises and the north bank of the Kansas River in the distance, 1947.

Senator Rock Chalk Jay Hawk, Rex, and Jerry.

"The Mists of Time," by Richard Gwin, Lawrence Journal-World.

Younger Schaakes: Frieda (third from left) and Irene (standing on the porch railing).

College days: Irene in the green Chevy coupe.

Irene Schaake: engagement portrait.

Schaake reunion, ca. 1920. Far left: William Schaake. Center right: Edward C. Schaake (with mustache) and Magdalena Schaake with one hand on Frieda and one on Irene.

Two Vogel brothers: Howard J. (John Deere dealer in Phillipsburg) and John H.

John at his desk, post 1951 flood

John at his desk in the state house

Nancy at an autograph party in Hays for the Frost book.

Nancy – Professor Emerita.

Irene and John.

Jerry's first day of school – ready for Bismarck (#79).

Facing the setting sun: Irene and John at "The Secret Garden," the opening of the Lied Center at KU.

Nancy, Frosty and Jerry.

Jerry at the ARRL Convention (American Radio Relay League).

A family with their William Phillips steam plow in Belpre, 1910.
Photo by W.O. Durstine courtesy of the Library of Congress.

FAMILY HISTORIES

ABSALN – James Absaln and Viola Hagman purchased the 160 acres in Smith County, north of Claudel, KS, on April 4, 1916 from F.M. Claudel for $5,500.00. J.A. raised mules and horses that were sold to Camp Funston (Fort Riley). Other animals on the farm were hogs, milk cows and a Hereford cow-calf herd. Chickens were raised by Viola and Bunny for eggs and fryers. Milk, eggs and cream were taken to Claudel and transported by train. Viola, Bunny and Melba all had vegetable and flower gardens.

Carrying on the family legacy

The barn was added in 1922. Raymond Dow and Bernadine purchased more farm acres and continued farming with wheat, corn, milo, soybeans and alfalfa. The present farm home was built in 1959 after the original house was torn down. Jerry Raymond and Melba joined the operation and added more land and increased the amount of irrigation. Irrigation water is from the Kirwin Irrigation District and wells.

Jerry graduated from Fort Hays State University with a B.S. degree in Agriculture and Melba graduated with a B.S. degree in Home Economics. She taught in the Kensington School District. They have two daughters who are both Kansas State University graduates and one granddaughter enrolled at K-State now. Pets included parakeets, farm cats and one special German Shepherd.

J.A. in front of his 1922 barn

Jerry was busy in his Hagman Repair business, working on customer's cars, trucks and farm equipment that needed repairs. It was a gathering place for local farms, to visit. He built a 22 T-roadster, metal hoods to cover irrigation motors and other projects created from metal.

Walking or riding through the pasture up the hill (where a black metal buffalo stands now) provides a view of the valley and sunsets. Picnics, water skiing at Kirwin Dam, Go-carts and fireworks were very popular. *Submitted by Jerry and Melba Hagman.*

ACKERLY – In 1854, at age 17, my great grandfather, Gottelop Ackerly, homesteaded the 160 acres across the road to the north. When my grandfather, John David Ackerly, was married, Gottelop purchased these 160 acres and gave it to John D. as a wedding gift. In 1930, when my mother and father, Kittie and Wm. Wilson, were married, John D. was blind and an invalid. He gave them this farm if they would care for him the rest of his life, which they did. In 1966, when I returned from combat in Vietnam, Barbara and I purchased the farm from my parents and began farming. We farmed here until 1983, when I began work in seed sales. Our son, Jamie, was born in Hawaii while we were

Painting of the original farmstead

stationed there. After we returned we had another son, Jon David and two daughters, Denise and Michelle. We have continued to live here and raised our children here. They are now adults with families of their own. Our grandchildren number 17.

We raise corn, soybeans, wheat and cattle. Our son, Jon, runs the operation here and on another half section we own. *Submitted by David Wilson.*

ADAMS – Robert E. Adams Jr. is the third owner of the 160-acre century farm located in Labette County, KS. On April 10, 1869 Eugene Journot purchased this farm located in Township 32, Range 31, Section 19. The farm was originally acquired by Eugene Journot on April 10, 1869.

This farm originally produced wheat, corn and livestock. The 160 acres were retained and now produces wheat, corn, beans and hay. Robert Sr. planted the first wheat crops, pulling a drill with a dump truck.

There are still signs of three homes and one barn that were located here. The barn was added on.

Robert Jr. and Carolyn have one daughter, Dana Jean. Their grandchildren are Justin Carl, Levi and Eugene Willems.

Robert Jr. has helped on the farm all his life and is still working it at 78 years old. *Submitted by Robert E. Adams Jr.*

ALLEN – In 1904, the Allen homestead in Sugar Creek Township in Miami County, KS was purchased by Sydney and Minnie Allen. The accompanying picture is my great-grandparents, Sydney and Minnie and their two sons, John and Ed Allen. They worked the farm to make a living raising cattle, pigs, chickens, sheep and goats. At one time they even had a sawmill business.

Sydney and Minnie Allen, with John and Ed

The Allen Century Farm is currently owned by Jake and Pat Allen, who raise cattle on the farm. They are the fourth generation of Allen's to live and work on the farm.

The original house was a small one level home. With lumber cut from the timber of the property, it was later made into a two-story home. A detached garage, which is still standing and in use, was also made from that same lumber. Electricity was added to the home in 1948. In 1962 the old house was town down and a new one was built with inside toilets and running water.

One tidbit about the farm is that a country western singer/song writer worked on the farm. There is a picture of him with my Uncle John, butchering a goat in 1915. The entertainer was Tex Owens, best known for the song "Cattle Call" which Eddie Arnold made famous. Mr. Owens married a local girl from Drexel, MO.

For many years the farm was work with a horse and plow. Then in 1942 Ed Allen bought a new John Deere B Tractor for $975. The tractor was the first modern equipment on the farm.

With great-grandfather Sydney Allen, grandfather Ed Allen, father Bill Allen and brothers, Rick, Jake and Bob Allen, we have a combination of over 300 years of farming in Miami County, KS.

With all the generations of Allen's on this Century Farm, you cannot take a step without stepping on the footprint of an Allen. *Submitted by Jake and Pat Allen.*

ALTWEGG – The south one-half of 6-14-4 in Dickinson County, KS was purchased in 1893 by Conrad and Rosetta Altwegg for almost $24.00/acre. The land was deeded to J.J. Altwegg (born in Bern, Switzerland) in 1909. Herman Robert Altwegg Sr. purchased and inherited the land, who then passed the land to his son, Herman Robert "Bob" Altwegg Jr., in 1999. The farm's main crops have been wheat, milo, soybeans and alfalfa. It is presently in the CRP Program.

The farm house, which still stands, was built in 1917. It was sold along with five acres to the current owners.

Jacob and Louise Altwegg had four daughters (Louise, Lillian, Bertha and Marvel) and two sons (Lawrence and Herman). Family life on the Altwegg farm included long hours of hard work, however, there was always time in the evenings to invite family and friends for an evening of playing cards, games and singing and playing the 1917 Baldwin upright grand player piano. The most memorable winter activity was to watch Herm pull taffy. Those calloused hands could handle the hottest of taffy and everyone wanted their share of the tasty licorice flavored treat! On Sunday the family found their places in the pews of the Alida St. John's Evangelical Church.

Bob and Pat have been married 49 years, on Dec. 21, 1968. They have one daughter, Kristin Ann, a doctoral student at the University of Texas, San Antonio, Health Sciences Program. Her focus of study is Cancer Research. *Submitted by Herman Robert and Patricia Altwegg.*

ANDERSON – The Homestead Act certificate on my office wall is dated 1880. Since five years were required to "prove up," that means 1875 was when our people arrived from Sweden. (Can you imagine what they thought of their first Kansas summer?) The Cheyennes were still raiding in north central Kansas then.

Peter Anderson, my paternal great grandfather, homestead two 80-acre parcels running east and west.

The original home—probably a sod house—was somewhere on the west "80." When the two story, bay window Victorian house was built on the east 80 acres, I don't know. However, the rock smokehouse was 1898 chiseled into one rock.

The 80 acres adjoining the east parcel was purchased in early 1942, immediately after the Feds for Camp Phillips virtually confiscated 800 of the 880 acres owned by my father, Chester Peterson Sr. and grandmother Alphild.

This homeplace "Quarter" was immediately enrolled in a Soil Conservation Service plan. This involved constructing waterways and building terraces after grass establishment.

I recall SCS Work Unit Conservationist Jay Payne, all five feet four inches of him, standing on our International M's drawbar continually adjusting the three-bottom plow's levers as the terraces were built—the first terraces to be completed in Saline County!

My father won conservation awards and helped form and was Saline County Conservation District Chairman for 20 years. He was also, for almost 20 years, the Saline County Chairman of what's now known as the Farm Service Agency.

Three years after WW II ended and the land had been raped and pillaged by big lease operators, my father and grandmother were able to buy back the two nearby 80-acre tracts to the west, including the original west "80." This increased the Falun farm to 320 acres. The remaining 720 acres remain in the Air Force bombing range.

NOTE: This farm is owned by the family's Smoky View Farms & Ranch L.L.C. *Submitted by Chester Peterson Jr.*

ANDERSON – The Anderson family farm located on the banks of Turkey Creek, in Banner and Hope Townships in Southern Dickinson County, KS, was purchased by Adolph Bertschinger (also known as Robert A. Bertschinger) of Davis County (now Dickinson County), KS in December of 1865 for the sum of $500.00.

Adolph Bertschinger and his wife, Anna, probably constructed the home and outbuildings during their marriage, although it is known that Edward Bertschinger, the youngest of their five children, was born in a log cabin on the land. Edward A. Bertschinger purchased the farm from his mother and his siblings in 1914. He and his wife, Cora, lived and farmed on the homestead and were the parents of one child, Orpha Florence Bertschinger. Edward Bertschinger's grandson, Glee An-

Members of the Altwegg family working with their tractor and old threshing machine.

Pictured l-r: Eric, Clema, Francis, and Rodney Anderson

derson and family moved to the farm in 1952, farming the land until the early 1970s when Glee's brother, Francis Anderson, look over the farming operation and still farms the Turkey Creek farm today.

Orpha Bertschinger Anderson inherited the land upon the death of Edward A. Bertschinger and passed the land to her two sons upon her death in 2000. In February of 2001 Glee Anderson's interest in the land was purchased by his brother, Francis Anderson and wife, Clema Anderson. The Turkey Creek farm, containing 160 acres, has been continuously farmed by members of the Bertschinger family for 150 years. The original house has been demolished but a large barn still stands on the property.

Francis and Clema Anderson are the parents of two sons, Rodney Anderson and Eric Anderson and have six grandchildren. They are proud they can continue the tradition of the family farm and plan for the land to stay in the family for generations to come. *Submitted by Francis and Clema Anderson.*

ARENSMAN – Louis H. (known as L.H.) Arensman purchased the SE 1/4 of section 32-23-19 in Edwards County on April 29, 1912. L.H. moved to this area from Huntingburg, IN. He was of German descent and a lifelong farmer, born in 1874.

He was married to Sarah Hemmer until her death in 1931, due to cancer.

L.H. had five sons: Henry (born in 1903 in Rice County, KS), Sam, Ben, Leo and August. Sam died as a young adult, after being kicked in the head by a horse.

Wheat, alfalfa, feed, as well as Hereford cows and calves were raised on this property for many, many years.

Son Henry and wife of one year, Elsie (Lippoldt) built their new home on this farm in 1928. Upon L.H.'s death in 1964, they inherited this property.

On Feb. 14, 1986 Marvin Arensman, L.H.'s grandson, became the owner of this farm, of which he had already been farming for many years.

An old story told to L.H.'s great-great-granddaughter in the late 1970s, when working as a nurse, by an elderly George Hetzel I goes like this – I wouldn't have survived if it weren't for L.H. I moved out here as a teenager and lived in a dugout that first winter. L.H. brought me food and checked on me from time to time. He even gave me seed wheat for my first planting, depending and risking that I would have a crop to repay him. For an elderly man remembered as crotchety, this was quite a story. *Submitted by Jerri (Arensman) Segman.*

L.H. and Sarah Arensman

Four Arensman sons – back row, Ben and Henry; front row, Sam and Leo.

ARENTS – John Arents was born in Germany in 1845. When he came to America he resided on his farm in the early 1870s, west of Leonardville. In 1871 he married Elizabeth Nanninga. She died one year later, leaving an infant daughter. Mr. Arents was married again, May 23, 1873 to Wobina Visser. She was also born in Germany and came to America in 1871. To this union 14 children were born, four of whom died in infancy. Mr. Arents was a charter member of the Fairview Presbyterian Church north of Riley.

John E. Arents, age 15, with his dad, Groen J. Arents, 1941.

Around 1900 Mr. Arents' oldest son, Groen J., took over the farm. He was married to Mary Jahnke, April 10, 1900 in a double wedding with her sister, Lucy, marrying Albert Heller. John Ervin, their only child, was born Oct. 8, 1905. He farmed with his parents. John attended K-State for a six-week course in agriculture. Aug. 31, 1932 he married Inez Peterson at Walsburg Lutheran Church. Groen and Mary farmed until 1934 when Groen was killed by a runaway, farming with his horses. He died at the KU Med Center with a blood clot. John and Inez had a diversified farm. Terraces were built and the crops were rotated. They were very successful. A new home was built in 1948. They farmed until 1993 and put most of their land in the CRP program. John E. passed in 1999. Inez passed in 2001.

Their only child, Nyla, was born Sept. 20, 1936. She now owns the farm with her husband, F.L. Young. They plan for their children to inherit the farm. *Submitted by Nyla J. (Arents) Young.*

ARMOUR – This Century farm has been in the Armour family for four generations: Josiah Armour, John Allen Armour, Joseph M. Armour and now the Armour children: Janet (Martin), Joe A., Jim, Jack and Jean (Thome).

Josiah Armour was born in Hardin County, Kentucky on March 8, 1836 to Dempsey and Rutha Jane Armour. Between 1860 and 1865, he moved to Montgomery County Illinois where he bought 160 acres in Bois D'Arc township. His father and mother died there in 1871 and 1872. Shortly thereafter, Josiah who was single, headed to Kansas.

Josiah & Mary Hall Range Armour

Josiah & Mary House In Kansas (Sec 29)

On December 26 and 27, 1873, Josiah Armour paid $2.00 and filed his Declaratory Statement, # 9130, in the Wichita, Sedgwick County, Land Office (they spelled his name "Armar") for the South East Quarter of Section 20, Township 28 South (Illinois Township), Range 2 West containing 160 acres.

By June 17, 1874, Josiah had built and moved into a 10 foot by 12 foot single story house with one window and one door. By the end of the year, he had "forty eight acres broke and under cultivation, a well 24 feet walled and plenty of water, a stable – 15 by 15 feet built of sod and planted 100 forest trees". These improvements were valued at $200 according to the affidavits signed by his two adjacent neighbors, Isaac Barnett and Alexander Gard. Josiah paid $1.25 an acre for the 160 acres, which was recorded in the Register's Office on December 21, 1874. Afterwards, certificate number 6743 was signed by U. S. Grant, President of the United States, (actually signed by his secretary) and issued to Josiah. The family still possesses this Certificate.

Being a single man, Josiah found it difficult to "plow all day and then come in and get my own meales" as he had someone write for him to a Miss Mary Range in Illinois in June 1874. He also promised that if she married him he would "try and do the best by you that I no how". They were married on March 5, 1876, at her mother's home south of Palmyra, Illinois. He was 40 and she was 38. On July 2, 1877, they had a son, John Allen, and in 1879 a daughter, Susan Elizabeth, who died in March, 1880 while in Illinois. In Kansas, Josiah and Mary lived in the original "homestead" house and later expanded it to two rooms. In 1881, they bought the adjacent SW quarter of section 20. In 1883, they purchased the NE quarter in section 29 where they built a two-story house. They also bought 240 acres in section 33 in 1892 – 1893.

While homesteading, Josiah experienced grasshoppers, droughts and all the difficulties of pioneers. Yet, he and Mary were able to expand their holdings so that at the time of his death in 1913, they owned 720 acres. Josiah was a tall man, probably over six feet tall while Mary was short. She had a specially built rocking chair since she was so small.

Their son, John Allen ("Al") helped on the farm. As a young man of 22, he took out a three-year loan to buy the SW quarter in section 34. The seller from Wichita taunted him saying, "I hope you ain't like the last guy who couldn't pay it off". As he was fond of telling his grandchildren, he paid it off in *one year* to the surprise of the seller. Al went to a local one-room school and received some schooling in Illinois and attended business school in Chillicothe, Missouri.

In 1902, Josiah and Mary rented out their farm, sold their equipment and moved to El Cajon, California where they bought a 40-acre orange and lemon orchard. Al accompanied his parents to California. Josiah and Mary brought along their maid, Ella Dugan, who grew up on a nearby home in Kansas. Some family members say that his parents were playing matchmaker. It worked. On November 4, 1903, Al and Ella were married in El Cajon, San Diego County, CA. They had six children, Mary, Grace, Julia, Edna, Joe M. and Catherine.

After his father's death in 1913, Al moved his young family and mother to Virginia where he bought a 198-acre farm ("The Refuge") and ran a country store near Charlottesville. He later moved to Winchester where he dabbled in Real Estate. In 1926, he moved his family back to the home place in Kansas, which he had been renting to neighbors and relatives. On their trip back to Kansas, they stopped in Washington, D.C. where they toured the White House and shook hands with the President, Calvin Coolidge. During the dry hot year of 1936, their two-story house burned down. They had a new two-story house built on the same spot. Outside of some remodeling done in the 1970's, this two story white house still stands and serves as a landmark in the surrounding area. In the 1939, Al sold the land in section 33.

His son Joe M., born in 1914, began farm work and later took over. He married Evelyn Jansen on April 3, 1940 in St. Mark's, Kansas. They built a two-bedroom home on section 20 near the site of Josiah's original house. They raised five children, Janet, Joe A., Jim, Jack, and Jean. Joe M. farmed, growing mainly wheat and milo, alfalfa, corn, atlas and other crops for hay and silage, and milked cows. Later, he switched to raising beef cattle. In 1937, he bought a new car by selling a thousand bushels of wheat. As he would later tell his children, "You sure can't buy a new car for a thousand bushels of wheat nowadays". When times were tough, Joe worked in town to supplement the farm income. Evelyn canned much of the garden produce. She also sold eggs to customers in town and used the proceeds to help buy groceries. Like most farm families, Joe and Evelyn worked long hours along with their children. They all helped with the plowing, planting and harvesting, tending the cattle, raising chickens and working in the garden where they would find stones which were part of the foundation of Josiah's original house. Joe remained active in farming until two years before his death in 2001. Evelyn continued working and living on the farm until she passed away in 2009.

For the past few decades, Jack, who received his agricultural degree from Kansas State University, has employed conservation methods and been farming and raising beef cattle where his great grandfather, Josiah, first homesteaded and later expanded the Armour Farm. *Article submitted by the Armour children.*

ARNOLD – The Arnold Century Family farm is currently owned by Allen & Sharon Arnold and their daughter, Rochelle Falk. The farm, consisting of 163 acres located in Jackson County, Kansas, was purchased for $800 in 1864 by Rochelle's three-great Grandfather, John Arnold from Andrew Riederer. John was born in Germany, came to the United States and received citizenship in 1855, then lived in Pennsylvania prior to coming to Kansas. Upon John's death in 1899, ownership passed to his son, Franklin Arnold, who lived there for many years before moving to town. Extended family lived in the house at various times until Frank's great nephew and his wife, William & Kathryn Arnold, were married in 1942 and moved into the house sometime after. Frank died at the end of 1961, so they purchased the farm, but William passed away unexpectedly in June 1962. Kathryn owned the farm until her death in 2010. Her four children inherited the farm and her oldest son, Allen, purchased the farm from Allen's siblings. In 2013, Allen's eldest child, Rochelle, gutted the interior of the house. In 2014,

The Arnold Century Farm

she purchased the house with 3 acres and remodeled the interior. She lives there currently.

Of the original 163 acres, 117.29 acres remain. The only buildings left are the barn, corn crib, chicken house and the home. At various times, the farm also had a silo, scale house, oil house, garage, brooder house, farrowing house and hay barn. John planted a large apple orchard and traded apples with Indians. Other fruit trees were added over time but only one apricot and one pear tree remain. Frank installed a cave sometime after 1900 close to the west side of the wash house (south porch). This is a cement structure that was partially buried and is still being used today as a storm shelter and vegetable storage. At some point, a grove of catalpa trees was planted west of the house. Many of the trees remain today, providing a wind break and shade.

The house, already on site when the farm was purchased, started as a 2-room cabin. It was enlarged prior to 1900, adding a half-story upstairs that contained 3 bedrooms and a bathroom along with 2 additional rooms downstairs. A bathroom /laundry room was added to the west side of the house in 2001.

Livestock have always been a part of the farm. Kathryn had a large flock of sheep, providing 4-H lambs to her grandchildren and other 4-H kids. There has also always been a portion of the farm planted to crops. Currently Allen farms the land, rotating between corn and soybeans.

Allen likes to tell people that the Arnolds didn't get far (in distance) in life. Allen's great-grandfather, Grant (John's son), had a farm 1 mile east of John's place. Allen's grandfather, Clarence, had a farm across the road from John's place and Allen's father lived in and farmed John's farmstead. Allen settled across the road and less than a quarter of a mile east. And now, his daughter has restored and moved into the original home.

ATHERTON-ETTER FARMS sit between Independence and Sycamore, on both sides of Hwy. 75, on a knoll visible to drivers climbing north from Elk River. The integrated cattle ranch produced award winning Registered Hereford breeding stock into the late 1960s.

The first 160 acres of the Atherton Farm was bought in 1894 by Samuel Etter and his wife, Josie, for $7.03 an acre, by paying the mortgage within a year (in gold dollars). Samuel was born in Indiana. In 1865 he came to Kansas in a 16-member wagon train at age 3, with his widowed mother and grandparents. At 21 he earned his way with a team of mules, by breaking sod for other farmers.

Samuel became a substantial cattleman on his own agricultural operation. With his four sons, he sold fresh beef wrapped in brown paper on the streets of early Independence. A 1918 charter member of the Radical Grange, Samuel's five children became known as the Etter Grange family.

Samuel's son, William, was a committeeman bringing rural water to their district. He owned the Farmer's Supply and Exchange and several farms. The 1919 barn still stands near the farmhouse, where in 1990 William and Lula Etter celebrated their 70th anniversary. Warren, Sam's grandson, participated in 4-H, winning awards for his bulls. Prizes from the Kansas City Royal Cattle show financed his way through college.

Today's principle owner, Anita Etter Atherton, a Business and Accounting graduate, continues the farm business. The farm was en-

Samuel Etter, a substantial farmer of Sycamore Valley, and his family, 1903

larged and diversified after her husband retired from engineering. Corn, soybeans, wheat and hay are grown, along with cattle. The new generation – William, James and Elizabeth – have ownerships in the farming enterprises. Noblestone Housing rents out country homes. Heritage Honey traps endangered bees for pollination of crops on area farms. After 2003 her children purchased two adjacent farms, renaming all farming operations as the Atherton Farms to honor their father, James Atherton. *Submitted by Anita Etter Atherton.*

ATKINSON – The Atkinson Century Farm in Cherokee County is currently owned by Atkinson Farms, LLC, which is owned by Marion W. Atkinson, his son Charles W. Atkinson, and grandson Mathew C. Atkinson.

The Atkinson story reflects a long history in the area. The farm was originally settled by Albert Rhoads, who came from Suffolk, MA, and settled on 80 acres. The farm raised corn, wheat, potatoes and vegetables as well as cattle, hogs and chickens. At that time the farm raised food to feed the family as well as travelers that passed through on the stagecoach. Albert ran a livery stable that kept horses for the stagecoach and switched out the horses so they could continue on their journey. During this rest period is when Albert and his wife provided meals for the travelers and the cowboys.

In 1917, the farm was purchased by Oscar Rhoads, son of Albert, who continued to farm and expanded the acres to 240 acres. The farm now consisted of wheat, corn, sorghum, cattle and hogs. During the depression, 80 acres had to be sold off in order to keep the farm and remain in business.

In 1951, the farm was purchased by Nellie (Rhoads) and Clarence Atkinson. They continue to farm and raised their children on the homestead and continued to expand the farm. Soybeans were now added to the crop rotation. They purchased another 160 acres to add to their crop and livestock business. In the early 1970s their son, Marion E. Atkinson, came back to help them farm until their passing.

James W. Atherton stands in front of the barn at the family farm north of Independence.

Matthew, Charles, and Marion Atkinson

At that time the farm was divided between Marion and his two sisters; Marion kept the original 80 acres and the house in 1974.

The original house is still standing and up until 2016 was occupied by a family member. The original home began as a sod house and latter a two-story home was built alongside the sod house. Later an addition was built on in the 1950s over the sod home, which became a cellar to store food and provide protection from the storms.

In 1998, the farm was purchased by Marion W. Atkinson from his father. The farm was soon moved in to a No-Till cropping system of soybeans, wheat and sorghum, along with Brahma and Brangus cattle. Marion W. continued to expand the farm that also included the addition of Bermuda Clover and Native hay. On other acres of the farm, a turkey operation was added and for several years growing commercial turkeys, which allowed Marion W. to purchase back the 80 acres that was sold during the depression.

In 2014 Marion, Charles and Matthew Atkinson formed Atkinson Farms, LLC now owns the homestead that consists of 160 acres located in the SE4 of section 6-34-24 in Cherokee County, KS. Turkeys are no longer part of the operation and now consist of soybeans, corn, sorghum, wheat, cover crops and hay as well as Brahma and Brangus cattle. Matthew and his family reside on the farm as the seventh generation.

Marion W. Atkinson is widowed and has one child (Charles) and three grandchildren: Chelsey Rziha (Sam), Lynette Hofer (Tim) and Matthew Atkinson (Marry Ann). He is currently still farming and not retired. *Submitted by Marion W. Atkinson.*

BAIR – The Bair Century farm in Minneola, Clark County, Kansas was acquired by Thomas and Alice Bair in 1905, and has been farmed by four generations of the family. The original 1907 farm house has been restored by a great-grandson of TL Bair, Mark Mahieu and his wife Michelle Reinert Mahieu.

Henry B. Bair Family c. 1887

Tom Bair and Alice Fairley, ca. 1895

Thomas L. Bair Family ca. 1912

Tom Bair with seed cleaner, ca. 1929

The Bairs brought old-world farming expertise from Germany and Switzerland, and two centuries of experience in Western Pennsylvania. Tom's parents, Henry Brinker Bair and Margaret Martha Latta, brought their family to Kansas by flatboat from Mount Pleasant, Westmoreland County in April 1879. This river passage down the Ohio, up the Mississippi to the Missouri, then by wagon from Fort Leavenworth to Jackson County, made quite an impression on eight-year-old Tom, who would relate stories of the journey to his grandchildren. Henry and Martha's substantial farm in Holton raised flax, corn, oats, wheat and hay, as well as cattle and hogs, and boasted an orchard.

Thomas Latta Bair, b. Sep. 12, 1871, married Alice Almira Fairley on Feb. 20, 1895 in Soldier, Jackson Kansas. Alice was born in Circleville, Jackson, Kansas on May 6, 1873. Her parents, William Fairley and Sarah Frances Proctor, were also farmers, having arrived in Kansas by 1870 from Indiana. In 1904 Tom Bair moved with his family from Soldier to be foreman on the Denton cattle ranch, southeast of Minneola. The family built a farmhouse on the original quarter section, one mile east of Minneola, and moved in on January 1, 1907.

Situated south of Dodge City, and along the Rock Island railroad, Minneola enjoyed excellent cropland, hospitable grazing pastures, and a critical mass of commerce. It was a thriving community for Tom and Alice's children- Bernice Ella b. 24 Apr. 1896, Ross Thomas b. 5 Sep. 1899, Otis Harry b. 21 Jun. 1902, and William Henry b. 23 May 1911.

Tom Bair embraced both time-honored and novel approaches to dry-land farming of wheat, oats, and sorghum: crop rotation, seed cleaning, summer-fallowing, and disease management. His wheat often was best in Clark County competitions, and in 1929 he was chosen Wheat Champion for all of Kansas.

The *Kansas City Star* on Feb. 7, 1930 described the silver trophy and $300 cash prize for Kansas' Wheat King. Tom Bair's success was attributed to, "use of recommended tillage methods, good seed wheat, planting after the fly free date, and studying the market to sell advantageously." His wheat sample had a test weight of 58.5 pounds with average yield of 42.5 bushels per acre.

Kansas State Agricultural College sponsored a 56-stop Santa Fe education tour that summer to "promote profitable wheat farming" featuring Tom Bair speaking atop a flat car, with amplifiers, so that, "visitors remaining in their cars will be able to hear perfectly."

Demonstration Train
Santa Fe All the way
July 21 to August 2 1930
Operated Jointly by the Kansas Agricultural College and the Santa Fe Railway
The Santa Fe is furnishing an eight-car demonstration train. The College is furnishing the Speakers and Exhibits.
PURPOSE
To Present a Plan for Higher Profits on Wheat Belt Farms
MEET TOM L. BAIR
STATE'S CHAMPION WHEAT GROWER
TRAIN WILL STAND NEAR SANTA FE STATION. LECTURERS TALKING FROM A FLAT CAR EQUIPPED WITH ELECTRICALLY CONTROLLED VOICE AMPLIFIERS. VISITORS REMAINING IN THEIR AUTOMOBILES WILL BE ABLE TO HEAR PERFECTLY.

Tom continued his farming into the 1940s, with the assistance of his children, particularly Bill and Ross. In time Tom and Alice moved into Minneola; Ross and his wife Nellie moved their family to the home place and took over farm management.

Ross married Nellie Dale Moore on 26 Jul. 1925. Nellie was born in Protection, Comanche, Kansas to John Lloyd Moore and Jessie Dale. John, born in Wellington, Sumner, Kansas, had driven cattle for the Dalton Brothers from Texas to Dodge City before taking up farming. Jessie Dale, b. Aug. 3, 1876 in Industry, McDonough, Illinois, came with her family to Protection by 1885.

Ross & Nellie, ca. 1970

Ross and Nellie raised 5 children in Minneola: Ross Jr., Mary Lou, Alice Anne, Linda Lee, and John Lynn. Building on Tom L's disciplined operation, they added cropland and pastures for cattle, and introduced modern approaches including terracing, dammer plows, and mechanization. With wheat harvest and field preparation, summers were always busy and Ross's kids, his nephew Norman Bair, and nearly all the grandchildren were pressed into service over the years to help with the fun.

In the 1970s Ross turned over management of farm operations to his daughter Mary Lou Sibley and her husband Don Sibley, who leveraged earlier successes by investing in modern technology for automation enabling no-till farming. Their family, particularly their son Kent Sibley, carried the farm into the new millennium. Today the fields remain cooperatively farmed with ownership divided among Ross's children and grandchildren.

Ross T. Bair Family c. 1959

BAKER – Morris T. Baker purchased 80 acres in Labette County, Kansas, Township 32S, Section 27, Range 19E, in November 1882. The farm was later sold to his son, Willford Baker, and then to Willford's son and daughter-in-law, Howard (deceased in Feb. 2007) and Lois.

Howard, Lois, and Bruce Baker, 1973

Howard and Lois have three children: William J. Baker (deceased July 2017), Brenda J. Baker (deceased Dec. 1969), and Bruce A. Baker, and two grandchildren: Bryce A. Baker and Hannah J. Baker (children of Bruce and Connie).

As the land lies, half of the farm runs into Pumpkin Creek (into the Verdigris River) and the other half runs into the Hackberry Watershed (into the Neosho River). Wheat, soybeans and corn are still raised in 2017.

BALLET RANCH – Countless century farms started in the 19th century when someone had a vision of what he or she wanted to achieve. The vision always focused on a unique combination of people and the land. This one involves railroads, although there was never a railroad within 18 miles of the Ballet Ranch. It begins with William Henry Brownback, who, after serving in the Civil War, returned home to Tower Hill, Illinois, and became a merchant. This was in 1865, and he was 20 years old. He could see opportunity in the railroad that passed through the town. William Henry began with a grain elevator, buying grain and shipping it to the Illinois River by train. Then he bought hogs and shipped them as well. The first grain elevator became a string of elevators stretching to the Illinois River. The wealth that the railroad provided allowed William to pursue his dream.

In 1884 William Henry, now 39 years old, and his younger brother Joe traveled to Kansas along with a group of men from central Illinois, lured by inexpensive government land in the Cherokee Strip. The Cherokee Strip was a two and one-half mile piece of land along the southern border of Kansas that had been incorrectly surveyed as belonging in Indian Territory. In April William Henry bought his first parcel. The following year the *Medicine Lodge Cresset* reported "W. M.

Top row, l-r: Jesse Peach, Michelle Reinert Mahieu, Jeff Sibley, Jack O'Brien, Nancy O'Brien, Mark Mahieu, Jake Peach, Jed Bair, Paula Bair, Charles Russell, Kent Sibley, Joe O'Brien. Middle row: Norman Bair, John Bair, Linda Bair Mahieu, John O'Brien, Alice Bair O'Brien, Ed Mahieu, Susan Sibley Russell, Sarah Russell, Kasey Russell. Bottom row: Susan Vera Peach, Brad Callison, Brook Peach Alvarez, Maximus Alvarez, Annika Mahieu, Colleen Mahieu, Karri Mahieu, Allison Mahieu, Deborah Mahieu, Branden Alvarez, Quincy Bair, Elizabeth O'Brien, Emma O'Brien, Nancy Anne O'Brien, Henry Bair, Thomas O'Brien, Jamie Vera, John O'Brien, July 2012.

Ballet Ranch, 2008

(sic) Brownback and A. M. Craddick of Tower Hill, Ill., arrived here last week. These gentlemen own quite a large tract of land on the neutral strip, which they will probably fence and stock this spring." The ranch story almost ended before it started when William was caught in a blizzard, rode his horse until it foundered, and walked into an Eldred cowboy camp in "a perishing condition."

Aetna, Kansas, was a small town in the southwest corner of Barber County, north of the Salt Fork River. Like many other communities of the period, it developed on the hope that the railroad would come through the town and enrich its early citizens. The newspaper, *The Aetna Clarion*, existed for only two years. It extolled the virtues of the surrounding area, a siren song to lure settlers and railroads. In an article titled "The Salt Fork Valley," the paper's boosterism records: "The land is rolling, but not broken, being a gradual, gentle slope of valley land. The soil is a red, sandy loam, which has proven itself to be very rich and productive. Salt Fork has quite a large body of timber along its banks, while Cave creek has much more. Every quarter section is taken and a family is living on nearly every one. We do not believe there is a prettier or a richer body of valley land in the State of Kansas than the Salt Fork valley south of Aetna."

William Henry Brownback continued expansion of his ranch for the next quarter century, purchasing and trading parcels of land. One of his last enterprises was involvement in the creation of the Kiowa, Hardtner & Pacific Railroad. Spearheaded by Hardtner founder Jacob Achenbach, William Henry Brownback and several others, with Peter Ballet providing his expertise and acting as secretary and treasurer, the railroad built a spur line from Kiowa to Hardtner. Construction began in 1909 with the dream of extending the railroad to Trinidad, Colorado. All of this ended on October 30, 1910, when a cowboy sent to roust him for breakfast found William Henry dead in his bed, the victim of a brain aneurism. He was then 65 years old. This led to another situation common to century farms--the fracturing of the family farm. The Brownback Ranch was divided among the three children, Del, Gertrude and Margaret. The obituaries and reports of William Henry's death listed his ranch as being between 30,000 and 50,000 acres. In reality it was between 25,000 and 30,000 acres. The acreage was split into three parcels, with the land closest to Hardtner going to his son, Del. The next parcel went to his daughter, Gertrude. The furthest west parcel straddled the Barber and Comanche County Boundary along the Oklahoma border and went to the oldest daughter Margaret. This is the present land forming our century farm. It was 9,310 acres of butte and mesa topography. The Cave Creek drainage reached from the Oklahoma border east and then north to empty into the Salt Fork of the Arkansas. Most of the Cave Creek drainage and about 1.5 miles of the Salt Fork are on this parcel.

The youngest daughter, Gertrude, married James "Rube" Harrington on September 22, 1897. He appeared to be the most eligible bachelor in Edinburg, Illinois. Unfortunately, Rube was never able to find financial success. In 1914, they traded their part of the Brownback Ranch for an apartment building in Kansas City. Two years later, the apartment building was gone. Deserted by Rube Harrington, Gertrude filed for bankruptcy. She was alone with two children, Don and George Harrington.

The east and best end of the Brownback ranch was willed to Del Brownback. He had a wife and three children. And he had a vision: he would introduce Hereford cattle to the region. He would sell high quality cattle in a show barn that people would come from great distances to purchase. He began by making trips to the ranch from the family home in St. Louis where they had moved after his father's death. In 1915, the family moved to the ranch. He immediately spent $10,000 buying a Hereford bull, Sir Donald Britton. Over the next eight years he made other expenditures to support his vision, culminating in the construction of a show barn in 1919. This barn was round, situated six miles west of Hardtner, and still stands in 2017. He held sales for his SilverCrest Herefords in the new round barn. But the profligate spending was more than the cattle sales could support. In the fall of 1923 Del Brownback sold his ranch and declared bankruptcy. A receiver's sale for the other ranch assets followed in early 1924.

The oldest daughter Margaret married Peter Edward Ballet (pronounced like Ballot). He did not appear to be the "catch" that Rube Harrington had been. He came to Edinburg, Illinois, the home of the William H. Brownback family, to work for his older brother, Jacob Ballet. Peter had become fatherless at under two months of age. His mother had remarried when Peter was three. Three years later his stepfather, Jacob Gehm, died, leaving his widow with five children. His mother died 11 years later, when Peter was 17. The family was scattered after this. One of the five children, Peter's half-brother, Henry Gehm, was placed in the home of Thomas Fox, a railroad freight agent and later station manager. Henry followed the railroad profession of his foster father. Eventually he became the manager, general manager and then president of the Venice Transportation Company, a company that repaired and later manufactured railroad cars. Peter Ballet joined his stepbrother, Henry Gehm, in the business. The company changed its name to the St. Louis Rail & Equipment Company. Henry Gehm was president and Peter Ballet was vice-president. The company introduced a low-height flat car that was easier to load. This innovative railcar resulted in a close relationship between the St. Louis Rail & Equipment Company and many of America's circuses. The circuses bought and leased flat cars, animal cars and sleeping cars from the company, and stored their cars at the company railyard during the off season. When Henry Gehm died, his obituary appeared in the show business daily newspaper, *Variety*, because of that relationship.

Margaret and Peter Ballet were a childless couple. When Gertrude was left bankrupt, the Ballets took in first Don Harrington and then George Harrington, allowing their nephews to obtain college educations. Don would use that education to later become a multi-millionaire benefactor of Amarillo, Texas. When Del Brownback lost his part of the ranch, the couple took in Del's son, Wilber, and later Del's daughter, Marion. Again, both were helped with their college educations.

Peter and Margaret Ballet were tied to St. Louis by his employment. But they made frequent and long trips to their part of the Brownback Ranch, now called the Ballet Ranch. Ida May Bird, in her book *Raised On Methodist Pie* (1975) remembered Peter Ballet. "He lived in St. Louis and came once or twice a year to check on his holdings. He would spend a week or two visiting, with headquarters set up in Hardtner. I remember him as a small, dark man who always wore a moustache, and moustaches were quite unusual at that time. He was always well dressed. As young children we received a roll of funny papers from him every week, mailed from St. Louis. They came for years without fail. Pete Ballott (sic) was Dad's friend." The couple stayed in a hotel at the northwest corner of Hackberry Road and Main Street in Hardtner and drove back and forth to the ranch. Their visits ended September 3, 1934. Peter was driving Margaret back to Hardtner from the ranch when he suffered a stroke and drove the automobile into the ditch. Margaret extracted the car and drove Peter to Hardtner where he was rushed to Alva, Oklahoma. He died the following day.

Although Margaret Ballet was only 60, she felt inadequate to the task she faced following Peter Ballet's death. Her niece, Leanore Brownback, left her high school teaching position in Kansas City and moved to St. Louis to help her aunt and manage the ranch. For the next 34 years Leanore lived with her aunt and drove to the ranch at least twice a year. Leanore would stay with her friends, Lois Platt and her mother, Alma Olson, in Alma's house just east of Hardtner. She would always arrive for the wheat harvest. Days were spent on the ranch with the evening reserved for listening to the St. Louis Cardinal baseball games on the radio. If the weather was threatening, the listening, card playing and reading were relegated to the basement, which functioned as a storm shelter. Margaret died in 1968 at age 94.

The ranch passed to another generation. The five nieces and nephews became owners. They were a scattered group. Don Harrington was in Amarillo, Texas. His younger brother George was in Wilton, Connecticut. The Brownbacks were also dispersed. Wilber was living in New York City. Marion Brownback Miller was in Decatur, Illinois. Leanore remained in St. Louis. She badgered and cajoled the other four with her single-minded vision: the ranch should not be sold or split.

Leanore moved to Hutchinson, Kansas, in 1973 to be closer to the ranch. She lived there until January 18, 1985, when she died after four years of increasing ill health. Her ashes are scattered on the ranch.

Today, the ranch is owned by the only living descendants of William Henry Brownback. They are the son of Marion Brownback Miller, Michael, and his wife Barbara Miller and their three children, Chris, Craig and Charles. In 2007, the ranch was recognized by the Barber County Conservation District as a Banker's Award winner. The Ballet Ranch is now Ballet Ranch LLC to reduce the likelihood of its being sold or split. The windmills, once a fixture on the ranch, are being replaced by solar-pumped wells, and a rural water system extends across the ranch. The galvanized stock tanks are gone, with fiberglass tanks and wildlife protectors in their stead. There is less cross fencing, with the ranch now operating a short-season, rotational grazing system. Along with increasing use of cover crops, the cropland is being no-till farmed. The 133-year-old ranch is now a 21st century ranch working on its second century.

BARTZ/ANSTAETT CENTURY FARM – In 1900 Herman Bartz and his wife, Amelia, settled in Osage County and paid $2,500.00 to purchase a house and 160 acres in Sections 3-16-16 and 4-16-16. Herman was born in Germany in 1865, immigrated to the US in 1890 and first lived near Clearfield in Douglas County, KS. He and his wife Amelia had five children.

The Bartz farm was in the Smartsville Community, which consisted of a general store, a blacksmith shop and a ball diamond. The blacksmith shop was on a corner of the property and an acre of land was set aside for Neill School District #56.

In 1915 Herman died of complications following a broken leg. Amelia and her sons continued to farm. One son, Walter and his wife, Grace, eventually purchased the farm from his mother and siblings.

The current owner of the farm, Garlena Anstaett, was born on the farm in 1922. Walter, Grace and Garlena saw many changes on the farm. Tractors replaced horses, trucks replaced wagons, combines replaced threshing machines and electricity replaced lanterns.

Garlena married Roger Anstaett in the farm house in December 1942 while he was on a short Army leave. Before Roger was shipped overseas, Garlena spent the summer with him in South Carolina. Those few months mark the only time Garlena has not lived in Osage County, KS.

From 1940 until 1946, Garlena taught in one-room school houses, including the Neill School near her home. Garlena and her first son lived with her parents on the farm until Roger returned in 1945 to

Aerial view of the Bartz/Anstaett farm in 1989

farm with Walter. Garlena and Roger eventually purchased a house up the hill from the Bartz farm and Garlena took over doing the farm bookwork as she raised six children.

Walter Bartz died in 1982 and ownership of the farm passed to Roger and Garlena. They continued to farm with their son, Jerry, until Roger's retirement in 1985. The land has been leased to the Sturdy Brothers since that time. In 2003 the farm sustained substantial damage by a tornado and most of the buildings were damaged or destroyed. Garlena became the sole owner of the farm in 2006 when Roger died.

THE BATES FARM PROPERTIES – The 40-acre "Century Farm" in Section 9 on SW Dike Road west of Augusta, KS has been owned by the Bates family since it was purchased IN the late 1800s by Henry Clay Bates (1837-1920) as additional acreage for his 160-acre farm located directly west in Section 10, Township 77, Range 4E. Henry, a Civil War veteran, came to Kansas from his native Michigan in 1872. He first constructed a log cabin as a home for himself and his wife, Jeannette Negus Bates and later built a two-story frame home, a substantial barn and other outbuildings.

The 1880 Nonpopulation US Agricultural census described Henry's farm as 125 acres of tilled land, with 25 acres of woodland and 10 acres of "other". In 1879 he had planted 40 acres in Indian corn yielding 1,462 bushels, 10 acres in oats, 60 acres in wheat and a 1/2 acre in Irish potatoes. He had six working animals – a cow, two calves and four other. He had hired an employee for 52 weeks of the year. The 1875 Census for Augusta Township, Butler County, KS reported the value of Henry's property as $2,000.

The Bates' adopted son, William, grew to adulthood on this farm. The five children of William and his wife, Maude Beaver, grew up on the farm as well. Henry and Jeanette moved to the city of Augusta soon after the 11910 Census. William inherited the farm upon Henry's death in 1920 and lived there some 30 more years with his second

Aerial view of the Bartz/Anstaett farm in 1949

Jeanette (Negus) Bates, William Donaldson Bates, and Henry C. Bates at the original log cabin on the Henry C. Bates farm, Augusta, Butler County, ca. 1890.

wife, Edith Boggs until they moved into Augusta. He continued to manage his rural holdings until his death in 1966, after which William's heirs sold the 150-acre farm. The smaller plot (dotted with working oil wells since the early 1900s) is now owned by Henry's great-granddaughter, Beverly (Roblin) Goodson Baker. *Submitted by Beverly (Roblin) Goodson Baker.*

BECK – The Beck Century Farm in Reno County, KS started with the birth of Jacob John Beck in Dettingen, Germany July 10, 1858. July 25, 1878 he arrived in New York. After stopping in Iowa to visit a brother, he made his way to Kansas and a job helping build what became the Santa Fe Railroad through southern Reno County. Another brother, Joe, living north of Plevna offered to sell Jake the north 80 of a 1/2 section he owned. Jake declined because of his job at the time and worked in Arizona where the railroad job turned into logging timber for a short time. In 1896 Jake returned to Reno County, KS and bought the N2 NW 4 Sec 24 T235 R9W from brother Joe. Jake moved a small house on site and lived alone until 1900 when he married Ottilia Kucton from Nashville, KS. To this union four children grew to maturity. The youngest was Edward J. Beck, born 1918, who started farming about 1932. Edward married Elner Zimmerman in 1940 and they subsidized the farm with a binding and baling business. A son, Dale, was born in 1942 and a daughter, Kay, was born in 1945.

J.J. Beck and saddle horse at the original barn, 1909

After Ed's heart attack and surgery in 1983, his son, Dale and wife, Diane, assumed management of the farm, merging it with their own. To Diane and Dale were born Renae, Carma and Seth. After the death of Edward and Elner, settlement of their estate saw Seth receiving title to the property and living in a new house built in 1973 by Ed and Elner. The original house was bought and moved north of Hutchinson and remains in the Rayl family. The original barn built in 1909 remains in use on the farm.

Crops grown over the years include wheat, corn, milo, alfalfa, soybeans and sunflowers. Livestock over the years included hogs, cattle, a few sheep and chickens. Horses and mules were the original sources of power. *Submitted by Dale Beck.*

BECK – Martin Beck, great-grandfather, purchased 80 acres on March 22, 1883, S 1/2 of SE 1/4 of Section 18, Township 22, Range 14 in Coffey County, KS and is still owned by great-granddaughter, Frances and husband, Jim Force. This farm was purchased from cousins, George and Elisabeth Storrer at a price of $1400. ($17.50 per acre)

Tile barn built by Frank Povenmire, dad to Frances Force, in the late 1940s to feed his cattle indoors. Native lumber was cut locally for use in the barn at about 3 cents per foot.

Martin Beck also purchased the north half of Section 19, Township 22, Range 14 in Coffey County on Oct. 31, 1893, of which 80 acres is still owned by Frances and Jim Force. Two sisters still own land in this section.

Left to right in front of family home: Retha (7), Ruth (7), Mary (9), Wilma (3), Frances (1), Povenmire on their horse, Smoky, won by their mother in a "Yes" facial tissue contest on why she liked that brand of tissues. The top two prizes were either $100. cash or a pony. You can see what Esther Povenmire chose (maybe influenced by her five girls).

Martin and Christina (Hauser) Beck were born in Berkfelden, Germany, and the family of Martin and children came to America. (Christina passed away at age 37 in Germany.) Martin did not want his sons to enter the military service in Germany. The family arrived in New York, then traveled by train to Burlington, KS, where Preacher Joe Huber brought them directly to this farm by box wagon. The Apostolic Church was built across the road from this farm and the cemetery for this church is still 1/2 mile west of this farm house. The family was affiliated with this church and Chris Beck and a neighbor dug grave sites by hand for many years.

This farm has a homestead, certificate #994 by James B. McKinley on Sept. 16, 1874 and signed by Ulysses S. Grant.

All the buildings on this farm were built by the four generations who owned this farm. The horse barn and milk barn were built in the early 1900s and still stand. The stucco home was built by Chris Beck from 1932-1934 and is still used as a home.

Horses, cattle, hogs, mules, sheep, goals, chickens have been raised on this farm. Crops raised include prairie hay, alfalfa hay, red clover, soybeans, milo, wheat, kafir corn, popcorn, oats, several types of fruit in an orchard, as well as a lot of garden produce. *Submitted by Frances Force.*

BEIGHTEL – In 1907 C.W. Beightel purchased 320 acres located in Section 21, Township 7, Range 13, Grant Township, Jackson County, KS. Native pasture crops were raised to support the livestock on the farm.

The farm was passed from C.W. Beightel to his grandchildren – Cathleen Reed, Elizabeth Howells and Joseph Kennedy and in 1979 passed to Joseph Kennedy.

Although no original buildings remain on the land, the farm retains the original 320 acres. Now the farm produces native grass pasture. In 2016, the farm was awarded the Jackson County Grassland Award.

Joseph and wife Kathy have five children: Kelly, Carmen, Kalena, Kaycee and Jeanie. They have 16 grandchildren: Joseph, John, Joel, Corey, Breanna, Kaden, Savanna, Kyler, Conner, Caroline, Koy, Kennedy, Cole, Luke, Lisa and Andrew. *Submitted by Joseph L. Kennedy.*

THE BELL CENTURY FARM – Clemence and Annerberry Bell homesteaded their farm in Clark County, KS (south of Bucklin, KS) in 1886. Clemence fought with General Sherman in the Civil War in 1864. Their son J.E. "Ed" Bell and his wife, Edith, returned to farm this land with his father in 1892. They raised their nine children on this farm. Ed's and Edith's sons, Charley (and wife Mabel) and Ross (and wife Leona) Bell, took over the farm when their father retired. After Charley and Ross divided the farm, Charley's grandson-in-law, Ed Chamberlin, and wife, Linda, farmed half of the land and Ross's great nephew, Larry Feikert and wife, Kathy, farmed the other half. Today Ed still farms Charley's land and Larry and Kathy's sons, Lance and Kyle Feikert, farm Ross's land.

Mostly wheat and milo have been grown on the Bell farm along with a cow-calf herd. A beautiful landmark on this farm was the huge barn that was built around 1915. You could see this red barn for miles around. Unfortunately, the barn had to be demolished in 2015 because of a crumbling foundation.

Clemence Bell and family members in the field

Farming practices have changed tremendously through the years... from horses pulling plows to new and bigger tractors and combines (with GPS guidance) to no-till farming (a way of farming the ground from year to year without tilling the soil).

The huge Starbuck fire that burned through Clark County in March of 2017 came within 1-1/2 miles of burning the farm. This huge fire is officially the largest single wildfire in Kansas history, burning over 502,000 acres in Kansas. The majority of this is in Clark County where 85% of the county burned.

THE BERGER CENTURY FARM – Abner Berger homesteaded a claim of 160 acres four miles south and one-half mile east of Ford, KS in 1884. He brought his wife, Lucinda, and oldest child, Edna, from Howe, IN to Newton, KS where they had relatives. From there they came to Ford County by covered wagon in 1885.

During these first years in Kansas, they experienced many hardships common to pioneer life. But they persevered when others became discouraged and returned to the eastern states. Hot winds, drought, grasshoppers, prairie fires and blizzards made crops uncertain. After some good crops, they managed to add to their original claim. Abner helped build the railroad track from Pratt to Bucklin.

Abner and Lucinda retired in 1912 and moved to Bucklin. Abner died in a farm accident in 1914 when a team of horses trampled him to death. Their son Albert and wife, Elizabeth, moved to the farm. After WW II Albert's son, Kenneth and wife, Wilma, returned to farm. When Kenneth retired in 1977, his son-in-law, Larry Feikert and wife, Kathy (Berger), started farming the land. Today Larry and Kathy's sons, Lance and Kyle Feikert (5th generation farmers), continue to farm this same land with the help of their young sons and daughters.

Farming practices have changed a lot since this land was homesteaded. Besides wheat, they have grown corn, soybeans, milo and alfalfa. Kenneth started irrigating in the 1960s and they continue this today. They also have a cow-calf herd. Probably the biggest change in their farming practice is no-till farming...a way of growing crops from year to year without tilling the soil. In 1990 the Kansas Board of Agriculture recognized our farm as a Kansas Tradition Farm for keeping this land in the same family for over 100 years.

BIGGS – On Nov. 1, 2013 Rhonda K. (Thompson) Biggs became the owner of 25 acres of the original 160-acre farm purchased from Daniel Brown. On Feb. 22, 1900 William W. Thompson purchased the land for $15.00 an acre, located in Labette County, Section 6, Township 32, Range 20 from Brown.

The farm originally produced horses, cows, hogs, wheat, oats, corn, soybeans, milo and prairie hay. Now the farm produces prairie hay, horses, cows, chickens and rabbits.

No original buildings are on the farm as they have been torn down.

The farm has remained in the Thompson family, being passed from William W. to his son, Fred R. Then to Irvin A., to his widow Cora Edna and to Donald L. Thompson. Donald promised his father, Irvin, he would keep the farm in the family for 100 years, which he achieved Feb. 22, 2000. Irvin passed away one day later on Feb. 23, 2000, having

The Berger farm, ca. 1950

The Thompson homestead in the early days

kept his promise. This farm was divided between Rhonda K., William, Anthony and Donald L. Jr., the children of Donald and Patricia Thompson. *Submitted by Rhonda K. (Thompson) Biggs.*

BLANKENSHIP – The Blankenship Century Farm in Sumner County, KS, is currently owned by the Edith C. (Blankenship) Peck Trust. The Blankenship history reflects 140 years of farming in Kansas. Edith's grandmother, Sara Ann (Lawless) Blankenship, came to Kansas with parents, James W. "Jimmie" and Nancy (Cook) Lawless, along with three siblings. The Lawless family left Russell County, KY in 1871, settling in Sumner County, KS in Harmon Township.

Pictured l-r: Felix, Nick, and Sarah (Lawless) Blankenship

John "Nick" Blankenship was born in Russell County, KY in 1863 to Cornelius and Belinda (Wilkerson) Blankenship. Nick came to Sumner County in 1883 and worked for friends who migrated earlier. His parents came to Kansas in 1903. Nick married Sarah in 1885. To this union one son, Felix C. Blankenship, was born.

Felix and Maude Bender were married in 1916 and lived on the family homestead. To this union two daughters were born, Edith C. and Patricia A.

The 160-acre farm has been owned by the descendants of James "Jimmy" Lawless and Nick Blankenship from 18171 to the present. Edith (Blankenship) Peck became the owner in 1984.

They raised turkeys, chickens, beef cattle, swine and horses. A few dairy cows were on hand to provide milk, cream and butter for the family. Wheat, milo, corn, soybeans and alfalfa were grown. A large family garden and orchard was planted, much was canned for the winter. With the farm being on the Ninnescah River, fish was commonly eaten. During the summer they often hosted fish fries. On Saturday night you could find family and friends parked on Main Street in Belle Plaine, just to visit and get the latest news.

Edith married John Peck in 1947. They have four children, four grandchildren and one great-grandchild. *Submitted by Jeanette O'Hair.*

THE BOEHME FARM – The Boehme farm, located one-half mile east of Offerle, KS on Highway 50/56, came into the family in September 1905 when William J. and Sarah C. Lightcap purchased the property from Lillie M. Baxter and D.C. Baxter. The land then went to John and Luella Boehme in 1935, as Luella was a daughter of the Lightcap couple. John and Luella's son and wife, Elmer and Joyce Boehme, began farming the land in 1942 and managed the farm until Elmer's passing in 2009. At that time, Elmer and Joyce's sons, Galen and Jary (and wife Alice), began operating the farm.

The Boehme home, 2017

The farm now has acreage in three Kansas counties – Ford, Edwards and Kiowa. Through their tenure, Elmer and Joyce also managed a Shorthorn cattle operation. Elmer was one of the first operators of the area to incorporate irrigation into his farming practices. The farm now basically produces grain, while Jary and his family do custom hay work in Kiowa County also.

The Boehme farm was declared a "Farm Bureau Century Farm" in 2009 with Jary and Alice Boehme receiving the plaque. The two-story house on the property was originally built in 1917 and has seen at least two major remodelings. *Submitted by Galen R. Boehme, Ph.D.*

BOLLER – Philip William Boller immigrated from Germany in 1850. Philip married and farmed in Pennsylvania. The family of children arrived and a son, William P., was born. William P. Boller grew to manhood and married and had a family. William P. served in the Civil War. After the war was over in 1866, he desired to move on to another area, perhaps for more space or a quiet area. The decision to come to Kansas with his family was made. The family had grown up and married and had families of their own.

Welcome sign to Boller Ranch

Fertile ground and water beckoned the family to locate 2500 acres of farmland in S.W. Riley County and N.E. Geary County. Clarks Creek flows into the Kansas River and is accessible to the land. Through the years wheat, milo, alfalfa, prairie pastures, beef cattle and swine were raised.

Church, 4-H and the rural community were important to the families. Limestone homes were built and some or remnants of them still remain. A stone barn was built in 1855. The stones were secured with wooden pegs. The barn is still sturdily standing and usable. One of the first Grist Mills in Kansas was on Clarks Creek by the farm.

Today, eight families live in the area. The age ranges from 92 to 5 years old. This includes fifth, sixth, seventh and eighth descendants of the original Wm. P. Boller legacy. Present owners of the farmland are Phillip C. Boller, Darryl Boller, Daniel P. and Dorothy Boller, Peggy Boller, Patricia Regelman and Chad and Amy (Boller) Blockcolsky. The family celebrated 150 years in Kansas September 2016. 150 relatives gathered for a reunion then. *Submitted by Peggy Boller.*

BRIGGEMAN – Ida Dorothy Louise Scheele was born in Haven, KS in the late 1800s. She came to Pratt County as a teenager by train to visit her cousin, whom was married to the Lutheran minister at Natrona. She was known for being a good seamstress. As the word spread about her sewing skills, she moved from home to home to sew wardrobes for families in the Natrona and Preston area. One such home was the Henry and Gustena Briggeman home. It was here she met their son, Henry William "H.W." Briggeman. After a proper courtship the two were married. At that time Ida sold 80 acres of land she inherited around Haven. With that money she purchased the SW/4 of 28-26-12 in Pratt County. Henry "H.W." then built for his bride a new four-bedroom house in 1915. He had the home plumbed for water, electricity and steam heat. Unto this union four daughters were born: Ilene, Florene, Ethel and Velma. Ilene passed shortly after birth. Henry and Ida farmed the land until his early death of pneumonia in late De-

The Briggeman Farm, March 3, 1918. The house was built in 1915 and the barn was destroyed by tornado in 1917.

cember 1928. This left Ida now a young widow with three girls to raise. When the Great Depression hit in 1929 banks closed, times were hard and some people lost everything. Ida was able to hang on and raise the girls by renting out the farm land. They had chickens for eggs, a cow for milk and cream. The cream that they separated was taken to Pratt to sell on Saturday's.

Ida stayed on the farm till Velma married Merle Trinkle from the Preston area. Ida and Florene then moved to a house she had built in Pratt. Ethel had already married and moved to Chicago with her husband she met when he was stationed at the Pratt Army Airbase. Florene never married and was devoted to caring for her Mom.

Ida passed in 1974. At that time the land and home was inherited by Velma and Merle. They had one son, Morgan. Velma and Merle lived in the house until shortly before Merle's death in 2011. In 2012, Velma decided to deed the property to Morgan and his wife, Deb. They still own the land today. In late 2011, their daughter, Sheila and her two young sons, Tyler and Dylan, moved into the house. They still live there today. That makes them the fifth and sixth generations to live in the house. They are very proud to be carrying on the family legacy. *Submitted by Morgan and Deb Trinkle.*

BROCK – In 1879 Walker C. Brock and his wife, Amy (Mercer), came to Kansas from Morristown, OH in a covered wagon along with their two youngest children, Louisia and John W. They settled on an 80-acre farm in southeast Woodson County (Perry Township W 36 S).

This sign on the old red barn reminds visitors where they are.

On Dec. 31, 1891 John W. Brock married Eliza, daughter of David and Emily Crist who lived on the adjoining farm. In 1870, at 6 months of age, Eliza, with her parents had traveled with a large covered wagon caravan from Vinton County, OH to Kansas. The Crists settled on the farm that joined the Brock farm to the east.

John W. and Eliza purchased the Brock farm in 1898. While living on the farm they built a new house in 1907 and a new barn in 1911. After John's death in 1918, Eliza moved to Chanute, KS leaving Lawrence, the youngest of the boys, on the family farm.

Lawrence Brock married Charlotte "Lottie" Beeman on Jan. 1, 1931. After his mother's death, Lawrence and Lottie purchased the farm in 1961. Before their deaths, Merrill, the oldest son and his wife, Nancy, purchased the farm on Sept. 17, 1981.

Merrill married Nancy Coonrod Dec. 23, 1955.

They are the fourth generation of the Walker C. Brock family to live on the farm. Along with their family, they have hosted a three-day Brock reunion at the farm every five years on Memorial weekend.

Wheat, corn, soybeans are among the crops that have been grown over the years on the farm. Many animals have been raised on the farm - cows, hogs, sheep and chickens. At one time Lawrence and Lottie ran a dairy.

Nancy continues to live on the farm as Merrill passed away March 14, 2016. She stays active in church and many other organizations. *Submitted by Nancy L. Brock.*

BROWN – In 1880, Francis and Louisa Brown bought 160 acres located in Section 16, Township 33, Range 13 East. Wheat, milo, alfalfa and corn were the grain crops raised.

This farm passed from the Browns to son-in-law Henry and daughter Lela (Brown) Aiken, then to their son Glen and daughter-in-law Dola Aiken and to their son Kenneth and Beverly Aiken in 1997.

The farm still retains 160 acres and now produces pasture to support the beef cattle.

Kenneth and Beverly have three children: Deborah Gregoire, Rebecca Arellano and Edwin Aiken. Audrey, Betty and Cavya Gregoire, Ashron and Scarlet Arellano and Finley and Stella Aiken are their grandchildren. *Submitted by Kenneth and Beverly Aiken.*

BROWN-TODD, LLC - Joshua Brown came to Comanche County, KS, looking for land for expansion at a price lower than could be found in his home county of Pawnee, KS. In 1910 he purchased one section four miles south of Wilmore. Joshua Brown was national president of the Farmer's Cooperative Association in 1926, at a time when it was having a struggle for survival. He was among those received in the White House by President Calvin Coolidge. He gave 52 years of service to the Co-op movement.

His son, Mark, came to Wilmore in September 1915 to farm the land owned by his father in the New Eden neighborhood, traveling overland with a team of horses his father had given him. He and Esther Libbey were married in January 1916.

Together Mark and Esther raised Polled Hereford cattle, the start of which was a wedding gift from her father, who was a chartered member of the American Hereford Association. New Eden Polled Herefords were sold to breeders in 21 states from Maryland to California.

Wendell Brown was born on the farm in 1916. He and Lorene Todd, of Nescatunga Township, were married in 1941. When Mark died in 1983, Wendell and Lorene had farmed and raised cattle for nearly 40 years and were established as agricultural leaders in Comanche County. They inherited the original acreage.

An eight-year county commissioner, Wendell served on the local Soil Conservation Board 40+ years and as county representative 25 years. He initiated the RE&D Range Land Improvement program, was early to terrace his land, was instrumental in acquiring financing for Lake Coldwater and was an early proponent of Rural Electric.

Upon Wendell's death in 2008, daughters Judith and Marilyn inherited the family farm. Three grandsons spent summers on the farm and return annually as adults for hunting.

BRUMBAUGH FARM #1 – Marjorie M. Brumbaugh Trust and Harry L. Brumbaugh Trust own a farm in Athelstane Township of Clay County. The land was original purchased by Marjorie's great uncle, Frank Burgess, in August 1890 from Franklin and Sarah J. Smith. Frank Burgess was the proprietor of the hotel at Longford at the time of his death in 1905.

George and Lizzie Simpson, with children George and Mary Ann

Frank and Anne Burgess sold 262 1/2 acres to his sister, Lizzie and her husband, George Simpson, in 1902. George Simpson was born in Scotland in 1853 and emigrated to America. George crossed the ocean three times and settled in Dickinson County, where a second cousin lived. George then moved to Colorado for 14 years, where he was naturalized in 1894. Lizzie Burgess Simpson was born in La Salle County, IL in 1869. Lizzie moved to Kansas in 1879 with her parents. George and Lizzie married in 1901 in Oakhill, KS. To this union was born four children, one pair of twin girls, both of which died in infancy, Mary Ann and George Theodore.

George Simpson suffered with rheumatism and was an invalid for four years prior to his death in 1915. George left the land to his wife, Lizzie, to have and hold until her death (1961). After her death, the land was to pass to his two children, since his wish was that the farm shall not be sold. Thus, Lizzie became a shrewd businesswoman, as she raised her two children and managed the farm. Family stories are told of her taking her yardstick out to measure the grain, to confirm she received her share of the harvest.

The original farm was evenly split in December 1962 according to George's will with their two children, Mary Ann and George, inheriting the farm. Marjorie (Simpson) and Harry Brumbaugh inherited 220 acres of the 262 1/2 acre farm from George Theodore Simpson and an aunt, Mary Ann (Simpson) Rush. The farmhouse, built by Lizzie in 1929, is a fun location for family to gather, while attending the Longford parade and rodeo. *Submitted by daughter, Judy Curtis.*

BRUMBAUGH FARM #2 – Jacob Brust was born in Germany in 1842. In the year 1868, Jacob sailed on the ship *Hammonia* from Hamburg to New York. He arrived in New York on July 14, 1868, at the age of 25. He listed his occupation on the passenger list as a Hacker (maker of hoes). He made his home in Pennsylvania, where he met and wed Miss Margueritte "Margaret" Siegle in 1870. Margueritte had emigrated from Switzerland to Pennsylvania. Jacob and Margaret came westward, homesteading 80 acres near Industry. They had five children: Wilhelminia "Minnie," Matilda "Tillie," William, Amelia and Jacob Jr.

Twila (standing) and Marjorie Simpson

The Brust family purchased land surrounding the farm that they homestead, including 160 acres from the Land Department of the Union Pacific Railroad in 1881. Today, their great grandchild and her husband own 117 acres of the land purchased from the railroad.

The Brust family was successful and helped many in the neighborhood to get homes for themselves. The Brust farm grew to 570 acres in 1885 and eventually to 640 acres in 1895, according to the Kansas Censuses. The Brusts grew corn, oats, and Irish potatoes. Their livestock consisted of horses, milk cows, other cattle and swine.

Jacob suffered from Bright's disease and died in 1910. His estate was divided between his wife and their children, one of whom was Matilda (Brust) Lewis. Margueritte died in 1924 of pneumonia.

The Joint Last Will and Testament of Jesse C. Lewis and Matilda "Tillie" Lewis left land to their daughter, Twila M. Simpson. Matilda and Jesse died in 1939 and 1942, respectively.

Over the next 59 years, Twila Simpson utilized the property mainly as pastureland, with approximately 80 acres in farmland. She mainly grew wheat and a few row crops. In addition, she had prairie and alfalfa. In the early years, the Simpsons grazed their herd of Hereford on the pastureland. Eventually the Simpsons' herd included Angus. Twila put in waterways and terraces. She installed a large farm pond, which she stocked with catfish since she loved to fish.

Upon her death in 2001, Twila passed 117 acres to one of her daughters, Marjorie Brumbaugh. The Marjorie M. Brumbaugh Trust and Harry L. Brumbaugh Trust own 117 acres that the Brusts purchased in 1881. Today, the land is in pastureland and some wheat ground.

BRUNING – Jere and Patty Bruning are the present owners of this farm. His grandpa bought the farm for $3,000 in 1902 and lived on it till 1927, when Jere's dad, Omar Bruning and Pearl Schuler were married. Jere's grandparents moved to Hiawatha, KS. Jere's parents moved on the farm in 1927. They had two children, Ann and Jere. Ann married and they live in Raleigh, NC. Jere and Patty have four girls – Luann, Jane, Mary and Sara. They had seven grandchildren and two great grandsons. Jere celebrated his 87th birthday Sept. 17, 2017. He retired in 1992 when he was 62. They moved to Hiawatha then.

The Bruning homestead

Jere started to County School District 8 for eight years. Then he went to Hiawatha High School for four years and graduated in 1948. He took FFA for four years so he could learn how to farm, which he did. He farmed for 45 years. Going to college wasn't an option. He was farming 80 acres in 1992 by himself except for what his Dad helped for essentials.

Jere married in 1951. They have been married 66 years.

Jere's dad did buy a tractor, which was a 1942H. Jere helped shuck corn by hand for two or three years with a pair of mules. Omar had five mules with which he did all of his work. Back in the '40s, we raised corn, oats, wheat and red clover. When Jere quit farming, corn and soybeans were raised, each 50%. He remembers that back in the '40s they had chickens, beef and hogs. Omar would cut wood every winter so that he would have it to burn the following winter.

Omar and Jere were life members of the Zion United Methodist Church.

Family birthdays were important. Caring for the land was important. They have all of it in conservation now – terraces and waterways, fence rows cleaned out, etc.

Our neighbors meant a lot to us. We exchanged labor and work. Land has really increased in value.

This farm is worth 1-1/2 million today for 16 acres. This is located in the southwest quarter of 25-1-18, Brown County, KS. The original deed was filed July 17, 1860 at Iowa Point, KS by James Buchanan, President to James S. McDowell.

Jere was born and raised in that community. His grandpa, father and he attended the Zion United Methodist Church their entire lives and held many offices in it. Jere has been on 10 or 12 organizations and held offices in Kansas. *Submitted by Jere Bruning.*

BUBB – The first of our family to own these 80 acres was Louis Bubb. He was the second child of immigrants, Antone and Katherine Vosse Bubb, from Baden, Germany. Louis was born on Nov. 21, 1856 in Weston, MO, two years after Kansas became a territory in 1854.

The windmill and concrete water storage tank, May 4, 2014. The house in the background was built in 1959 after their original house burnt from a crack in the chimney.

Antone, a shoemaker, moved his family to Kansas in 1857 and built the fifth house in Atchison. Antone taught Louis to be a hard worker and at 26 years old, he bought this farm on Feb. 21, 1882 for $29.06 an acre from John M. and Lucy Cain. This farm was along the Military Road that angled from Fort Leavenworth to Fort Kearney, NE Territory. Louis farmed, raised cattle, hogs and broke horses for the military.

He married his next door neighbor, Rosa Walz. They had five children on this farm. Herman Bubb, the second son and the second Bubb to own the 80 acres. He never married but bought this and several farms in the surrounding area totaling about 1,000 acres. Herman enlisted his nephew, my father, to work for him.

My father, Roy Roloff Sr., bought the farm in 1967 and continued farming and raising cattle. In 1975 my father passed away from a cancerous brain tumor. My mother, Hazel Rupert, continued farming it with help. She had cattle and crops of corn, soybeans, and hay. In 2001 she passed away and we settled the estate with my husband, Jerry and I getting the original 80 acres. My mother did conservation work

Louis and Rosa Walz Bubb with children (back row) Julius, Emma and Herman, and (front row) Theodore and Bertha.

of terraces and waterways. We continue to improve the farm by plowing in ditches and keeping the waterways clean. The three original barns have given away to age. The hand-dug well and windmill still stands and could pump water if needed and was used until Dec. 30, 2010. *Submitted by Jerry and Janis Falk.*

BURGESS – William Burgess was born in Ireland to parents who were land owners/farmers. William, being a second son, was unable to own land. During the Great Potato Famine in 1850 he decided to come to the US to pursue his dreams.

William Burgess and Rebecca (Wilson) Burgess

William, his wife Rebecca Wilson and their three daughters arrived in New Orleans, LA in October of 1850 after a six-week journey. The family moved to St. Louis, MO where they lived for 11 years. William supported his family by working as a custodian at the Methodist church where they were devout members.

In 1862 the Civil War was in full force. While William was in sympathy of the North, he was too old to serve. He was given the opportunity to purchase a 160-acre farm from Merriman Cahill, located in Pottawatomie County, KS. William traveled by boat to Leavenworth and walked nearly 100 miles on the Oregon Trail to the homestead 10 miles north of Wamego in 1863. After he purchased the land, he was held by slave state supporters and accused of being a spy. After being found innocent he walked back to Leavenworth and brought his family to their newly purchased farm. They had two oxen, two wagons, two good saddle horses cows and $600.

He planted an orchard and vineyard. This orchard bore fruit for over half a century, and in the early days was like an open community hall for the surrounding farmers when they all gathered to conduct their Fourth of July celebration, picnic and birthday get togethers.

He made friends with the Indians who raced horses past his home. When the renegades of the Southern Army came through after the war, they would steal livestock and cold-bloodedly shoot the owners who objected. These Indians guarded the Burgess home for three days and nights until after the renegades left.

The original farm expanded after WW II when land became more available. This expansion continues today with land used for pasturing cattle, crop growing and feed lot production.

The farm has proudly existed for five generations: William Burgess, 1860-1893; William Burgess, early 1900s; James Burgess, 1925-1945; brothers William V. Burgess and Roscoe J. Burgess, 1945-1978 and current owner, Gary J. Burgess, 1978-present.

Gary's son, Rustin Burgess and children, Liam and Nealy and daughter, Taryn Westenberg and son, Nolan, will be the next generations to preserve the Burgess family legacy of "custodians/stewards of the land."

BURK – The Burk family has been involved in farming since Edward Burk arrived from Ireland before the Revolutionary War. The Burk family arrived in Clifton Township, Washington County in 1871. My grandfather, Walker, purchased his first 80 acres from neighbors in 1880 (1137 years ago) in Brantford Township.

The Burk farm, East ½ of the SW ¼ of 27-4-1, Bradford Township, Washington County

My grandfather was a hard worker and added acreage to the 80 as adjacent land became available.

He married Margaret Craig in 1882. As their family grew, so did the house. In later years, barns, granary, corn crib, chicken coops, smoke, wash and ice houses were among buildings added to the farmstead.

The Burks were members of the United Brethren in Christ Church and took an active part in church and community affairs. Walker and Maggie stayed on the farm until 1927.

My parents, Clarence and Delena Burk, moved on to the farm in 1931, living in the house and farming the land until 1947.

In 1955 Shirley, granddaughter of Walker, married Arlan Benteman and we returned to the area. Our three children, four grandchildren and their spouses now have a connection to the original 80.

Farming has changed over the years. The farmstead is gone, building moved off or sold, the pasture is now used for hay, hedge trees for windbreaks and fence lines are mostly gone. Terraces, stubble mulch tillage and land in conservation would surprise Walker.

The land acquired by Walker and Maggie is in the Clarence Burk Trust. The Trust is owned by great-grandchildren: Gary (farmer of the land), Glen and Linda. Shirley and Arlan, retired in 1998, are the beneficiaries. The Benteman's grandchildren are: Logan and Ryan Benteman and Brishen and McKinley Hawkins.

This family is proud of Kansas and pleased to live in or near Kansas! *Submitted by Shirley Benteman.*

BUTEL – Our Century Farm story began with my grandparents, Charles D. Butel and Gertrude M. (Dyer) Butel. However, The Butel lineage is traced back to 1695 with the birth of Antoine Butel in Palante, Haute-Saône, Franche-Comté, France. My great-great grandfather Charles CONSTANT Butel "Constant" (1829 - 1871) left France in 1853 or 1854 settling in Illinois. By the Fall 1855 he had settled in Prairie City, Kansas. My great-great grandmother Genereuse Stiquel "Rose" (1828-1912) left France with 3 children including my great-grandfa-

The Weigel family

ther Emile Joseph Butel to join Constant in 1855. Then my great-great-great grandparents Claude Joseph Butel "Joseph: (1800 - 1868) and Marie Josephe Guchemand (1802-1856, died at sea en route to the United States) left France to join Constant and Rose in Kansas. Family history shows that the Butel family were tile makers in France and that Marie was a farmer.

One of the Butel children and dog at the family barn

Charles and Gertrude Butel purchased 160 acres, Southwest ¼ Section 36, Township 16 Range 22, in Miami County Kansas on August 8, 1910. They moved into the home on that land with their oldest child, Charles, in 1911. They moved from Prairie City, Kansas (near Baldwin). They had 6 children, Charles, Henry, Herman, Mary, Joseph, and Anthony. Anthony planted an Oak tree near the house when he was a small boy.

After being honorably discharged from the Navy in 1946, Anthony (Tony) returned to the farm to help his parents. On December 16, 1946 Tony purchased an adjoining 80-acre tract, the South ½ of the Northwest ¼ Section 36 Township 16 South Range 22. The first tractor Tony purchased was an International Farmall "M".

Tony married Annetta Elizabeth Wolken of Scipio, Kansas on October 8, 1949. They moved onto the farm to care for the land and animals. Charles and Gertrude moved into the town of Paola, Ks. Over the years they raised wheat, corn, soybeans, oats, milo, dairy cattle, chickens, beef cattle, sheep, pigs, brome & alfalfa hay, prairie grass, and horses, as well as a large garden from which much produce was enjoyed fresh from the garden and the rest was canned, frozen, or shared with friends and family.

The Butel-Weigel home

Tony and Annetta raised 13 children in the original farm house. Those children, Mary, Tony, Theresa, Margie, Rita, Angela, Marcella, Diane, David, Phyllis, Dolores, Roberta, and Karen – and our families – 31 grandchildren, 32 great-grandchildren are living throughout several states within the United States. Annetta passed away August 30, 2012 and Tony passed away January 31, 2014. His hope was that someone from his family would take ownership of the farm.

Church, school, and 4-H leaderships were important to our family. We were raised with a deep abiding faith and Mom singing. Birthdays were celebrated with hand-cranked, homemade ice cream and homemade chocolate cake with chocolate frosting. The 4th of July was a fun time of celebrating with the Seck cousins who traveled to the farm from the city.

Karen (Butel) Weigel & her husband Derrick purchased the family farm June 6, 2014 from Karen's siblings. Corn and soybeans are currently grown on the farm. Continuing to care for the land and heritage of the Butel - Weigel family farm is our dream.

The Butel family

Charles and Gertrude Butel

Tony Butel

CANTY – The Canty family farm in Buffalo, KS, is currently owned by Reginald Glenn Payne, son of third generation Canty, Mary Lou Canty Payne. In 1870, Steven Canty of Winchester, VA, homesteaded the 160-acre farm in Wilson County, KS, after serving in the Civil War. Steven married Jane Ellen Steele in 1876. They had ten children: Miles Everett Canty, William Canty, Steven Canty, Bryan Lee Canty, Jesse Margaret Canty, Kate Canty, Jane Ellen "Ella" Canty, Teresa Canty, Virginia Canty, and Mary Canty.

The farm was passed down and owned by the Canty siblings, and additional acreage was acquired during that time. On the farm they raised cattle, sheep, and chickens, as well as building the haybarn, chicken house, and two silos that still stand today.

Gardens provided fresh vegetables and living off the land was a way of life. Crops included corn, milo, oats, silage, and soybeans.

The last Canty sibling to own the farm was Bryan Lee Canty, his wife, May Stewart Canty, and brother Steven Canty. Brian and May had two girls, Mary Lou Canty and Ruth Ann Canty. Bryan continued in his dad's footprints and farmed the land from 1950-76. In 1986, Bryan, Steven, and May all passed away.

At that time 500-plus acres were left to daughters Mary Lou Canty Payne and Ruth Ann Canty Pringle, and their spouses Glenn Payne and Lauren Pringle, respectively. The original homestead of 160 acres was retained by Mary Lou Canty Payne, and the additional acreage by Ruth Ann Canty Pringle.

Glenn and Mary Lou Canty Payne retired and returned to the farm in 1994. The farmland continued to be leased for farming. The legacy of the farmhouse continued to be the site of many family celebrations and holidays.

Aerial view of the Canty farm

In 2012, Mary Lou Canty Payne passed away, survived by her husband Glenn, who continued to reside in the farmhouse until his passing in 2016. At that time it was handed down to their son, current owner, Reginald Glenn Payne and family.

In the last 147 years, the many landscapes of the farm have changed, but the heart of the farm that Steven Canty built is still standing strong after all this time.

CARLS – The Carls Century Farm in Shawnee County, KS is currently owned by William D. Carls and his three daughters. Part of the original 160 acres purchased by William Frederick Carls in October 1900 belongs to his brother, Dean Carls. William D. Carls manages all the farm ground, which still includes the original acreage. He works every summer to maintain healthy pastures and rents out the ground for corn, soybeans, feeder cattle and hay.

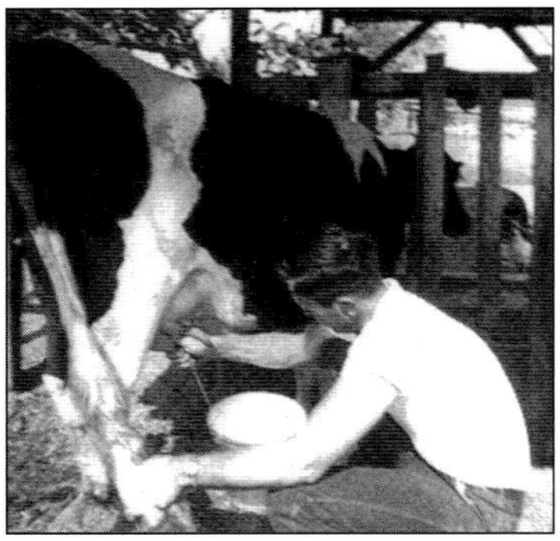

William D. Carls on milking duty

Like many Kansas farmers, the Carls family immigrated from Germany. The Carls Family Dairy became Carls and Sons, then Carls Bros. to Carls Family Farm. Selling eggs in town, riding a horse to school, hard work, neighbors helping neighbors, hired hands, homemade bread and pies and garden vegetables are part of the history. Recent memories include 4-H, your choice of homemade ice cream and cake flavor every birthday, raising hogs and chickens, horseback riding, hunting, fishing, children playing outside, numerous agriculture and community organization memberships, and faith in God.

Today, each formerly widowed, the William D. and Shirley Carls family includes four grandchildren from five children on his side and two grandchildren from one child on Shirley's side. Life events continue to increase the gift of family all around. *Submitted by William D. Carls.*

CHRISTERSON – The first owners of the 1,525 acres in Ottawa County, KS were Chris and Mary Christerson. The farm is located in Section 03 Township Range 05 West (2008), SW 1/4, EX TR DOB SW 1/4, NE to NW 137.

On April 10, 1901 this property transferred to Carl and Ruth Christerson. On April 8, 1920, The Christerson Trust was set up. The property transferred from Carl to Clyde and Alma Christerson.

Wheat, soybeans and milo have always been produced on this farm and still are to this date.

Clyde (deceased) and Alma L. had four children: Chris, Connie, Carma and Cathy (deceased). They have eight grandchildren and four great-grandchildren. *Submitted by Alma L. Christerson.*

CHRISTIANSEN – Our Century farm in Cherokee County was purchased in 1872 by our great grandfather, Ernst Christiansen. He purchased the 160 acres from the Kansas City, Fort Scott, and Gulf Railroad for $420.00. The farmhouse built in 1905 is in amazing condition today. The ownership has passed down from Ernst, to Casper, to Arthur and to daughters, Doris Christiansen and Edna Christiansen Lassman.

My siblings, Thelma, Doris, Galen, Erma, and I loved growing up on the farm. Our dad, Arthur, had shorthorn cattle, Shropshire sheep, Poland China hogs, chickens and harvested corn, wheat, soybeans and a large garden. When our grandfather Casper was 95 years old, he rode Galen's horse five miles to the fair in Mineral.

Our dad and our mother, Pauline, were baptized and confirmed at St. John Lutheran Church in Pittsburg, KS, as were all of us children. Our faith and life in the church was central to our family.

In 1941, Pauline and Arthur received the Kansas Master Farmer Award. In addition, Arthur began a 21-year term on the SEKAN Electric Cooperative Board of Directors, a position he held until his death in 1962. As one of the organizers, he solicited $5.00 memberships from neighbors, needing three people in each mile in order to establish

The Arthur Christiansen farm (minus the hay barn)

electricity. In 1941, when I was 4, we got electricity to the farm which made life more enjoyable for us all and easier for our mother.

Our parents loved to have people stay with us—farm helpers, two displaced Latvian families, foster children, and International Farm Youth Exchange (IFYE) students from Italy, Switzerland, Chile, Algiers and India to learn about Kansas farm life. During WW II, Mother invited soldiers from Camp Crowder to spend weekends with us. She never complained about cooking extra meals. Music, 4-H, sports, church, and school were all important to us. Mother served over 50 years with Extension Homemakers and other community groups. She had blankets made from the wool of our sheep.

This year, Galen's daughter, Lisa Christiansen Howard, will become the new owner of the farm. We are thrilled the farm is staying in the family. *Submitted by Edna Christiansen Lassman.*

CHRISTY – The Christy Century Farm in Labette County is currently owned by Merrill and Norma Lee Wiggins Christy. Elisha L. and Clara May Christy purchased what we refer to as the Christy Farm in 1892. The original farmstead was 110 acres and is north and west of Oswego. They moved from Indiana, first settling on a farm just east and north of the current farm.

Merrill and Norma Lee Wiggins Christy

Modern dairy barn on the Will Christy farm near Labette

Some of Will Christy's high producing Jersey cows

Aerial view of the Christy Century Farm

Cynthia Christy and her horse, Ginger

Merrill Christy and sons Stephen and David

By 1905 their oldest, Daisy, had married and there were four additional children. Tragedy struck in August of 1905 when their oldest still living at home, Lizzie, died of typhoid fever followed by the death of Elisha a month later. This left Clara May with six children at home, the oldest being William, age 18 and the youngest Mabel, 6 months. Times were tough and neighbors told Clara she should put the kids out in neighboring homes. She and her son, William, worked to keep the family together.

Further tragedy struck Valentine's Day 1906, when the house burned. At this point they moved into a shack, with a dirt floor, on a neighboring farm just west of their farm. They then moved into the chicken house on their farm and put up a tent for the four boys to sleep in. One night a high wind blew the tent away and they were all forced to stay in the small chicken house.

By the fall of 1906 William, who was now 19, with help from the neighbors, built a new house. The family moved into it in October 1906. There were no windows installed in the house, but they had a roof over their head and a floor under their feet. Clara later married and moved into the village of Labette City, two miles away. William stayed on the farm and later purchased it from his mother.

In 1916 William and Esther Hallock were married, she was 23 and he was 29. Esther's parents were concerned about her marrying William as they felt she would have such a hard farm life. A daughter, Ruth, was born in 1920 and a son, Merrill, in 1922.

William and Esther worked hard and in 1927 William built a large Grade A dairy barn with a gothic roof design. The Depression hit and they started purchasing farms. By 1940 the original 110 acres had grown to 880 acres and they had hired help to assist with the farming operation. The farm consisted of registered Jersey's, Hereford, pigs and chickens. Crops were wheat, soybeans, milo and corn. In 1948 with rural electric they added more acreage and built a new dairy barn to transition to machine milking. At this same time, they hired a young man, Don Pickens, who along with his wife, Betty and their two children, moved from Oklahoma to Kansas. During the '50s and '60s the dairy farm operation continued expanding.

Today the Christy farm is enjoyed by Merrill and Norma Lee Christy and is treasured by their children, grandchildren and great-grandchildren.

CLOGSTON-BERTENSHAW-DIKLICH FARM – The farm is owned by siblings, Alleen (Diklich) VanBebber, Richard Diklich, Kitty (Diklich) Shinkle, and Neva Diklich. It began in 1871 with 40 acres of Osage Indian Trust Land. By 1881, it had grown to 196.5 acres owned by our great-great grandfather, Lewis E. Clogston. The farm then came to our great-grandparents, Kit (Clogston) Bertenshaw and her husband, James Edward (Ed) Bertenshaw. Their son, Lewis Jeffrey (Pat) Bertenshaw, left it to his only child, Lillian Grace (Bertenshaw) Diklich, and it came to us from our parents, Lillian and Donald Diklich.

Edward and Kit Bertenshaw, with sons Lewis Jeffrey and Charles Richard.

Under care of prior generations, this was a typical Kansas family farm. The families raised beef cattle and hogs, along with corn, soybeans, oats, alfalfa and other types of hay. Milk cows, chickens, a small orchard and large vegetable gardens fed the families. Kit was known for traveling with a medical kit to help those in need. Pat's wife, Grace, died in the 1918 flu epidemic, leaving newborn Lillian in the care of Pat, Ed, and her maternal grandmother, Dora Palmer. Toddler Lillian spent many hours parked at the end of a row while Pat planted or harvested it. When she was twelve, Pat remarried.

Pat and Amy Bertenshaw were devoted to public service. One of the pastures became the Fall River High School baseball field for a time. Amy determined that the City needed a golf course, and saw that one was built. Lillian worked in Manhattan during World War II and married Dick, a city-boy soldier from Pennsylvania who wanted to be a farmer. Pat and Dick farmed the land together for many years, acquiring more property and watching us grow up to love the land as they did.

COLSTROM FAMILY FARM – The Colstrom Family Century Farm is currently owned by Larry and Jeannine Colstrom, third generation farmers of the land.

The land, which consisted of 160 acres was purchased on July 14, 1902 by grandparents Emil and Selma (Fagerstrom) Colstrom. Emil, along with his family, immigrated from Sweden to America in the late 1800s. They were the descendants of Gustaf Colstrom. After arriving in America, some of the brothers changed their last name from Colstrom to Gustafson to be known as the sons of Gustaf or Gustafsons. Our grandfather and two of the brothers, Alfred and John, kept the last name of Colstrom.

Emil and Selma Colstrom on their wedding day.

It appears that in 1867 the land was originally owned by Seyfert, McManus & Co. from Berks County, PA.

A deed dated Jan. 15, 1869 conveys that Seyfert, McManus & Co. also known as the Reading Iron Works, was granted permission by the US Government to sell 19,028.78 acres at a consideration price of $57,000.00, which contained the 160 acres.

In June of 1869 John and Mary Wetherell purchased the land for a sum of $60,000. In July of 1874 the land was sold to Thomas Peter at a price of $114,864.42. On Oct. 6, 1884 the 150 acres, which is the original land of the Colstrom farm, was purchased by Caroline Selix, who then sold the land to Emil and Selma Colstrom at a price of $2,300.00 or $14.38 an acre.

In the early 1940s, the 160 acres was purchased by Clarence and Martha Colstrom, son and daughter-in-law of Emil and Selma. In 2003 Larry and Jeannine Colstrom, son and daughter-in-law of Clarence and Martha, purchased the land.

150 acres of the original land still remains in the Colstrom family name.

The Colstrom family has always been very active citizens of their community and church.

COMBES – Obadiah and Rebecca Combes moved from Terre Haute, IN to Kansas in 1870 by covered wagon with their several children. They settled on a farm two miles south of Lebo. They moved in 1876 to the present day Combes farm, located one mile west and 4-1/2 miles south of Lebo and lived in a little stone house.

Original home lived in by first homesteaders in the Combes family. Obadiah on the right, Rebecca sitting, and an aunt is standing.

Obadiah lost his eyesight so his son, Walter and wife, Lillie Belle, lived on the farm with his parents. They did the farm work and took care of his parents. Walter was deeded the farm on Nov. 16, 1907. Walter and Lillie Belle raised their nine children - Alvin, James (father of Gale), Harvey, Thelma, Nadine, Clinton, Cecil and Wayne - on this farm. They had a huge garden to produce the food needed to feed such a large family. They also raised hogs, cattle, chickens, corn, milo and wheat. When Walter passed away, his son, James, continued to farm the land until we purchased it in 1972.

We continue to live in the house built by Walter and a neighbor. Our children - Andrew, Matthew, Jennifer and Julie - were raised here. We have 10 grandchildren who enjoy spending time on the farm.

Gale farmed until 2016, growing milo, corn, wheat and beans. We raised sheep, cattle, hogs, chickens, rabbits and goats. He continues to help son, Andrew and nephew, James, farm. Gale spends his spare time fishing, hunting and trapping. I enjoy raising miniature horses.

At times of family reunions, the older generation would often talk of the Oklahoma reservation Indians stopping to camp in the field west of the house. They traded for chicken and pork with Gale's grandmother. His uncles tell of having various contests with the young Indian boys. Many arrowheads were found in this field thru the generations. *Submitted by Gale Combes.*

CORNELISON – In 1867 Jeremiah Croley purchased 185 acres located in Jefferson County, Rock Creek Township. Crops raised on the farm were corn and sorghum. Shorthorn cattle were also raised here.

This farm was received in 2006 by Jerry L. Cornelison. Through the generations, it passed from his great-grandfather, Jeremiah Croley, to his grandfather, W.F. Croley, to his mother, A.J. Cornelison.

The Cornelison home and barn

The farm still retains the original 185 acres and today produces corn, soybeans and hay. The original house and barn remain on the farm.

The Century Farm Award – Soil Conservation was presented to the owner.

Jerry L. and wife, Linda, have two children, Cary and Nick. Their grandchildren, Ashton, Nicole, Carly and Maura, are the sixth generation.

Jerry retired in 1993.

COX – William Cox of Orange County, IN came to Kansas Territory in 1857, looking for land to settle. However, he died in November of that year before he could make any purchase. (His wife, Elizabeth (Cloud), had remained in Indiana.)

His eldest son, John C. Cox, then came to the Territory and purchased the NE quarter, Section 32 in Township 20, Range 22 (160 acres) in what is now Linn County on Feb. 19, 1858.

In June of 1861 John returned to Indiana and helped organize Co. K of the Fourteenth Infantry Regiment. He died on Nov. 3, 1861, at Huttonsville, VA, leaving the Kansas land to his mother and 10 brothers and sisters.

Joel, the next oldest brother, moved to Kansas to make a new home on that land. Joel married Rachel Ellen Lindley March 20, 1879, in Montgomery County, IN. They purchased another 100 acres adjacent to their Kansas farm, so there were about 245 acres. (The railroad had taken some acreage.)

Joel and Rachel's son, Howard Cox, inherited the property. He married Blanche Bearly and they had six children. Howard died in 1960 and after that Blanche passed on the original property, plus other land, to her children.

Joseph and J. Frank inherited the original land so they had 245 acres. J. Frank later bought out Joseph's half.

Following J. Frank's death in 1996, the land went to his two sons, Larry J. and Terry J., who have been caretakers since and have made numerous improvements. It is pasture with some wooded area and is rented to a local rancher. The house (not the original) is also rented.

Joel and Rachel Cox at their Kansas farm home. Their daughter-in-law, Blanche Cox, and her children, Lester and Frank, are also in the photo.

Larry and Terry are the fifth Cox men in a row to own the land, but are the fourth generation since early owners John and then Joel were brothers.

It is the sincere intention to keep the property in the family in the future. *Submitted by Larry J. Cox.*

CRAIG – Homesteaded May 20, 1883 by Wm. Shaffer, sold E 1/2 to J.S. Bennett on Sept. 6, 1880, then through marriages the W 1/2 was purchased by James Johnson who was married to J.S. Bennett's daughter. In October 1892 the land once again became a quarter. Bennett's and Johnson's were family through marriage. Emma Johnson Cousins was Joann Cousins, Craig, Walker mother and JoAnn is the mother of Robert Craig Jr. and Grandmother of Robert Craig III. Emma passed the farm to JoAnn; Robert Jr. purchased the land from JoAnn April 1998.

Currently through the forming of Double Rockin R Ranch LLC, Robert Jr. is passing the farm on to Robert III. Thus keeping this farm in the same family for 137 years as of 2017. JoAnn is still active in farming along with her son, Robert Jr. and grandson, Robert III (sixth generation). She lives on the farm just east of this farm.

The current operation has grown to farm ground 550 acres, with an active Feedyard business, thus the formation of the LLC. The farm ground supports the active cattle business for the LLC. Members of the LLC are Robert and Rebecca (Strand) Craig Jr., son Robert Craig III who manages the LLC., daughter Charlsie Craig member of LLC, Nashville, TN , she works as a producer for RFD-TV. These families have been devoted to agriculture for over 100 years. *Submitted by Robert Craig Jr.*

CRESSLER – In 1950 Larry bought the SW /4 of the SW/4 of 25-7-27 in Bow Creek Township, Sheridan County, KS, totaling 40 acres, from his grandfather, Herbert E. Cressler. H.E. had purchased the homestead rights from a man who was unable to marry the woman he wanted to and to make the farm pay. H.E. finished the requirements of the homestead and received the document from the US government. At the time H.E. purchased the ground, he and Lillian (Russell) Cressler had five children. Three more were born while they lived there. Howard and June Cressler, children Judy and Bob, lived on the property until Judy was to start school in Hoxie in 1945. The house was moved to just east of Hoxie, KS and is still there today.

Memorial tree planted on the Cressler farm in honor of Keller

Once Larry purchased the land, it was broke out and dry land corn was planted and harvested with a one-row picker. Larry added irrigation and planted a variety of crops on the land. Flood, dragline sprinkler pipe and underground pipe were installed. Finally, a pivot system was installed and is still in use today. In 2015 the land was given to his son and daughter-in-law, Herbert E. and Marita Cressler.

Larry and Irene Cressler had six children, 10 grandchildren and 15 great-grandchildren.

On the century farm property, a memorial tree was planted for Keller, an infant grandson, who will always be remembered as extended family watches lovingly over the tree as a part of the Cressler legacy. *Submitted by Larry Cressler.*

CURLESS – The Curless Century Farm in Sedgwick County, Kansas, is currently owned by me, Robert Curless, and my wife of 67 years, Ruth. The Curless-Long family history tells of farming as their occupations back for several

Current owners of the Curless Century Farm, Robert Curless (age 92) and Ruth Curless (age 88).

Ivel Marie (Long) Curless feeding her chickens, one of her many daily chores.

farming daily and driving my tractor and combine.

Draft horses, cattle, hogs, chickens, wheat, native grass and corn were grown. Today we raise beef cattle, wheat and alfalfa. Wives and daughters planted large gardens. As we grew it, we ate it fresh and preserved the excess. We worked from early morning until after the sunset, caring for our livestock and our crops. Planting and harvest are the most important times of the year still today. Some of the best times on the farm never change.

Favorite family memories include family gatherings, church, ball games, peach picking and going on Sunday drives.

Pictured l-r: Nory, Steve, and Robert Curless

Today the Robert and Ruth Curless family includes two children, Anne and Rick and their families, three grandchildren and five great-grandchildren, located in Kansas and Oklahoma. *Submitted by Robert Curless.*

CURTIS – The Curtis Century Farm in Stafford County, KS is currently owned by Lee Eldean Curtis. The quarter of land, the NW 1/4, 17 -25-13, was bought in 1891 by my great grandparents, Thomas and Evaline Curtis. They paid $1,000.00 to the widow of the original homesteader, Fannie M. Brockman. She moved back east when her husband, James C. Brockman, died. This quarter is directly across the road from my great grandparents' quarter. Thomas and Evaline Curtis moved to Rose Valley Township from Illinois in 1878 with two children. Four more children were born and raised in a one-room house. The Curtis who came to Kansas to homestead belonged to a family of six brothers. All six brothers and their father served in the Civil War and all survived. Three brothers stayed in Illinois to farm and three brothers

Field on the Curtis farm

came to Kansas to farm. The two other brothers settled nearby and those homesteads are still owned by their heirs. The story is told in the family that the great grandparents lived in a dugout with two small children their first winter. The winter was so severe their oxen froze to death.

My great aunt, Pearl Curtis Lyon, inherited this quarter on the death of her parents, Thomas and Evaline Curtis. She owned the land until her death in 1987. She passed it to her six children, mentioning in her will that if it was sold she hoped it could be sold to a Curtis descendant. We purchased the land in 1988 and it will be passed on to my children.

There are no old building sites on the land, but there is a small grove of cottonwood trees in one corner that could be part of the original tree planting. The land raises wheat, corn, rye and milo today.

I am pleased to own this land as a fourth generation Curtis. *Submitted by Lee Eldean Curtis.*

DEDRICK AND BARTLETT – This year marks 130 years that members of the Dedrick family and their immediate descendants have farmed in Union Township, Sheridan County, KS. Current landowners of the original 160 acres purchased by W.H. and Mary Walker Dedrick on March 12, 1887 are great-grandson, Steven Launchbaugh and wife Kathy and great granddaughter, Vivian Herrmann. It has also been the location for the Dedrick-Launchbaugh family home for five generations and the headquarters for their farming operation.

Mary and W.H. Dedrick, ca. 1880s, prior to Kansas relocation

Landowner Carrie Griffith Bartlett and children, Ava Bartlett and Blanche Bartlett Leist, ca. 1930s.

A Civil War veteran, W.H. Dedrick was raised in Pennsylvania, married in Ohio and had farmed in Iowa by the time he and Mary decided to move their family to Kansas in 1887. Within 10 years, they purchased 150 adjoining acres, which belongs to Vivian and in 1901they filed a tree claim on an adjacent 80-acre parcel the Launchbaughs own. These three pieces of ground became the core of the Dedrick farm. While W.H. farmed a small portion of the land he owned, much of it remained in native grass and was used for grazing. Service related ill health prompted him to rely on help from his future son-in-law, Tallis Launchbaugh, whose father had homesteaded nearby in 1888.

When W.H. died in 1901, Mary continued to live with the young family until her death in 1938. Home on the Dedrick place was a soddy until several years after Tallis and his wife, Ella Dedrick Launchbaugh, had welcomed their first child, Ura, in 1903. This sod home stood about 200 yards from the current farmhouse. Later, a new frame home was built. Two sons were also born to Tallis and Ella. The oldest, Vernon, died as a teenager, while the youngest child, George William, began farming all of the Dedrick land soon after returning home from service in the US Navy during WW II. George continued in this role when the farm passed to him and his sister, Ura, upon the death of their last surviving parent, Ella, in 1962. Ura's daughter, Vivian, now has ownership of her mother's former interest.

George married Carolyn Leist in 1944 and in 1968 she inherited nearby farm ground that had recently belonged to her uncle and aunt. In 1901, Wirt and Caroline Griffith Bartlett left their 1889 homestead in neighboring Decatur County and purchased 230 acres of land within two miles of the Dedrick-Launchbaugh farm. This ground along the North Fork of the Solomon River was some of the earliest settled in Sheridan County when a Receiver's Receipt on 40 acres of - it was taken out in June 1880. Wirt raised cattle, turkeys and alfalfa. Their son, Alva, and wife, Neta Matthews Bartlett, took over the operation on Wirt's death in 1934 and owned it from 1943-1968. With no children of their own, the Bartletts deeded the land to their nephew and nieces. Niece Carolyn Launchbaugh and husband George farmed the Bartlett place until 1974 along with raising their children, Steven and Carol Lee.

In 1974 Steven purchased the entire property his Great-Grandfather Bartlett first bought. Four years later he purchased the balance of his parents' farm ground, combining the land of W.H. Dedrick and Wirt Bartlett into one farm. He and Kathy are the current owners of this property. When they applied for the Farm Bureau Century Farm status in 2009, the Bartlett-Launchbaugh land was included as part of their heritage farm because of its distinction of also having been farmed by members of the Bartlett family and their immediate descendants for over 100 years. This year marks 116 years and four generations of landownership. Their two sons, George Gilbert and Karl, and Karl's daughters, will make the fifth and sixth generations of family when the responsibility passes to them.

Under the stewardship of the numerous caretakers of the land, these two properties have raised a cornucopia of crops and livestock. Land primarily once used for grazing has seen the expansion into modern, productive farming and ranching. This small part of God's creation along the North Solomon River bottoms has sustained generations of our family. *Submitted by Kathy Launchbaugh.*

W.H. DEDRICK CENTURY FARM – The history of the Dedrick farm begins in 1887, when W.H. Dedrick purchased 160 acres in Union Township. Sheridan County, KS. He was born in Pennsylvania, moved to Ohio where he married Mary Walker, and moved on to Iowa after serving in the Civil War. The purchase of the Kansas acreage completed his westward trek.

William and Mary Dedrick, ca. 1901

He and his wife and two youngest daughters (Ida and Ella) settled on the land where they built a sod house and sod outbuildings. Some years later they purchased 160 adjoining acres. This piece of land, along with 40 acres of the earlier parcel, is now owned by great-granddaughter Vivian (Long) Herrmann. William farmed only a small portion of the land to wheat and corn, leaving most of it in native grass to graze his cattle. He and Mary also raised chickens, milk cows and horses used for both farming and transportation.

As ill health from military service injuries worsened, W.H. relied on help from his future son-in-law, Tallis Launchbaugh, who lived nearby. Tallis and Ella were married in 1901 and continued to farm the Dedrick land after William's death. In 1903 their daughter, Ura, was born in the sod house. They later welcomed to their new frame house two sons, Vernon T. and George and grieved as a sudden illness claimed Vernon's life at age 16. Ella's mother, Mary, lived with the family until her death in 1938. This was a time of much transition. In addition to the house, Tallis and Ella built a new barn and numerous outbuildings. Horse drawn implements and transportation gave way to mechanization. Telephone and electricity were introduced. Their farming operations were primarily milk cows, beef cattle, some pigs, wheat, alfalfa and milo. Ella also raised chickens and always had a large vegetable garden. There were hardships as well with the dirty '30s, the Great Depression and war.

When son George returned from WW II service, Tallis and Ella gradually turned the farming operation over to him. After the death of Tallis in 1955, the old house was torn down for lumber. Ella lived with George and his wife. Carolyn. At Ella's death in 1962,the original Dedrick claim passed to Ura and George. George farmed Ura's land and continued the expansion of mechanization and tilled land. He leveled fields and implemented irrigation with new wells. He and Carolyn phased out the milk cows and transitioned to beef cattle. Their crops included wheat, alfalfa, corn, milo and cane.

Tallis and Ella Launchbaugh, ca. 1941

Eventually their son, George Steven, purchased his parents' share of the Dedrick land. He and his wife, Kathy, have implemented new conservation methods with no till crops, improved crop rotation, terraces and water conserving irrigation methods. Their primary crops are beef cattle, corn, alfalfa, wheat, and milo.

In 1969 after the death of her husband, John Long, Ura returned to the home place where she lived until her death in 1979. The land passed to her children, Vernon and Vivian, who had both been born in the original frame house of their grandparents. Vivian and her husband, John Herrmann, purchased Vernon's share and in 1996 built their home on the 40 acres of the original claim. George Steven and Kathy farm Vivian's 200 acres. Vivian has two children, David and Kelly and five grandchildren who are the fifth and sixth generations to enjoy the land. It has been 130 years since W.H. Dedrick purchased the original parcel. Thus far, four generations of the family have nurtured and respected their heritage as will those who follow. *Submitted by Vivian Herrmann.*

DeGEER FAMILY RANCH – The story of the DeGeer Ranch begins in the 1850s when James Winterstein DeGeer and his brother, Lyman Winterstein DeGeer, restless and determined to make their marks, left the Ontario, Canada home their grandfather established in 1801 and headed across the border to the US. First attending college in Michigan, the two arrived in Cameron, MO. There they became smitten with two sisters, Emily Jane and Mary Ellen Renfrew. James married Emily and Lyman married Mary and the brothers soon had families to support. Seeking greater opportunity, the brothers decided to move their respective broods to Wyoming, where they would raise sheep. On the way to Wyoming, they stopped in Coffeyville, KS to visit their older brother, John Winterstein DeGeer. Continuing on, they chanced upon Barber County, KS where land was available for homesteading. They decided there was no need to go to Wyoming, that this was the perfect place to build a life.

The area they chose had been part of the fabled Comanche Pool and was now open for settlement. Their claims were in the southwestern part of the county, where neighbors were few and far between. In the spring of 1886, they began farming the red clay by Mule Creek, building two dugout homes and bringing their families from Missouri. By 1893, when the Cherokee Strip opened, they were well established with stick

J.W. DeGeer homestead, 1895

Roundup on the DeGeer Ranch

Vaughn and Hazel on their way to church

built homes, corrals and outbuildings. Lyman, by then the Barber County District Attorney (he won on the Populist ticket), decided to move his family to Alva, OK, making "The Run" to claim city lots. James stayed in the Deerhead Community, buying acreage from homesteaders anxious to move on, gradually accumulating a sizable property.

In the early 1900s, anxious to see more of the country, James and Emily decided to continue west, eventually settling in Chula Vista, CA with their daughter, Eva. Their son, Vaughn Eugene DeGeer, became the next steward of the ranch. Vaughn, his wife Hazel and their two sons, Vaughn Eugene Jr. "Gene" and Myron Winterstein DeGeer worked hard, raising pedigreed Shorthorn cattle, wheat and chickens. The chickens helped finance the boys 'college during the Depression. After receiving his Master's Degree in Engineering from Kansas State University, managing a CCC camp in Wyoming and serving in the Army during WW II, Gene returned to the ranch with his wife, Mary, where the couple raised their five children. The Depression years and the accompanying drought were difficult and Gene became keenly interested in soil conservation and best practices for land management. The ideas in his K-State Master's Thesis, *A Study of Terracing as a Method of Erosion Control*, formed the backbone of his farming methods and he was actively involved in efforts to protect and improve the land.

This interest in conservation and planned management of the environment continues today with Gene's family. In 1990 after Gene's death, the ranch became the DeGeer Ranch, LLC. His four surviving children oversee those same acres their Great-Grandfather first claimed, but in a much more complex and hostile world of economic risk and regulation, coupled with age-old adversaries such as weather and natural disaster. In 2016 the Anderson Creek Fire swept through the ranch and burned all but 20 acres, consuming fences, animals and the heritage buildings constructed so long ago by James. Looking at the ranch headquarters, it is not apparent that four generations lived and worked there—all is gone. What remains is the determination to rebuild and the love and appreciation for the preceding generations who chose such a magnificent spot to establish a home and raise a family. *Submitted by Mary J. DeGeer Benefiel.*

DeLONG – Earl DeLong purchased 160 acres in Lyon County, Fremont Township, Section 25, Range 111-12 from his father, Elvin DeLong, around 1911. Over the years, 40 additional acres were purchased, adding to the original farm. Earl DeLong built the house in November of 1914 and over the years added a barn, granary and a chicken house. Earl married Josephine Pearson and they lived on the farm for over

Aerial view of the DeLong farm today

Aerial view of the DeLong ca. 1950s.

50 years raising wheat, kafir, alfalfa and livestock. Earl and Josephine were the parents of three children: Joe, Betty and Don DeLong.

Earl passed away in 1965 and Joe and Don continued the farming and feeding the livestock. Joe DeLong passed away in 1972. Betty (DeLong) Stewart passed away in 1992. Josephine DeLong lived to be 100, spending her life living on the farm. Don and Mary Lou moved to the farm in 1990 and remodeled, adding a family room and garage.

The granary and the chicken house have been replaced by metal buildings and the original big red barn is still in use today. Don and Mary Lou are the parents of four sons: Dean and Nan DeLong, Dave and Julie DeLong, Dale and Bobbi DeLong and Darryl and Brelana DeLong. They also have nine grandchildren. In 2016 and 20117 two of the grandchildren married, adding to the family. All of the DeLongs have been involved in FFA and Vocational Agriculture and this will continue with the grandsons.

Currently Don and Mary Lou live on the farm. A feedlot has been established by Dave DeLong and is operated by him with his two sons, Dylan and Devin DeLong.

This farm will continue for many years to come. *Submitted by Mary Lou DeLong.*

DEXTER – In 1874 Oscar N. and Celestia Peck Dexter homesteaded the southwest 160 acres of 26-19-10 in Pioneer Township of Rice County. Their daughter, Elma Dexter and her husband, Levi Flora, bought the farm from her father's estate in 1911. In 1949 their sons, Elton, Ora, Irving and Vernon Flora each inherited 40 acres from their mother's estate. That same year Ora and Irving sold their holdings to their cousin, Bertha Dexter and her husband, Fred Hathaway. In 1947 Fred and Bertha Hathaway bought Elton's 40 acres from his daughter, Janette Flora Hume, following the death of her mother, Elizabeth Flora. Fred Hathaway deeded the 120 acres to his son, Ralph Hathaway, in 1967, who in 1980 bought the remaining 40 acres from Vernon's widow, Mabel Flora. This purchase by Ralph, the great-grandson

Oscar and Celestia Dexter with children Edith, Edwin, and Elroy in front of their stone house.

of the original homesteaders, reunited the original 160 acres, which had been divided by his Great-Aunt Elma. Ralph's children - Marlene Hathaway Carson, Martin Hathaway, and Carolyn Hathaway Kern - inherited the quarter from his estate in 2009.

Oscar and Celestia survived the grasshopper plague in 1874 Kansas. In addition to being a farmer, Oscar was a stonemason and built a small sandstone house on his farm. The stone was from a quarry at Raymond six miles away. Oscar also built sandstone basement walls for the house of his neighbors, John and Mary Hathaway, where three decades later his granddaughter, Bertha Dexter, would live after marrying Fred Hathaway. The farm's present owners, Oscar and Celestia's great-great-grandchildren, are in the process of excavating the ruins of his house, which collapsed following a fire several decades ago that destroyed the roof and floor joists. *Submitted by Carolyn Hathaway Kern.*

DILLER – On June 8, 1872 Doratha Diller acquired the760 acres located in Butler County, Sycamore Township, Section 36, Range 6 and Section 2, Range 6. Corn and wheat were crops raised on the farm to support cattle and pigs.

This farm has remained in the Diller family since that time, being passed down through the generations. The farm passed from Doratha to her son, John III. Then the farm passed to John III's children – Lula, Anna and Leonard. Leonard's son, Robert, was the next owner, finally passing to Roy L. Diller, Roberta (Diller) Rose and Zita (Diller) Wallace on Nov. 13, 2009.

The farm still retains 360 acres that Roy L. leases for pasture, corn, wheat and soybeans.

Roy and Jackie Sue have four children – April, Jessica, Brian and Marc and seven grandchildren. *Submitted by Roy L. Diller.*

DIRKS – My great-grandpa, Gerherd Dirks, was born in Poland, Russia and moved to the US with his parents and settled in Pawnee Rock, Kansas in 1874.

In 1908 Gerhard and Julia Dirks purchased the West Half of Section 34, in Township 28 South, of Range 19 West of the Sixth Principal Meriden for $7,000.00.

Harvesting on Gerhard Dirks' farm

Lincoln and Leah Dirks farm

My grandpa, Lincoln Dirks, first son of Gerhard and Julia Dirks, batched in a four-room house one and one half miles south of Joy and farmed the land. At age 18, Lincoln owned his first horse and buggy, he was very fond of his horse. In 1916 he purchased a new model T Ford for $397.00.

In 1917 Lincoln married Leah Peaster. That same year Gerherd Dirks gave the land to Lincoln and his siblings. Harvesting was done in two operations, cutting and stacking and then threshing. The first harvest of 1,000 bushels sold for $1.00 per bushel and barely covered cost of cutting and stacking. Lincoln sold the car to cover threshing expenses. The second crop cleared $150.

In 1926, Lincoln bought his first combine, a Case and a Twin City tractor. In 1934 Lincoln and Leah bought the half section from his sibling for $12,000.

The half section was given to Lois Dirks Koehn and her siblings in 1975. In 1987, Lois Koehn bought the half section from her siblings.

The farmstead was sold to Mike and Jan Unruh in 1989.

In 2010, the half section was purchased by Paul and Shirley Koehn Unruh from her sister, Beverly Koehn Seyfert.

The farms crops are wheat, milo, corn, and soybeans. *Submitted by Paul and Shirley Unruh.*

DIXON – Thomas Dixon, born Nov. 5, 1846 in Silsden, England, arrived in North America via the St. Lawrence Seaway in October 1869. On Oct. 26, 1907 he and Sarah C. purchased 80 acres located in the east 1/2, northeast 1/4, 11-27-6 in Little Walnut Township, Butler County, KS. Purchase price was $3,450. Corn and cattle were raised on the farm.

Charles R. Dixon is the third generation to own these 80 acres. The farm passed from Thomas and Sarah C. to their son and wife, Charles T. and Mary E. to Charles R. on Aug. 6, 1997. The farm now produces cattle, brome and prairie hay.

Historic buildings include the Quito School House, built in 1872 and the farmhouse, built in 1907, is still occupied.

Many registered Hazlett Herefords from this farm have been sold to various regions in the US, sometimes sight unseen. In 1986, the farm was awarded the Kansas Bankers Award for soil conservation.

Charles is still ranching, now owning an additional 800 acres and renting 320 acres. *Submitted by Charles R. Dixon.*

DONAHY – On June 18, 1884 80 acres owned by W.R. and Helen Stebbens was purchased in Clear Fork Township, Marshall County, KS by John M. Dowd. The legal description was the south 1/2 of the NW 1/4 of Section 26, Township 5, Range 9 East. Oats, milo, corn, soybeans, hay, alfalfa and prairie hay were the grain crops grown.

Upon the death of John M. Dowd, the farm passed to his wife, Mary E. Dowd. The farm then passed to Michael J. Boyle, uncle of Mary E. and Dan Donahy. The farm then passed to Gerald "Jerry" and Irene L. Donahy and Dec. 15, 2001 to widow Irene. The farm is now in the Gerald and Irene Donahy Trust.

Gerald Joseph Donahy and Irene Hochard married June 16, 1951 and celebrated their 50th anniversary before Jerry passed away Dec. 15, 2001.

Over the 50 years they raised chickens and hogs, fed out fat cattle and had a herd of cattle and pet dog. They raised wheat, oats, corn, milo, soybeans, prairie hay and alfalfa. Irene now raises corn and soybeans on the 80 acres. *Submitted by Irene L. Donahy.*

DOYEN – Gerard and Marie Theresa L'Heureux Doyen came to the United States by ship from Seraing, Belgium, in 1869. They settled first in Braidwood, Illinois, with two small children. After losing their daughter to diphtheria, they moved to Rice, Kansas, in 1877 and purchased 160 acres in Cloud County for $420 from William C. Bramwell. Money to pay for the farm came from their relatives in Belgium. They had six more children and four of the eight siblings formed a Doyen Brothers farming partnership. Simon Anthony's (and Andelina Robin Doyen's) son Orville also joined the partnership. Orville purchased his first row crop DC Case tractor in 1941. Prior to that, he farmed with horses and mules. Eventually Orville's (and Millie Derby Doyen's) sons Ray and Ross Doyen worked the farm and sons Lee and Mark became teachers. Orville and Millie were both active in 4-H and received state awards for soil conservation. Orville enjoyed working a large vegetable and flower garden. Millie was active in Farm Bureau and served as safety chair. She was also a wonderful cook and baker. They fed out about 200 head of cattle and hogs each year. Orville had a 200 hen laying flock at one time. Today the Doyen farm is still owned and managed by the surviving family members of Lee, Ray, and Ross Doyen.

Orville and Millie Doyen along with sons Lee, Ray, Ross and Mark in front of the home that was built by Simon Anthony and Andelina Doyen, which still stands today.

DUNBAR – The Dunbar Century Farm began with a purchase of 80 acres in Franklin County, KS in Section 28, Township 18, Range 20E on Nov. 26, 1908 by Charles Atron Dunbar. While Charles was a boy, his father, Leander, had rented property across the road and a bit east. Leander had walked to Lane, KS from North Carolina during the Civil War so as not to be conscripted into the southern army. Leander was a Quaker and didn't believe in war. His grandfather had come from Scotland in the 1700s. After joining a sister's family in Lane, Leander

The Dunbar family

worked for a time carrying mail on horseback from Osawatomie to Leroy. Once he was surprised and surrounded by Indians while carrying the mail, but they let him continue on his way.

As a young man, Charles taught in rural schools in the county and also worked out of St. Joseph, MO as a railway mail clerk. But farming was his true love and he and his wife, Irene, moved to the farm fulltime in 1918. Besides farming, he had dairy cows, hogs and raised chickens. Farming was done with horses.

Following Charles Atron, the next owner was his son, Charles Albert and wife, Dorothy, who inherited the property when Irene passed. Their five sons were kept busy with the farming and raising hogs, cattle and dairying. One of the sons, Ronald and his wife, Elaine, are the current owners, purchasing the property in 2010. Their three children are Janna, Aaron (with wife, Cassie and children, Grayson and Winnie) and Ashley Hayden (with husband, Patrick and children, Pippa and Matilda). Aaron is the fifth generation to farm the property. The Dunbars have cow-calf and feeder cattle enterprises and raise feed grains, wheat and soybeans. The farm has received conservation and water quality awards.

DUNN – The Dunn farm has been heavily owned and operated by women over the past century. Duanna Dunn originally bought this 160 acres located in Hayes Township of Franklin County, Section 27:15:20, on September 7, 1907. With her family, Duanna raised sheep, pigs, cattle, corn and wheat. Dairy cattle were even a part of the operation in later years.

Robin Dunn

Duanna passed ownership to her son and his wife, Perry and Verna Dunn. Since 1990, the property now known as Dunn's Landing has been under the ownership and management of Duanna's great granddaughter, Robin Dunn. Thanks to blood, sweat, and John Deere tractors, Robin farms not only this 160-acre plat, but several hundred acres more. At Dunn's Landing, one can find corn, beans, oats, alfalfa, wheat, black Angus cattle, Percheron draft horses, golden retrievers, chickens, and the occasional donkey.

In addition to being a working farm, Dunn's Landing has also served as a draft horse breeding facility and currently operates as a wedding and event venue. With the exception of the 1800s barn that serves as the focal point of Dunn's Landing wedding receptions, every other structure has been built by a member of the Dunn family...every shed, stable, coop, and even of the house.

The efforts of the Dunn family have not gone unnoticed. Robin has been the recipient of a Kansas Bankers Conservation Award, as well as named one of Franklin County Farm Bureau's Farm Families of the Year. Beginning with Duanna over 110 years ago and continuing with Robin today, the Dunn family is extremely proud of the female-led legacy of their Century Farm in northeast Franklin County, Kansas."

EASTON – The Easton Kansas Century Farm in Leavenworth County is currently owned by Brian and Renee Potter. In 1907 Brian's great grandparents, Frederick John and Florentine Louise (Knollman) Timpe, purchased 200 acres in the northwest corner of the county. Approximately five acres were deeded to St. John's Lutheran Church to construct a church/school building and one acre to establish St. John's Lutheran Cemetery. Five generations of the family have been married in this church. Road frontage left 193 acres. The original house was razed and a new house built in 1912 and still remains today as the home of the Potters.

Other owners include: (1936) Frederick's daughter, Edna and her husband, Alfred Kruse, (1980) Alfred and Edna's daughter, LaVerne and her husband, Austin Potter, together with son, Brian and his wife, Renee Potter, (1983 to present) Brian and Renee Potter. Three generations of daughters/granddaughters slept in the same bedroom over a span of 100 years. In 2003 Brian and Renee's son, Jeff and his family built a house on the farm.

Frederick and Florentine Timpe, original owners of the farm, 1912

The Brian Potter family

Over the years, the farm grew corn, soybeans, wheat, milo, alfalfa, hay, gardens and raised cattle, hogs and chickens. Today the Potters grow mainly corn, soybeans, hay, and raise cattle. The first and second generations sold eggs to the once-thriving town of Potter. Caring for the land has been an important aspect of ownership for this family. Alfred Kruse received the Leavenworth County Soil Conservation Award in the 1950s and Brian and Renee Potter received the same award in the 1990s. Faith and family have always been important to the generations of this family and still is today.

The Brian and Renee Potter family includes two children (spouses) and seven grandchildren: Daughter, Amy Schwinn (husband Randy and children Elijah, Emma and Wyatt) and son, Jeff (wife Amy and children Colten, Caleb, Ava and Cooper). All live within five miles of the Century Farm.

ELLIS CENTURY FARM – A covered wagon carrying the Ellis family and their possessions arrived in Comanche County, KS September 1909 from Pawnee County. They had purchased a quarter of land from J.F. Richards, Kansas City. As they topped the hill, one half mile east of the headquarters, Roy looked across the valley, stopped the horses and said to his wife, Myrtle, "This will be the Valley View Ranch."

Pictured l-r: Roy Ellis, Beulah, and Myrtle, with Harold in front.

Roy Ellis married Myrtle Deyoe in 1906. Two children, Beulah Mae and Harold W., were born while living in Pawnee County.

In the Spring of 1909, Roy had made his way to Comanche County. His dream: a home in the valley with streams of water and Hereford cattle. Midday he stopped at the home of Walter McKinney on the Kiowa/Comanche County line. 22 years later Roy's son, Harold, married the McKinney's daughter, Naomi.

Roy and Myrtle's legacy continues to serve the community they loved. Their faith lives on in the church they served.

Roy was on the building committee for the Coldwater Methodist Church. Some family members living in Coldwater worship there today.

He had a helping hand in organizing the Comanche County Farm Bureau. Roy served as President for several years.

In 1933, the Farm Credit Act was passed and the Farm Credit System became active. A group of farmers and ranchers from six Kansas counties gathered in Greensburg to organize the Production Credit Association. Roy served as a Director representing Comanche County.

Farm Credit assisted our family over the years in building the operation to include 11,000 acres of pasture and crop land today.

Myrtle's life was built around her faith, family, gardening, Eastern Star, Extension work and helping care for the animals. "Garden dinners" were fondest memories of the grandchildren. *Submitted by Judith Kerstetter.*

ELLISON – The Ellison farm is located in extreme Southeast Kansas between Mound Valley and Cherryvale at 552 17500 Rd. The farm rests on a rise allowing a scenic view of the valley to the west. Thus came the title of Valley View, which was proudly lettered for many years on the front of the large white barn, which was the focal point of the farmstead. The barn remains today though the lettering has faded away.

Painting of the Ellison farm, 1977.

The farm was originally homesteaded in the spring of 1871 by Isaac D. Ellison, who was reared in Marlboro, OH and his wife, Sarah, at a cost of $400 for the 150-acre plot. Isaac became a highly respected farmer in the area and lived on the farm until his death in 1907. At that time the farm turned to Isaac's son, Walter O. Ellison and wife, Cynthia, who continued the farming operation until the most current family members, Elwyn and Mable Ellison, purchased the farm in 1948 and built the current house in the photo in 1951.

They continued the family farm and dairy, working side by side, for nearly 50 years. The work was hard but they always seemed to make time for fun and fellowship as the nearby two-room school house acted as a gathering place for church services and community box suppers. Both proved to be very talented craftsmen in woodworking and quilting.

Elwyn passed away in 1992 and Mable remained on the farm until her health caused her to move to Oswego the following year to be near family. During the time Elwyn and Mable lived on the farm, they enjoyed hosting several family reunions featuring hayrides and wiener roasts and included several descendants of Isaac and Sarah reminiscing about the happy years spent on the farm. *Submitted by Peggy Strickland, daughter of Elwyn and Mable.*

ELSTROM/ECLOUD FARMS – Gustaf passed away in 1880, just 11 years after settling. Two generations resided on the farm in two separate houses, approx. 1885 to 1985.

The Elstrom family

In 1943, during World War II, Ted and Emil hired German prisoners from the POW camp north of Concordia to build a cement water tower on the farm - still standing. Bertha always prepared a hearty meal for the prisoners, who were accompanied by an armed guard. The guard told Ted, in private, that there were no bullets in his gun.

On June 9, 1967 the Elstrom farm was hit by tornado. Both houses were spared, but the barn was leveled except for the oldest section, made of stone.

Along with standard crops wheat/milo/soybeans, Ted and Bertha grew fields of hybrid iris and sold the bulbs wholesale in Salina for several years. Concordia residents used to visit in the spring to see the iris; and in later years, to see their peonies.

Bertha and Ted began raising dogs in 1965. They started with Pekingese and expanded to other breeds. Dogs were raised until 1996.

Interestingly, ownership of this farm skipped a generation - Steve's parents Allan and Janice and Angela's parents Barbara and Howard. Allan owned nearby land and lived on this farm for many years.

General Information: In 2007, Steve Elstrom and his wife, April, moved on to the farm raising their nine children on the homestead.

ENGELLAND – Jurgen Engelland and brother, David, came to Kansas in 1871. They acquired land in Rice County, KS, some by homesteading and some was "timber claim" land. The US government encouraged settlement of Kansas by giving a quarter section of land to an individual who would make improvements on the land, live and farm it for a period of time. After the agreements had been fulfilled, the settler was given title to the land. It was in this way that Jurgen Engelland became the owner of the south 80 of the Engelland home place. Jurgen then purchased the 80 on the north of the township road, the land on which the house and buildings are located.

Jurgen and Anna (Johannsen) Engelland were married and lived on the home farm all of their married life. Their eight children were

Eva Louise, Anna, George Albert (on Anna's lap), Jacob Anthony, Jurgen, and John Frederick, with the original home and barn in the background

Rex, Julie and Kirk on the family homestead

born in the farmhouse. The two 80s then were owned by Roy Engelland, youngest son of Jurgen and Anna. Ruth Engelland Cooper, daughter of Roy, later inherited the land. Rex and Julie purchased the land and original home (built around 1880) from Dale and Ruth Cooper in 1985. Jacob Anthony "Tony," oldest son of Jurgen and Anna, was Rex's great-grandfather. Rex and Julie have a son, Kirk, who is the 6th generation to farm this land.

Today the Engelland family, Rex, Julie and Kirk, reside on the family homestead land in a home that was constructed by Rex in 2016. The new house is northeast of the old homestead house site in Wilson Township/Rice County (33-20-7) at 2185 Ave. S., Sterling, KS.

In addition to this land, Rex and Kirk farm 2000 acres, growing alfalfa, corn, milo, soybeans and wheat.

ESTES – Caleb and Mary (Moody) Estes originally staked a claim to a farm at Bellville, KS. While they were at Ft. Riley buying cattle, they received word that their claim had been jumped and decided to stay in the Ft. Riley area for protection from claim jumpers and Indians. In 1866 they homestead a 160 acre farm in 7 Mile Township which became known as Estes Gate. Vegetables, apples, grain and hay were raised. Livestock was cattle and horses. Caleb was postmaster of 7 Mile Township and raised horses he sold to the US Army. He met Edward Bevins Huston, a blacksmith in the 7th Cavalry, who took Caleb home and the rest is history. Caleb became a highly respected and wealthy member of the community, with a large house and approximately 1,000 acres of farmland.

At least one of Caleb's brothers, Larkin, followed him to Kansas as well as his father-in-law and several sisters and brothers-in-law. Caleb fought in a battle at Gettysburg against his daughter's future father-in-law, John Huston.

The farm passed from the Estes family to the Huston family and then to the Deatherage family.

The farm has retained 338 acres, which produces hay. A stone barn and the homestead house remain, although both are in need of repair. *Submitted by Patricia H. Deatherage.*

EWBANK – John and Ann Ewbank, with family of 12 children, came to America in 1807 from the Parish of Ampleforth in Yorkshire, England and to the Territory of Indiana in 1811 from the state of New Jersey.

A son, John Smith Ewbank, hearing of opportunities West, moved to Kansas. He married Marie Antoinette Reed. He served in the Civil War, Co. E, 2nd Kansas Cavalry. They acquired land by Homestead in 1885, 160 acres in SW/4 11-27-14, Ninnescah Township, Pratt County, KS. Deed recorded Jan. 27, 1886 for $200 from US of America by President Benjamin Harrison. Approved Seal April 25, 1887, a Patent in 1889.

A "dug out" house was where the family with 10 children called home. Wheat and milo were grown. Milk cows, hogs and a garden put meals on the table. A stone building with running water from the windmill kept milk and butter cool. This building is still standing.

Sons, Oliver and Alvin, maintained a Hog Business, butchering and curing pork at the farmstead. They purchased a small grocery and during the depression kept many families supplied with pork and groceries when they could not pay.

Ownership history includes: Son, Oliver Morton Ewbank; Son, Alvin Marquis Ewbank; Daughter, Mina Ewbank, Herbert Eugene Huffman, grandson of John and Marie Ewbank, Currently Dorothy Huffman, wife of Herbert Huffman, trustee of H and D Trust. The Trust includes as co-owners their daughters, Dee Marie Stewart, Laree Hiser, Lynnea Johnson, great granddaughters of John and Marie. Six great, great, grandchildren come often to visit and observe planting and harvesting of corn, soybeans or wheat. Family ties are strong with church and family celebrations. There is a long family history in agriculture and a deep appreciation for the value of rural life, church and friends.

EWERT – Heinrich Ewert emigrated to the US from Germany in 1880. He arrived in Leavenworth County, KS that same year. It's not clear how the location was selected since no other family was in the US or ever followed him. Upon his arrival, Heinrich took residence in a boarding house in the city of Leavenworth. He tried several jobs including that of a coal miner, but soon turned to his love of the land and outdoors.

In 1884 Heinrich moved to the Mount Olivet Community six miles west of Leavenworth and began his lifelong career as a tiller of the soil. In 1885 he married Ottilia Schroeder and the couple began acquiring their own land in 1888. Much of the land was cleared by hand and they originally lived in a small existing house on the property. A new two-story house, large barn and numerous outbuildings were built around the turn of the century. The couple had nine children, five girls and four boys. All worked on the farm at some time of their lives. Major crops included wheat and apples. At the time of Heinrich's death in 1918 the family farm consisted of 385 acres.

Following Heinrich's death and the death of Ottilia in 1919, the farm was eventually divided between two of the sons, Fred and Herman. With Fred holding 120 acres and Herman 265.

Herman married Caroline Kunz in 1940. The couple raised beef cattle, sheep, hogs and maintained a large egg production. They sup-

Heinrich and Ottilia Ewert and four of their children

The Ewert family, 2017

plied eggs to several Leavenworth locations, including a hospital, for over 30 years. Herman and Caroline had one son, Ronald. Upon Herman's death in 1958 the farm management and operation was taken over by Ronald.

In 1963 Ronald married Kay Jones and they were blessed with five children. The 120 acres once owned by Fred changed hands several times following his death and was purchased back by Ronald and Kay in 2013, which brought the original 385 acres together again. The Ewert's presently own and operate a 600 acre cattle and cropland operation under the name Diamond E Farms.

The house and farmstead built by Heinrich and Ottilia at the end of the 19th and into the 20th centuries continues to provide a happy and satisfying life for their descendants. The farm echoes years of family gatherings, festive holidays and frequent reunions. *Submitted by Ronald and Kay Ewert.*

FALER – When I think of the Faler homestead, I think of the stories that have been told and retold. Buffalo wallows, wagon trains coming and going, Indians coming to pay respect to a "Chief" long gone and during the heat of the summer, the towns' people coming up and setting up camps for the cool breezes that swept up the hill. Everyone in the town of Elk City knew about "initial rock". A place for young lovers to carve their names to document their existence for prosperity. During my childhood the "hill" was a place to explore. The old orchards, the "smoke house" and the ravines seemed to me to hold ancient mysteries. Christmas brought my family and many town families out to find just the right "tree." The ponds were a place of summer fun when my Grandfather took me fishing.

For many years, Faler homestead was known as the Faler dairy. The milk room always had cold water and milk, cream and of course butter in different stages. The barn was made out of stones that had been quarried on the hill. The homestead was a free range for turkeys, chickens, and guineas. There were fox, deer and all kind of wild life. Trees to cut down, rocks to quarry for the town. People made use of the sandstones rocks lying on the hill to build houses and such. Another memory is the times that the townspeople would all come out to help save the "hill" on the numerous times it would catch fire. My Father's stewardship saw the dairy ending and the beginning world class Hereford cattle. One of my favorite memories was either flying or four wheeling over the farm checking the cattle. The homestead served not only the Faler family but also the town in many, many ways. *Written by Terry Faler Hogan.*

FANKHAUSER – Now owned by Larry and Mary Fankhauser. The location of the land is at 3196 BB50

Rd., Madison, Ks. The legal description is NE. 1/4, SE- 1/4 and all that part of NE 1/4 lying west of the Verdigris River and NW. 1/4, SE. ¼ lying west of the Verdigris River. All in 2-23-12.

On May 9, 1891, Larry's great-grandfather (maternal) bought the land from John and Mary Gilman. He then died Jan. 1, 1909 and Larry's grandmother inherited her share of the farm. Then Emma and Fred Fankhauser (Larry's grandparents) bought out the remaining heirs' interest. In time Grandpa Fankhauser bought the land adjourning south of the original farm. They moved into the original home. It was the first frame house built in Shellrock Township.

In 1915-1916 Fred and Emma Fankhauser built a new home. They had their first home, that they were living in at the time, moved off of the foundation about 10 feet north of where it had sat. Then they built their new home on the old foundation.

Grandpa and Grandma had three sons - Edward, Chester and Fred. Edward lived in the home after his parents passed. After Edward passed, Chester's son, Larry Fankhauser and wife Mary, bought the home. We have lived in the family home for 46 years already.

We have two children, Terry and Shari. Terry and wife, Hidi, have three children - Emma, Tess and Grant. Shari has two girls, Shelby and Tori. Shari and husband, Robert, live outside of Eureka, KS. Terry and family live in Colorado. Terry's plans are to move home and be the next generation of Fankhausers to run the ranch.

No cash crops are raised now. Just hay is grown for the cattle operation. We love the land and are continuing the cattle tradition. We named our ranch Rimrock Ranch. We were named a century farm in 2016. *Submitted by Larry and Mary Fankhauser.*

FARRIS – Ronald G. and Margaret E. Farris are the fourth generation to own the farm purchased by his great-grandfather M.T. Farris in 1858. M.T. purchased 640 acres located in Jefferson County, Jefferson Township, 33-20S-20E where he raised livestock.

The farm then passed to E.S. Farris (grandfather), to E.E Farris (father) to Ronald in 2009. Retaining 160 acres the farm now produces soybeans, corn and cattle.

During the 1950s, the house was moved from 1-1/2 miles north. E.S. Farris raised mules and horses on this century farm.

Ronald and Margaret have four children: Tim, Richard, Cynthia and Melinda. Their four grandchildren are Samantha, Ellena, Wyatt and Donovan. *Submitted by Ronald Farris.*

FARVER FARM #1 – Jacob "Jake" Farver was born in Londonderry, Dauphin County, PA on July 5, 1866. On Oct. 1, 1901 Jacob "Jack" Farver acquired 160 acres located in Southeast Quarter of Section 16, Township 20 S. Range 1 West of 6th Principal Meridian, Canton, McPherson County, KS. Wheat, soybeans, milo, corn and hay were grain crops grown and livestock was raised here.

The farm passed from Jacob to his nieces and nephews: Abner and Ruth Farver, Newton Farver, Lillie Farver and Ida Farver Peed. Then it passed to John H. and Cora Farver, and in 1992 to Ione Janssen and Adrian Farver. In 2001 at the death of Ione, her portion passed to her children.

Richard Vogts, Canton, has farmed Jake's farmland since he returned from the Korean War, over half a century ago.

The 160-acre farm still produces wheat, milo, corn, soybeans, cows, pigs, chickens and hay. None of the historic buildings are functional in 2017.

In 2009 the Century Farm Award was presented to Jonell and Jerald Schenk. *Submitted by Jonell Schenk.*

Original Fankhauser home beside the one the families lives in today.

Left to right, back row: Ed Farver, Fred Farver, Newton Farver, Jacob Farver, General Harrison Farver. Front row: Dr. Ida Farver Peed, Abner Farver, John M. Farver, Lillian Farver.

FARVER FARM #2 – Newton Farver was born Aug. 2, 1879 in Canton, McPherson County, KS. In 1907, Newton heard of the opportunity to "prove up" a homestead and headed west to Stanton County, KS. Arriving there, he built a dugout shack for protection from the winter cold. Newton's 160 acres were located in the Southeast Quarter of Section 31, Township 27 South, Range 41, west of the 6th Principal Meridian. Newton did not raise livestock, only wheat, milo and corn. This homestead was "proved up" on May 24, 1909. Eventually he returned to Canton to live with his brother, Jake, but would travel out to western Kansas to see old friends and check on his property.

Pictured l-r: Dr. Ida (Farver) Peed, Abner Farver, Mae Farver (half-sister to John H. Farver, died at age 17), Lillian Farver, and Newton Farver, May 1904.

This farm passed from Newton to John and Cora Farver, then to Jonell and Ione Farver in 1992. In 2001 at her death, Ione's portion passed to her children – Cynthia, Royce and David Janssen.

There are no historic buildings on the 160 acres but there once was a dugout there. The farm still does not produce livestock, only grain crops of wheat, corn and milo.

In 2009, the Century Farm Award was issued to Jonell and Jerald Schenk. *Submitted by Jonell Schenk.*

FARVER FARM #3 – Lillian Farver was born in Canton, McPherson County, KS. In 1907 her brother, Newton heard of the opportunity to "prove up" a homestead and they both headed west in a covered wagon to Stanton County, KS. Arriving there, they each built a dugout shack for protection from the winter cold. Lillian's 160 acres were located in the Southeast Quarter of Section 35, Township 27 South, Range 42, west of the 6th Principal Meridian. Only grain crops of wheat, milo and corn were raised. This homestead was "proved up" on Sept. 16, 1909. Eventually she returned to Canton to live with her brother, Jake, but would travel out to western Kansas to see old friends and check on her property.

Jonell and Jerald Schenk and family

This farm passed from Lillian to her sister, Ida Peed. Then it passed to her nephew, John and then to Cora Farver. In 1992 the farm passed to Ione Janssen, Jonell Schenk and Adrian Farver. In 2001 at her death, Ione's portion passed to her children – Cynthia, Royce and David Janssen.

There are no historic buildings on the 160 acres and the dugout that was there is gone. The farm still produces only grain crops of wheat, milo and corn.

In 2009, the Century Farm Award was issued to Jonell and Jerald Schenk. *Submitted by Jonell Schenk.*

FASSE – Harvey and Nancy Jo Fasse live south of Effingham in Atchison County. The Fasses are only the second family to live in the century old farmhouse built in 1907 by Nancy's grandparents, Ed and Ida Cook. Grandpa Cook planted a windbreak before the house was built and plans for the house were finalized while Grandma Cook was completing her education at Kansas State College, Manhattan.

Harvey and Nancy Jo purchased the homestead in 1973 from her mother, Bertha Cook Hamon and did extensive remodeling—starting with a new foundation and basement. They added and remodeled rooms and did the wiring, plumbing and carpentry work themselves. In 2002, their home was damaged by a tornado and a new roof, windows and many trees were replaced.

Harvey and Nancy Jo Fasse

Their family farm corporation, Hamon Seed Farms, Inc., is owned and managed by Harvey and Nancy Jo Fasse and her brother and sister-in-law, Charles E. and Mary Jo Hamon. They are proud to have continued the growth of Hamon Seed Farms started by Nancy's parents, John and Bertha Hamon, in 1946. The Fasses were honored to be selected as Kansas Master Farmer and Master Homemaker in 2006.

Harvey and Nancy Jo have been very active in their local church, Farm Bureau, 4-H, Extension and community clubs. The Fasse family includes three children: Brenda Jo Davids, Kaye Lynn Kabus and James Fasse. They were the fourth generation to attend KSU, following in the footsteps of Cook-Hamon tradition of purple blood. Now the Fasse grandchildren are attending KSU.

FAWN CREEK FARM – In 1879, Robert Ellison purchased 160 acres in Montgomery County near Coffeyville, Fawn Creek Farm Township, and raised wheat, corn, oats, horses, cattle, sheep, and hogs. The farm initially included a stone house that had served as a winery. A barn was built in 1889, and then a house in 1899. This house, with numerous additions over the years, still serves as the homestead.

Entrance to the Ellison farm

The farm passed to Robert's son, John Ellison, who consequently was the citizen who brought the dead body of Dick Broadwell (one of the Dalton Gang) back into Coffeyville after the infamous Dalton raid in 1892.

John's son, Orrin, farmed the land and increased its size to over 400 acres in his lifetime. He raised a large family, as well as wheat, corn, oats, horses, cattle, sheep, and hogs. Irma, Orrin's wife, raised chickens and sold eggs.

Kenneth, Orrin's son, continued to farm and expanded the property to over 600 acres. The farm has been a gathering place for many generations of Ellisons and holds a special place in the heart of all that have visited, whether family or friend. The present owners, Kenneth's sister Reba Ellison Miller, and her children, Mary Jane Miller Futch (Donald), David Leslie Miller, Nancy Miller Linder (Scott), and Jeffrey

The Ellison homestead

Ellison Miller (Mary Jo) bought 275 acres, including the homestead in 1999, and continue to work and improve the property, enjoying land and making memories. *Courtesy of Jim Miller*

FLORELL FARM – The Florell Century Farm in Republic County, Kansas is currently owned by three cousins: Barbara Palecek, Janice (Florell) McCaulley and Victoria (Florell) Nylund, great-granddaughters of Dayton Woods. Janice's husband is Roger McCaulley and Victoria's husband is James Nylund.

Aerial view of the Florell farm

The legal description is: the North half of the Northeast quarter (N1/2 NE1/4) of Section 36, Township 3 South, of Range 5 West of the 6th P.M. containing 80 acres in Republic County, KS.

Peter Peterson was the first owner of this farm. Dayton Woods purchased the farm on March 31, 1906.

Dayton Woods gave each of his four daughters and one son land. The 80 acres given to Belle (Woods) Florell, our grandmother, was deeded to her children and now to their children. Belle and Conrad Florell's children were Frances (Florell) Palecek, Ethel Mae Florell, Loren (Maxine) Florell, Marvin (Martha) Florell.

Dayton Woods lived in the Smith County homestead prior to the death of his father, John Sears Woods Jr., who was sitting in the door of the dugout peeling potatoes when he was struck by a bolt of lightning and killed in 1875. At that time Dayton Woods moved to the homestead across the road from the Florell farm. His father, John Sears Woods Jr., had homesteaded in both Smith and Republic Counties.

The original crops grown were wheat and corn. Chickens and beef cattle were raised along with having a large garden. Most of the food you had for a meal came from what you grew or raised. Crops raised today are corn, popcorn, and milo.

The original homestead has been torn down. James Nylund, Victoria's husband, currently farms the land.

We have many memories of the celebrations that were hosted on the farm by our grandparents. *Submitted by Janice Florell McCaulley.*

FORSSBERG – Nels Peter Forssberg was born Nov. 22, 1854 in Stockholm, Sweden. In 1882 Nels sailed on *Romeo* to Pennsylvania, immigrating to Illinois. Jan. 5, 1885 Nels homesteaded our farm in Phillips County, KS. In 1868 he confirmed his Lutheran Faith. In 1893 Nels became a legal citizen of United States of America.

Nels lived in a dugout. He also had a dugout barn, a homestead rarity, for his horses and cattle.

On Oct. 5, 1890 Nels married Barbara Magdalena Veeh. They built a small frame home. In 1907 Nels built our elegant nine room, two story home, using only a hammer, hatchet and hand saw. The "House of the Seven Gables" was a landmark to travelers traveling across Kansas, telling them their location. A tornado on July 5, 1924 destroyed all the buildings except the house. Tornado winds drove straw into the north side of the house. Cattle were found with 2 by 4s sticking out their sides.

The Forssberg home

Nels started our corn, wheat, sorghum, soybean, alfalfa and cow-calf operation. Nels brought one Dutch-belted cow, two chickens and a rooster with him to America. The original Dutch-belt blood has been retained in our cattle herd, also the Roan blood line from Nels' son, Gustavus and wife, Mary Vogel's, milk cows.

Our Kansas Farm Bureau Century Farm is a Center for Historical Research "Kansas Tradition Farm," registered with the Kansas State Board of Agriculture and home of the Classic Big Red Barn. Keith Eugene, son of Gustavus and Mary and his wife, Patricia Ann Scollin and their children - Kevin, Susan and Steven - were privileged to acquire the homestead when Keith's parents retired on Feb. l, 1972. Fifth generation members are Logan, Faith, Maia, and Miles. Keith and Patricia are still honored to live in our landmark home. Our family re-shingled and painted our home in 2008. Our son, Kevin, is currently assuming management of the farm. *Submitted by Keith Forssberg.*

FORSYTH LAND & CATTLE – Forsyth Land & Cattle operates the original Rau Family Farmstead that was established July 31, 1900 when Karl & Barbara Rau purchased the north half section 24-10-2, Athelstane Township, Clay County, Kansas. Three hundred and twenty acres were purchased for $3,600. Karl, born in Germany in 1857, came to the United States in 1893. Barbara Sender was born in Pennsylvania in 1864. They were married 1895. Their son Paul was born on October 23, 1902. Paul was a blacksmith and a machinist. The shop, which still stands at the farmstead, had a forge, trip hammer, drill press and a pit in the floor. Paul was deeded the farm from his parents on June 14, 1926. Paul married Elizabeth McPheeters on June 28, 1928. They had four children, Imogene, Pauline, Karl and Roy. 4-H was important on this family farm. They raised sheep, chickens, pigs and Registered Horned Hereford Cattle. Some original seedstock Herefords were purchased from CK Ranch at Brookville, Kansas. The family raised bulls for sale, which were halter broke in the old rock barn on the farm. Imogene married John Forsyth on May 1, 1957 and they had three sons, Karl (still born), John W. and Roger. Upon Paul's death in 1967, Imogene, Pauline, Karl and Roy divided the farm equally. Roger married Marla Lefert in 1984 and they had Daniel and Elizabeth. In 2007 the Rau Family Farmstead received the Century Farm Award. Another achievement to the family descendants includes five American FFA Degrees that span three generations. JW, Roger and Marla purchased the farm in 2008, buying their Uncle's remaining share to continue the family farm. The family still operates the original 320 acres as part of the 1,050 acres in Clay, Dickinson and Ottawa counties. They raise cattle, wheat, alfalfa, soybeans and native grass. The 5th generation includes, Danny and Allison, who live on the family farmstead, and Elizabeth and Aaron Shepard.

FOSTER/ALLEN FARM – The Foster/Allen farm is located in section 6 of Fairview Township of Labette County just 2 ½ miles east of Altamont Kansas. Jeanne Carole Allen McKinley's farm has a resilient family history of deep seeded roots in the farming community to this present day. Jeanne remains very active in the daily management and oversight of the farm, not to mention her physical contributions to the up keep of the large garden plot located next to the farmhouse and spraying fence line as needed. The original farmhouse that once was home to Irvin and Mary Foster was replaced around 1928 with the present day farmhouse, which was Jeanne's childhood home.

Hay bales on the Foster/Allen farm

The initial owner in the family farm was Irvin W. and Mary J. (Wall) Foster who purchased the 82 acres in October of 1883. Irvin and Mary Foster engaged in general farming for that era. The citizens in the community commonly referred to this farm as the IW Foster Farm. In the late 1800's the hand-dug water well on the Foster Farm was told to aide the community in a drought. Mr. and Mrs. Foster allowed the neighbors and community members to retrieve water from the well and it is said the well never ran dry during that time. This well is still there to this very day and it's well marked by the windmill, which towers above it. IW Foster and Mary J. Foster were Jeanne's maternal Great Grandparents. Mr. and Mrs. Foster originally purchased the farm with 82 acres of land and donated 2 acres for a school district, which is now believed to be the site of Elston Cemetery where Mr. and Mrs. IW Foster are buried.

I.W. Foster and Mary J. Foster had several children; their eldest daughter Anna Bell Foster Cave and her husband Russell Cave purchased the farm from Mary J. Foster in 1928 and continued with farms legacy. Russell and Anna Cave's, eldest daughter, Mary Louise Cave remained on the family farm after the tragic loss of her parents, Russell, Anna and sister, Esther Jean, in 1940. All three are buried in the Elston Cemetery in the Foster Family plot. Mary's older bother Marvin Foster Cave lived in California at the time of the loss of their parents. Mr. Eddie R. Allen who helped the Cave family with the upkeep of the farm married Mary a few years later.

Mr. and Mrs. Eddie Allen remained on the Foster/Allen farm and continued to work the farmland, farming wheat, corn, beans and alfalfa for hay. Mr. and Mrs. Allen raised cattle, hogs and chickens besides farming the land. Mr. Allen remained on the farm until his death in 1995; Mrs. Allen moved to Oswego, Kansas a few years later until her death in 2003. The Allen's only child Jeanne Carole remains the sole owner of the Foster/Allen farm.

Mrs. Jeanne Carole Allen McKinley has one daughter, Dana C. McKinley, one grandson, Christopher L. Taylor and wife Amanda, two great-grand children, Christopher Lance II and Noemi Grace. *Courtesy of Jeanne Carole Allen McKinley*

FOWLES – The Fowles homestead in Clay County, KS is owned by great-granddaughter, Alice (Adams) Iseli and husband, Dallas. Ralph Fowles Sr., born in Whitchurch, Shropshire, England, came to America in 1871 observing his 21st birthday aboard a ship. He homesteaded 120 acres in Union Township. He worked the land and planted many trees, an orchard, built a building to live in and store his grain.

In 1876 he filed his homestead claim and later returned to England and married Marion Kidman in 1880. They returned to the homestead

The farmstead today

and built a house and other buildings. A barn built in 1906 is used today. Ralph died in 1923 tending cattle and Marion in 1934. His daughter, Mary E. Yarrow, inherited the 80 acres and his other children the rest of the land he had acquired. Mary's daughter and husband, Julia and Alwin Adams, purchased the 80 acres in 1942, where they lived and farmed until their deaths in a car accident in 1964. Their daughter, Alice and husband purchased it in 1965 from their estate. We also purchased from two of his children, Winifred Hammond and Ralph Fowles Jr., the original 40 acres and adjoining 40 acres that he purchased in 1903, in 1978. It and another 100 acres that he purchased in 1887 were inherited from his wife's estate. We acquired this in 1977.

We continue to farm wheat, milo, soybeans and hay, having given up our livestock in 2015. We have a daughter, Jeanise Pride and a son, Gary Iseli, whom we lost to cancer in 2009, four grandchildren, and one great-grandchild.

We have previously received the Bankers Conservation, Kansas Wildlife Habitat and Farm Bureau Century Farm Awards.

The family has been involved in church, school, 4-H and other organizations. Our family lives close so many holidays are observed.

GALLE – The Galle Century Farm in McPherson County Kansas is currently owned by four siblings: Nelson Galle; Dorothy (Galle) Claassen; Edie (Galle) Dahlsten; & June (Galle) Krehbiel. The Galle history reflects more than 400 years of farming: 100+ years in Switzerland, 165 years in Germany, 30 years in Iowa and 130 years in Kansas. The Galles immigrated from Germany to Iowa in 1850, with 700 others, sailing on the ship – the *William Tell.* Attracted by lower land prices and after excellent land was located they moved to Kansas in 1877. In 1894, the 160 acres of the Galle Century Farm was purchased in Kansas. Ownerships included: Great-great Uncle Jacob Galle, 1894-1900; Cousin of great-great uncle, P J Galle 1890-1930; Wife of P J, Anna (Galle) Crary 1930-1942. Our parents: Dale and Bertha Galle 1942-2005; we, the Galle children 2005 – present.

Dale and Bertha Galle, recognized as Kansas Master Farmer Homemaker. From them, we learned the caring strength of family.

The Galle Family held a deep abiding faith. Friendships with family, friends and community were strong. We raised chickens, beef cattle, sheep and swine. A large farm garden was a must. If you grew it or raised it – you probably had it for a meal. Prairie grasses, wheat, milo, oats and alfalfa were grown. Caring for the land and animals was of paramount importance. Even today, seed time and harvest continue to create the seasonal rhythm of the farm.

Church, 4-H and School leaderships were important to the family. Family Birthday celebrations with homemade ice cream, angel food cake or burnt-sugar frosting recall favorite memories.

Today the Dale & Bertha Galle family includes: four children – Nelson, Dorothy, Edie & June and our families: ten grandchildren; ten great-grandchildren and five great-great-grandchildren, located in Washington, California, Ohio, Iowa, Missouri and Kansas.

GILBERT – In April 2001 Roger and Sheila Gilbert acquired the 160-acre farm that originally belonged to his great grandfather, William H. Gilbert. On Feb. 27, 1902 William bought this acreage located in Coffey County, Section 3, Township 22, Range 16. Oats, milo and barley were originally raised on this farm.

This farm passed to Kenneth D. Gilbert (Roger's grandfather), then to Will Carl Gilbert (Roger's great uncle), on to Dean and Vanette Gilbert (Roger's cousins) and then to Roger.

Roger, Sheila, Chelsey, Monica, Reece, and son-in-law Richard

Soybeans, wheat and corn are grown on the farm now as well as a cow herd.

In the early 1960s Carl Gilbert purchased a one-room schoolhouse that had to be moved because of the construction of John Redmond Reservoir. He moved it to the farm and remodeled it into a three-bedroom home for his family. Roger and Sheila still live in that house today. They recently installed wood flooring in much of the home very similar to what was in the original schoolhouse. Roger was sitting in the one room schoolhouse he attended at the time and remembers seeing the schoolhouse going down the road to his great uncle's farm. He couldn't have imagined at that time that he would one day be living in that same house.

Roger and Sheila have raised two daughters on the farm, Monica and Chelsey. They have one grandson, Reece Gilbert Morgan, who enjoys helping out on the farm.

In 2004 Roger and Sheila received the Century Farm Award from Kansas Farm Bureau. In 2011 Roger and Sheila were awarded the Kansas Bankers Association Soil Conservation Award. *Submitted by Sheila Gilbert.*

GOSCH-FIESER FARM – John C. Gosch, wife Mary, and daughter Emma (born Jan 4, 1869) left Warsaw, Illinois moving to Kingman County early in 1878. On March 15, 1878, they purchased this quarter section Osage Trust Land from the government for $1.25 per acre.

George H. Fieser (Sept. 16, 1836 - March 27, 1895) and wife, Elizabeth, came to Kansas in 1886 from Germany by way of Pennsylvania and Illinois. Their fourth child, George W. Fieser (Sept. 30, 1864-Aug. 25, 1935) born near Pittsfield, IL, married Emma Gosch on Sept. 16, 1889. He was elected to the state legislature in 1904, Kingman County Treasurer 1914-1918, city of Kingman Mayor 1922-1928. They were members of the German Methodist Church and the Methodist Episcopal Church.

George and Emma had two sons: Walter H. Fieser (July 5, 1890 – Nov. 30, 1940) and John E. Fieser. Walter married Mary Pankratz. They had three children: Earl, Everett (born Aug. 11, 1920) now living in Dodge City and Lorine. Everett and Earl worked the farm after their father died.

Celebrating the Norwich, Kansas Centennial, October 12, 1985

Everett married Nora Krause on Aug. 16, 1944. They had five children: Carl, David (Sept. 7, 1946), Dean, Joyce and Margaret. The family moved to Plains, KS in 1950. Earl farmed the land until David moved to the farm with his wife, Karen Osterman Fieser, in March 1973.

David continues to farm grain crops: wheat, milo, soybeans, corn, oats and to raise shorthorn cattle. From 1980 - 1999, we grew you-pick strawberries with a record 12,900 pounds picked from one acre in 1981. The farm now is in no-tillage with cover crops to improve the soil health.

Karen and David have one daughter, Elizabeth Green (May 28, 1984), an optometrist in Liberal, KS. *Submitted by David W. Fieser.*

GRAVES-STEVENS FARM – The Graves-Stevens Farm, currently Quad View Ranch, in Cheyenne County, KS is the original Holliman family homestead in 1894 on Section SW 1/4 29-1-39.

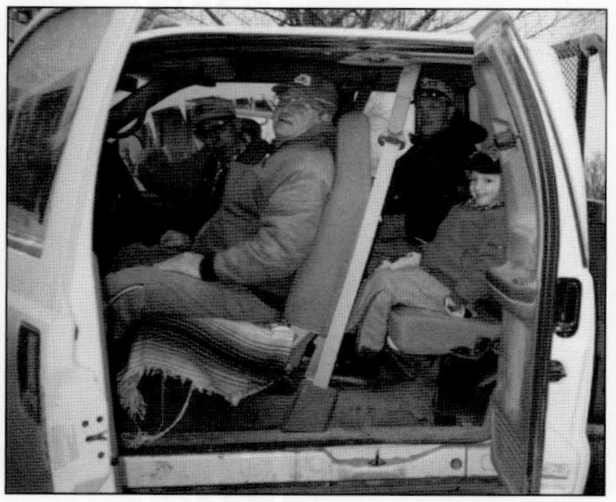
The Graves-Stevens men, 2001

The Holliman's came from Iowa when they became interested in homesteading in western Kansas, having heard about Cheyenne County, with its free land and many opportunities. The first quarter of land included trading a horse and $100 by Lola (Holliman) Graves' great-grandfather, David Holliman. A 16x24 foot dugout was built and then a sod house.

The Graves-Stevens family

Hurlbut and Lola (Holliman) Graves purchased the Holliman Family land in 1948. This was the home where Lola was born and raised.

The family was seeding wheat ground back in 1920 and currently continues the winter wheat tradition with approximately 900 acres. Irrigated ground was put into production between the years of 1970-1975.Crops raised through the years on this ground include, milo, oats, alfalfa, corn and sorghum feed. Today the irrigated ground is used mainly to raise corn for grain and silage and alfalfa, which is fed to the cattle in the feedlot. Besides the farm ground, the pasture ground is used to raise cattle. Currently there is approximately a 400 head cow calf operation running on 4,000 acres.

The steers are put into the ranch's own feedlot and retained through harvest. Some of the heifers are used for replacements while some are sold as replacements and the others are fed in the feedlot. Hurb had a Registered Shorthorn herd and through the years has transitioned from Limousin, Gelbviehs and now Red Angus.

During the '50s and '60s 'when times were tough, Hurb Graves raised and trained Greyhounds and used the money from this hobby project to purchase more pasture ground. There were a lot of fun stories of Hurb and his friend, Floyd Stull, training those greyhounds on jackrabbits out in the middle of fields.

The fourth, fifth and sixth generations are currently on the "family farm/ranch," Quad View Ranch. It is owned by sons-in-law Tom and Donna (Graves) Stevens and Shawn and Shawna (Stevens) Blanka and their two sons, Shakotah and Shadryon. The Blanka's live on the original Holliman homestead.

GRAY – The Gray Century farm, located in Cloud County, is currently owned by siblings Colleen Gray Larsen, Kane Gray, Kirk Gray, Robert E. Gray II and their cousin, Dennis Burdick.

Earl and Anna Gray

Our great-great-grandfather, Henry Gray, came with his family as a young child from Germany. His parents, Wilhelm and Mary Elizabeth Gray and their small sons settled in Pennsylvania around 1845. As a young man, Henry Gray served in the Civil War and was a member of the 3rd Battalion, Company G, 18th Regiment of the US Regulars. After the war he ran a general store in Franklin, PA and eventually, with his mother, followed his brother, Lewis, westward to find land. They homesteaded in Cloud County in 1872 on 160 acres located in Grant Township where Henry and Annie, his wife, raised three children.

When Henry died in 1899 his son, Frank Gray, our great grandfather, inherited the land. Frank raised corn and wheat, had a herd of grade cattle and also raised hogs. The farm was named the Walnut Row Stock Farm and walnut trees still dot the landscape around the farmstead.

Upon Frank's death in 1943 and his wife's death several years later, the farm passed to our grandfather, Earl F. Gray and his sister, Alice Gray Jones.

In 1981 the farm was deeded to our father, Robert Earl Gray, and his sister, Vera Frances Burdick. Wheat, corn, milo and alfalfa were still planted and cattle grazed the pasture.

When our parents passed the torch to our generation in 2001, we planted those same crops plus soybeans. Nature has also provided abundant wildlife and is a source of enjoyment for family.

We hope the Gray Century Farm continues to have ties to the Gray name and keeps this legacy alive for future generations. *Submitted by Colleen Gray Larsen.*

GREENE – Kansas Pacific Railway owned 160 acres located in Ottawa County, Henry Township, 11-28-5 and 11-33-5. In 1898 Roderic C. Greene purchased this land from the railway. He raised beef cattle and grew crops of wheat, alfalfa and sorghum.

Stone and frame barn on the original 160-acre farm of Kenneth & Muriel Greene. The man holding the two horses is Roderic C. Greene; the man on the saddle horse is Arthur C. Greene. Photo taken ca. early 1900s. The barn is still there.

This farm passed from Roderic to his son, Arthur C. Green. In 1979 Kenneth Greene acquired the farm from his father, Arthur. This farm has grown to 320 acres and now produces crops of wheat, milo and soybeans and raises beef cattle.

Historic buildings still on the farm are a stone house, stone/frame barn, a cattle shed and two silos. It also now has a chicken house.

In September 2003 this became a Century Farm.

Kenneth and Muriel have two children, Marcia and Marion Greene. *Submitted by Kenneth and Muriel Greene.*

GREENWAY HOMESTEAD – William Mottern Greenway was born in Carter County, TN on May 4, 1845 to Nathan Greenway and Rachel Smith. In 1873 he married Mildred Jane Keyes and they had 11 children: Cora Elizabeth (Swisher), Sarah Etta (Hill), Lillie Roberta (Newman), Alpheus K., Harlie B., Nathan Burt, Rachel Effie (Blake), William Jerry, John Jackson, Wilbur and Elmer.

The Greenway Homestead

The Greenways came to Kansas in 1880 from Missouri to file a homestead claim on a parcel of land six miles northwest of La Crosse in Rush County near the town of Hargrave. For many years, they lived in a modest two-room wood and limestone home on the homestead. From the time the first sod was broken until present day, wheat has been the primary crop raised on the land. William and Millie remained on the farm until their health necessitated a move to a larger city.

Mildred died in January 1923 and William died in December 1924 in Guthrie, OK. Shortly after, ownership of the land passed to their son, William "Jerry" Greenway. Jerry married Nellie Francis Speck on May 20, 1914 and they had two children, Dorothe and Kenneth "Bert." Bert was killed in a car accident in 1949. Jerry continued to farm the land being joined by his son-in-law, Raymond "Ray" Weiser. Ray and Dorothe had one daughter, Marjorie Raynell. Around the time of Jerry's passing in 1979, Ray and Dorothe retired and leased the land for several years. Following their death, Raynell and her husband, Leroy Penka, took over the farming operations with assistance from their son Bradley Penka, making Brad the fifth generation to work the land. Since being homesteaded in 1880 the land, affectionately referred to as "the old place," has been in the same family for well over a century.

GREER – The Greer Century Farm was homesteaded by George Robert Greer in 1869. The claim location was three and a half miles west

The Greer's in front of their home

of Altamont in the Mount Pleasant Township of Labette County. The original homestead was 160 acres. As time passed more acreage was added to the farm.

George met and courted his wife, Mary Ann (Mollie) Hill on the train coming to Kansas from Illinois. They were issued the first marriage license in the Mount Pleasant Township and were married July 12, 1870. After Mollie passed away in 1919, George remained on the farm for two more years. Six children were born to Mollie and George. Upon George's death the farm ownership was passed to their eldest son, John L. Greer.

George and Mollie lived the hardships of the early pioneer life. Their first year on the homestead was spent in a sod dugout. George first tilled the virgin soil of his farm with oxen. Back in those early days George and Mollie and their children went to town shopping in the wagon behind the yoke of oxen.

As a young man, George was converted and accepted the faith of Jesus, never faltering in the faith the many years that followed. That trust in Jesus has been passed down through the family to the present day.

John Lewis Greer was married to Catherine Alma (Callie) Bickham on Feb 14, 1900 and took ownership of the Century Farm upon the passing of George his father. Upon the death of John Greer the farm passed to his children, Euna and Vernon. Vernon planted the farm to fescue and Bermuda grass as a tame grass pasture. One of Vernon's most precious possessions was the Century Farm designation that was awarded in 1975.

GREGORY – NW Quarter Section 12 Township 7 South Range 16 west of the 6th pm in Rooks County, KS, was homesteaded in the 1880s by Charles and Martha Hall. They sold this land to Henry R. Raumaker for $1,000.00 April 18, 1898. John H. Gregory bought the land Sept. 11, 1901 for $1,350.

John H. Gregory, 22 years old and his wife, Mary Josephine, my grandparents, moved in to the two room limestone rock house on Sept. 12, 1901. He hired Mr. Robinson to build a two story modern house and Bob Britton to build a large barn,

The new house was modern with in-door plumbing, porches on all sides, a laundry room, a basement, coal furnace and carbide lights for the home and barn. The home had French doors and engraved woodwork with frosted windows.

John, Josie and Maggie

My grandparents were raised on farms and they were farmers all their lives. They developed the Lone Star Dairy. Twenty-four cows had to be milked by hand twice a day and the milk had to be taken care of without refrigeration. The crops were mainly alfalfa and grain. They sold whole milk for 8 cents a quart. Skimmed milk was sold for 4 cents a quart, delivered to the door.

John had many community interests. He had three loves beyond his Christian life, the love for children. They never had any of their own children, but raised several orphans from babies to adults. All of these children were thought-of as a part of the Gregory family. They raised two boys and three girls from babies. In 1915 they adopted a baby boy, Truman D. Gregory. In 1918 they adopted a baby girl, my mother, Junia Lorene Gregory.

John's second interest was a proponent of land conservation. He traveled to areas giving free lectures on natural fertilizer, summer fallowing, rotating crops, and saving land from washing away.

John's third interest was livestock nutrition. He was convinced of the value of chopped feed and grain. He believed in chopped corn and stored in upright silos. He developed a machine that would chop feed bundles into an up-right silo. This machine became known as the Blizzard Ensilage Blower. In 1936 he started selling Dodson Silos. For 12 years he was top salesman in Kansas, Colorado, New Mexico, and Nebraska. He was still selling Dodson Silos and buildings when he died.

John and Josie celebrated their 50th wedding anniversary April 21, 948. All the children and their families were in attendance.

John Gregory lived on his farm 50 years to the day. He moved to the farm May 12, 1901. Sept. 12, 195l. John Gregory dropped dead at the pulpit of the Woodston United Methodist Church.

The big house burned to the ground in the spring of 1952. Josie survived the disaster and she died May 2, 1959. Upon her death the two children inherited the land.

In 1960 Junia bought her brother's half interest in the farm, deeding to him where his new house set at the southwest corner of the quarter. At the death of Truman and his wife, Junia purchased the Truman Gregory home from his son, John.

March of 1981 Junia and husband, Willard Hale, leased all of their farm to Frank Dix. He still holds the lease 36 years later.

The land is still being owned by 99-year-old Junia, daughter of the Gregory's. In time will be passed to her son, Roger and grandchildren, Ryan and Dana Hale.

The 116-year-old family farm has many stories and memories for Hale's and John Gregory family.

GRIFFIN – The late Lyle and Ella Griffin, their daughters, Julia Griffin Fisher and Rita Griffin Shumate and their families were recognized as a Greeley County Farm Bureau Century Farm in 2015. Lyle's grandparents, Isaac Ezra and Mary Matilda Griffin, purchased four quarters of land in Greeley County for $2.50 in IRS stamps, which would have been between $1700-$2250 dollars. These land purchases were recorded in the Greeley County Register of Deeds office on Sept. 18, 1915. Of these quarters, only the Griffin home-place or "old place" remains in the family. The quarter (NE 10-18-41) is located northwest of Tribune and Horace, KS in Greeley County.

With deaths of Ezra and Mary, the quarter passed to their son, Elmer. At his death, the quarter passed to his wife, Ida Julia, who passed it to their oldest son, Emerson. Emerson sold the quarter to his younger brother, Lyle in 1953. When Lyle died in 1995, the quarter passed to his wife, Ella. Upon Ella's passing away in 2016, their oldest daughter, Julia Griffin Fisher inherited the quarter.

This quarter consists of 156 acres on which wheat, milo and sorghum have been planted in recent years. There are no buildings remaining save for a very large, old elm tree which shaded the original home.

This quarter is now being farmed by Julia's son, Ben Fisher. He is the fifth generation to farm this quarter and is very proud to continue the Griffin farming longevity and stewardship.

Today the Lyle and Ella Griffin family includes two daughters, three grandchildren and four great-grandchildren, all living in Kansas. *Submitted by Julia K. Fisher*

GRIMMETT/McMILLAN – A Homestead was granted to William Johnson and Martha Jane (McCollum) McMillan on November 20, 1884, southwest of Stafford. The family had come from Ohio in 1877 in a covered wagon bringing 50 head of cattle and 10 of their 12 children establishing a farm in Stafford County. A daughter, Martha Jane McMillan became the owner of the farm upon the death of her mother in 1909, farming with her husband, William T. Buckle. Martha Jane (McMillan) Buckle died in 1929 at which time ownership was passed to her son William R. H. "Bill" Buckle, born 1906. Bill married Stafford County teacher Naomi Shinkle in 1932. Bill and Naomi lived on and operated the farm until deeding it to daughter Barbara Jane (Buckle) and Clayton E. Grimmett in 1991, current residents. Children are Julie R. Grimmett and Joel M. Grimmett. Awards received: Kansas State Board of Agriculture, Tradition Farm, 1988, and Kansas Farm Bureau, Century Farm, 2002.

Bill Buckle

The farming was done with teams of mules and/or horses prior to using tractors to raise wheat, oats, sorghum and alfalfa. Livestock included horses, mules, pigs, chickens, beef and dairy cattle. As a young man, Bill Buckle raised thoroughbred Quarter horses. Earlier in the century his mother sold butter and eggs at the markets in Stafford. It is said that her butter was requested by many of the town's ladies. W.T. Buckle, her future husband, met Martha Jane McMillan as a widower while taking the orders for her desired butter at a local Mercantile. Son Bill lived on the farm his entire life, caring for the land and livestock by employing methods of organic farming and gardening with Naomi, who retired after 33 years of teaching. Their gardens yielded asparagus, onions, potatoes, corn, beans, strawberries, grapes and cherries.

GROFF – My great grandfather, Jeremiah Ellsworth Groff, came from Indiana with his parents and four brothers in 1884. In 1887 he homesteaded 160 acres south of Ellis, KS. On March 23, 1892 after the land was "proved up," he received homestead certificate #5190 for 160 acres in Ellis County, NE 1/4 section 32, township 14, range 20, west of the 6th principal meridian.

He married Petrea Peterson in 1887 and farmed until his retirement in 1924. He then sold the land to his son, Joe Groff. My grandfather Joe retired in 1949 and sold the farm to his youngest son, Clarence Groff, who sold it back to Joe in 1951. Then in 1952 my father, Leo Groff, traded his father Joe a parcel of land in Rush County for the homestead. After my father retired in 1984 I farmed the land. Upon the death of my parents, I inherited the homestead. The next descendant to inherit the farm will be my son, Travis J. Groff, who lives there now with his family, making my grandchildren the sixth generation of Groffs to live on the original homestead.

In 2002 the farm received a Farm Bureau Century Farm Award. *Submitted by Leo Groff Jr.*

GRONEWEG CENTURY FARM – The Groneweg Century Farm of Decatur County, KS, NE/4 13-5-30, Cook Township, bought by George H. Groneweg in 1902. He lived in a soddy for a few years. After his first marriage, a three-room house was built, which was added on to several times. George came to Kansas from Illinois, first buying land in Thomas County, KS. While in Thomas County, he enjoyed breaking wild horses, later to ride or farm with. Crops were wheat and corn, some oats for the horses.

Wilbur Dean, Mary, and George Groneweg

After the death of his first wife, he married Mary Demmer and she helped raise his son, Wilbur Dean. They bought more land, 1,200 acres. George and son farmed over 2,000 acres, caring for the land and animals – cattle, hogs and chickens were very important to them.

Church, 4-H and schooling was important, as George had very little schooling, as he did not speak English before going to school.

Family time in his later years was special, as he lived with the Wilbur and Dora Young family.

I, Dora E. Peggy Groneweg, am owner of the farm as I was widowed in 2011. Our children are Robert Dean, George Lynn and Joyce Coe. There are six grandchildren, nine great-grandchildren, living in Colorado, Kansas, Nebraska, Oklahoma and Illinois. *Submitted by Dora E. Peggy Groneweg.*

GUMBO HILL – Gumbo Hill, our century farm, is very cherished by our family! Lillie and Harmon Freesman settled the farm in 1900. Lillie came as a pioneer in a covered wagon from Ohio. Harmon was a German immigrant and his family lived in a dugout near Riley, KS. Harmon met Lillie and shortly after their marriage, they purchased 80 acres, now known as Gumbo Hill. The name of the farm, Gumbo Hill, came about because Harmon and Lillie took a farm wagon, pulled by two horses, hauling a crate of chickens and a cow in tow behind and parked it in the center of the 80 acres they had purchased four miles north of Keats, KS (now unincorporated). It rained that night and after sleeping under a tarp, Harmon got up in the morning, walked around, came back to the wagon and said, "Lillie, we have bought ourselves a Gumbo Hill and it won't raise a thing!"

Jeremiah Ellsworth Groff homestead farm, Ellis County, Ellis, Kansas

Donna's husband, Harry, and team haying on Gumbo Hill.

They built a one-room house and in 1910 adopted a daughter. Harmon then built a two-story addition to "keep his girl in." With the addition, the house had two bedrooms upstairs, with a parlor, one bedroom, a living room, kitchen and two porches downstairs. Erma was raised in the home and married Michael T. Sebring in the parlor. She became a writer and was known as "Erma - Girl of the Gumbo."

Erma and Michael had one daughter, Donna, also born in the house. She grew up and married a neighbor boy, Harry E. Whitney II. Donna and Harry built a house on the farm, a very short distance north of the original Gumbo Hill house. To this union was born two children: a daughter, Donita and a son, Harry E. Whitney III. Donita taught in a one-room country school and was a county superintendent in Nebraska. She married Dennis Bammerlin from the area and they had one daughter, Laurel "Lori" Bammerlin. After 10 years in Nebraska, Donita returned to Gumbo Hill with her daughter and Lori became the fifth generation to live on Gumbo Hill. Donita earned her Ph.D. and currently teaches at Kansas State University. Harry III helped on the family farm for a number of years and eventually created his own business as an equine clinician, "From The Horse's Point of View." Harry III continues to be an equine clinician, providing educational clinics around the US. Both grew up running in and out of the house at Gumbo Hill, playing on the back porch and enjoying a tree house in the backyard. Lori serves as an Extension Agent for Agriculture and 4-H in the Flint Hills District, Council Grove, KS.

In 2010 Donna and Harry repaired, restored and rebuilt the 100 year old Gumbo Hill farm house. Donita lives in the house and Lori has plans for Gumbo Hill to eventually be her farm, keeping it in the family for five generations. Her vision is to keep this piece of ground for future generations yet to come, continuing to have native hay crops, milo, wheat and soybeans along with raising horses, beef cattle, a few chickens and good kids! Yes, Harmon Freesman was wrong. Gumbo Hill can, and does, raise a good crop of livestock, grain and people. As the farm has been owned and occupied for 117 years by the same family, we manage by the motto: "God needed a caretaker for his creation, so therefore we farm."

GUTTERY FARMS – Charles Guttery was born in Washington, PA April 20, 1847, son of John and Elily Ellen (Teagarden) Guttery.

Showing patriotism at age of 15 years old, Charles enlisted in the Pennsylvania Volunteer Infantry and served in many battles including Gettysburg. After an honorable discharge Charles and his wife, Margaret Ely, traveled to Washington County, KS in 1869. In 1871 Charles was one of the first settlers to take up homestead in Rooks County. A few months later he relinquished his claims and set up a blacksmith shop in Bull City (Alton, KS). His wife passed away leaving him with two small children, Harry and Clara Guttery. In 1877 he married Victoria Doak and had three children: Charles Perry, who died 3 years old, Orville Grant and John Almarine. In 1879 Charles homestead 160 acres which he purchased for $100.

Brice, Porter, and Al

John Almarine was born on the homestead in 1887. After Charles' death, John Almarine continued to farm. He married Jennie Brown and had one son, Porter Guttery. Cattle, pigs, dairy, chickens, wheat and hay were raised on the farm.

Porter Guttery was born on the family homestead. Porter farmed with his father, Almarine, with a team of six horses up until they purchased first tractor in 1938. After an honorable discharge from the Army, he married Ruth Osborne from Bel Aire, MD. They returned to Kansas and continued to farm the original homestead. Four children were born and raised on the farm: Porter, John Alan, Bonita and Barbara. Porter and Ruth have 12 grandchildren with every one of them earning a college degree.

"Guttery Inn" on the Guttery Homestead, June 7, 2014

John Alan "Al" Guttery took interest in farming at a young age, buying his first self-propelled combine at age 15. Al married Debby Kendig. Three children were born: Brice, Jennie and Denise. Al was always reading and learning of new ways of farming and ranching. From implementing underground waterway systems, no till farming operations and artificially inseminating the heifers. Al passed without being able to teach his six grandchildren the love of the land and farming.

Brice Guttery now farms and ranches along with his wife, Shana (Preedy) Guttery. They raise three children: Doak, Clay and Stormy on the farm. Brice has expanded the Guttery Farm to the Swinging G Farm and Ranch from a few hundred acres his great-great grandparents farmed to several thousand acres. *Submitted by Debby Guttery.*

GUYER – W.H. Guyer acquired 320 acres of land located in Osborne County, 34-7-14 in 1902. Sorghum and grains were raised along with livestock in the pasture and chickens.

The Guyer farm

This farm passed from W.H. Guyer to Irvin Guyer and to his daughters, Linda O'Toole and Linda Rice in 2002.

The farm still contains 320 acres and now produces wheat, sorghum and grain crops.

The wooden house on that property where Irvin Guyer was born, is gone now.

Widow Janice O'Toole has two children, Shelly Poppelreiter and Darcy Pat O'Toole. Her grandchildren are Emily and Patrick M. O'Toole.

Linda and Dennis Rice have three children: Eric, Aaron and Denise Williams. Their grandchildren are: Caitlin and Gavin Rice, Callie and Jaren Williams and Maggie, Brooklyn and Mathew Rice. *Submitted by Janice O'Toole and Linda Rice.*

GWIN – On Oct. 9, 1906, Harry N. and Gertrude Gwin acquired 125 acres located in Anderson County, Lincoln Township, Sec. 28, 29 and 21, Township 21, Range 20. Three generations of Gwins have owned this farm. Harry N.

Harry N. and Gertrude Gwin with daughter, Ivy.

and Gertrude Gwin transferred this farm to their son, Raymond and his wife, Effie, and in turn it was transferred to their children: Charles, H. Floyd and Shirley (Gwin) Campbell Roeckers.

Originally this 125-acre farm produced grain crops of corn and grass to support livestock consisting of cattle, pigs, horses and chickens. The farm has grown to include 438.5 acres, presently producing corn, wheat, soybeans, pasture, cattle and hogs.

In 2006 this farm was designated a Farm Bureau Century Farm and received the Wildlife Habitat Award.

Shirley and husband, Richard H. Roeckers, have two children, Mark Campbell and Kim Gruner. Grandchildren are Jason and Jenna Gruner and Natalie Campbell. *Submitted by Shirley Roeckers.*

HAASE – In 1886 Frederick Haase paid $2,400.00 to Albert Banfield for 160 acres located in the southwest quarter of Section 27, Township 14 south, range 9 west, 6th PM. Frederick purchased this land for his son, Herman and wife, Emilie, who had married in 1889. The farm was turned over to Herman in 1893.

Frederick and his first three children, including son Herman Sr., were both born in Germany. The Haase family came to the US and first lived in Wisconsin. After the fourth son was born, Frederick and his family moved to Kansas. They lived in a dugout and Frederick would walk to town to help build Ellsworth, KS.

The original home and barn, which is still in use.

This farm then passed to Herman Jr. and wife, Elba after the death of his mother in 1962. Delvin Haase and wife, Jeanne now owns the farm.

Wheat, milo and cattle are now raised on the farm. The original home, limestone barn and smoke house remain and are being used.

Delvin and Jeanne Haase have three sons: Shane, Arrin and Ryan. Their grandchildren are: Logan, Ethan, Jenna, Nicole, Karlie and Marissa Haase.

Arrin will most likely continue running the farm and grandson Logan is now a hired hand. *Submitted by Delvin and Jeanne Haase.*

HALL – For 150 years, the Hall family has worked and cared for their ground in Atchison County, KS, in the Farmington Community.

In the spring of 1869, Durand C. Hall came to Kansas from Ohio, purchasing land for $18.92 per acre. The farm was known as Orchard Hill Farm and was described by historians as one of the largest, best equipped farms in the county, growing to 675 acres in one tract. The Halls erected an 80 x 54 foot barn, including a basement, which was reputed to be the largest in the county, to support their sizeable Herford cattle operation. The structure held 150 cattle and 200 tons of hay and grain.

Albert Hall and Bob Hall with work mule, 1941

In 1901, Albert Scott Hall inherited a portion of the land, which he named Walnut Grove Farm. His son, Albert Sparks Hall, inherited the farm in 1949.

In 1971, Albert's youngest son, Robert F. Hall and his wife, Winnie, then purchased the land. They are the current owners of the 325-acre tract known as Averie Acres. The Halls established one of the county's first confined farrow to finish hog operations, with 1,000 head of Duroc, Hampshire and Yorkshire breeds. In addition to raising cattle, the Halls have grown corn, wheat, oats and soybeans. The farm has survived numerous floods and a devastating plague of Armyworms.

Bob and his granddaughter, Averie Windsor, planning corn in April 1985

Albert (in overalls), Bob, and Scotty (on tractor), 1960

Bob has vivid memories of growing up on the farm, herding cattle down the county road every spring and fall, watching his father work the land with mules and the luxury of the family's first closed-cab tractor, complete with air conditioning and a radio!

Currently Robert and Winnie's family includes one daughter, Cyreesa Hall Windsor and one granddaughter, Averie Windsor Chapman, the farm's namesake and twin great-grandchildren, Miller and Reese Chapman, all residing in Kansas. *Submitted by Robert F. and Winnie Hall.*

HALLOCK – The Hallock Century Farm in Labette County is currently owned by Merrill and Norma Lee Christy. Merrill is the grandson of Thomas L. and Millie Steele Hallock, who purchased the 170-acre farm in 1904.

Peter Hallock came to the US from England in 1634, fleeing civil and religious oppression in England, settling on what is now long Island. Descendants of Peter Hallock, over the decades, moved to Pennsylvania and on to Illinois. Thomas Hallock's farm in Coles County, IL had been part of his parents' farm.

Thomas and Millie had five children, all born in Coles County, IL – four daughters and one son. They moved to Labette County, KS in 1904. Their son helped with farming, later marrying and moving to western Kansas. Three of the daughters became teachers, with one going on to become a dress designer in New York City. The fourth daughter, Esther and the youngest in the family, was an accomplished pianist and organist and in 1916 married William A. Christy, the red headed 29 year old struggling to make ends meet on his farm down the road from the Hallock farm.

Thomas and Millie Hallock's 50th Wedding Anniversary, December 22, 1931. Also pictured are their son and four daughters, plus granddaughter Ruth Christy and grandson Merrill Christy.

Hallock family home

Will and Esther's daughter, Ruth, was born in 1920 and their son, Merrill, in 1922. Will farmed the Hallock farm in addition to his own acreage. Will and Esther eventually purchased the farm from her siblings following her parents' death in the 1930s.

William and Esther Hallock Christy

Most of the acreage is tillable ground with some pasture. Merrill, as a kid, would move the registered Jersey's in the morning after milking to the pasture on the Hallock farm until time to bring them back to the Christy farm for milking.

Today, the Hallock farm is enjoyed by Thomas and Millie's grandson, Merrill; wife, Norma Lee and their three children, seven grandchildren, and the six little great-grandchildren.

HANDKINS – The Handkins Century Farm, now owned by Paul and Dorothy Handkins (NW 1/4 28-27-5 in Evan Township, Kingman County, KS), was first purchased by Paul's grandparents, Robert and Mary Ann Handkins, in 1898 and 1899.

Wedding picture of Robert S. and Mary Ann Handkins, February 21, 1889.

Robert Sheldon Handkins (Nov. 29, 1864 – Jan. 2, 1941) was born and raised in Crab Orchard, IL, but came to Kansas in 1882 when he was 18 years old after hearing about the wonderful farm land available in Kansas.

Five years earlier in 1877 Peter (1838-1909) and Magdalena Gemmer (1842-1922) and their six children left their home in Urbana, IL and made the trip to Kansas in a covered wagon. Along the way they sent a postcard home saying, "It is high noon and we have traveled 7 miles already today. We did not know it was such a long way to Kansas!"

Within a few years the Gemmer's oldest daughter, Mary Ann (July 28, 1866 – June 11, 1940) caught the eye of Robert Handkins and they were married on Feb. 21, 1889. The first few years they rented farm land but had their eye on a piece of property they wanted to buy. A letter dated Sept. 17, 1898 from Parks Real Estate in Hutchinson, KS read, "I have just heard from the owner of the 80 acres north of you, and the party will sell it to you for $600.00 cash." The offer was readily accepted and Robert and Mary Ann Handkins became the proud owners of the east 80 acres of the NW 1/4 of Section 28 in Evan Township.

The following year in 1899 they purchased the west 80 acres for $800 from the Hartford Steam Boiler Inspection and Insurance Co. They were required to pay $50 down, with an annual payment of $50

The family of Robert and Mary Ann Handkins, ca. 1915. Back: Edwin, Clarence, Earl and Wilbur. Front: Blanche, Robert, Mary Ann, and Stella.

plus 6% interest. Looking back Paul Handkins thinks his grandparents made very wise investments!

The 1905 census stated that the Handkins had 70 acres of winter wheat, 40 acres of corn, a quarter acre of potatoes, eight horses, 12 "milch" cows, 14 other cattle and four swine. They also now had six children: Clarence (1890-1955), Edwin (1892-1974), Stella (1894-1963), Earl (Paul's father) - (1897-1984), Wilbur (1899-1965) and Blanche (1903-1967). Earl and Clarence were the only ones who married, and the others stayed on the farm their entire lives.

After Robert and Mary Ann Handkins died, the quarter of land was divided among their six children and eventually passed on to the four children of Earl Handkins (Paul, Harold, Melvin and Edna Marie). Paul was the only farmer in the group so his three siblings sold their shares of the farm to Paul and Dorothy in 1993. Paul enjoyed his years of farming this land. He and Dorothy still own this quarter and their four daughters (Alice, Carolyn, Linda, and Paula), eight grandchildren, and seven great-grandchildren love hearing Paul, now 96 years of age, tell stories about "the olden days." All of the family is proud of the heritage and rich memories associated with this Century Farm!

Greg Reno is now doing a wonderful job of farming this original quarter of land for Paul and Dorothy. He is doing no-till farming and raises primarily wheat, soybeans and milo. *Submitted by Alice Bair, daughter of Paul and Dorothy.*

HANIGAN – On June 20, 1885 William S. Hyatt acquired 160 acres located in Labette County, Fairview Township, southwest quarter of Section 35, Township 33 South, Range 20 East.

Originally wheat, oats and corn were the crops raised along with cattle, hogs, chickens and turkeys. Today crops raised are wheat, oats, milo, soybeans along with cattle.

This farm passed from William S. Hyatt to L.G. and Chas. W. Felix. On Dec. 1, 1910 John E. and Ellen Hanigan acquired the 160 acres. Next owners were Clifford and Mae Hanigan. Their children were Alma, Robert, Melvin, Margaret, Sarah and Maybell. In January 1986, Maybell Hanigan became owner.

The original house and barn are still standing as well as a stone wall. *Submitted by Vera M. Hanigan.*

Taken May 1993.

HANSON – In 1903 Henry N. Hanson acquired 160 acres located in Saline County, Smokey View Township, 12-16-3. Wheat, corn and alfalfa were the grain crops grown and livestock was mules, horses, hogs and cattle.

The original Hanson homestead

This farm then passed from Henry N. to his son, Henry and in 1988 James W. Hanson acquired the century farm, which still retains 160 acres.

Now wheat, beans and alfalfa are the crops raised here. The original house and granary are still on the farm.

James has two children, James M. and Jesse Jore. His grandchildren are Will Jore, Laura Hanson and Harrison Hanson. *Submitted by James W. Hanson.*

HARDEN – Charles E. Harden came to Lexington Township in Clark County, KS in 1884 with his father, Nathan Harden. He bought the land for the farmstead in 1898 and built the barn in 1907 followed by the house in 1908. By 1905 he owned 1,760 acres of land and increased it to 2,000 acres by 1915. He was a wheat farmer and rancher. Three children were raised in the farmhouse.

Barn built in 1907 by Charles E. Harden

When Paul R. Harden, his son, came home from the Army in WW I, Charles leased him the homestead and land till Paul later purchased it from his father. Paul R. Harden raised seven children in the farmhouse. Paul was a wheat farmer, rancher, and operated a farm trucking company till 1936.

House built in 1908 by Charles E. Harden

Pictured back to front: Henry W. Hanson, Gertrude Hanson, Betty Sandquist, Lucille Hanson, Gladys Hanson, Freda Hanson, Clarence Hanson, Miriam Hanson, Henry Hanson (baby), Murtyle Hanson, and Cora Hanson.

His son, Willis M. Harden, took over the farmstead in 1941 and raised five children in the homestead. Willis was also a wheat farmer and rancher. Willis turned the farmstead over to son, James M. Harden, in 1987 where he has raised two children and now owns 4,000 acres of land. He is a farmer and rancher. He produces alfalfa, milo and wheat. He currently resides in the farmhouse. *Submitted by Jim Harden.*

HARPER – The Harper Centennial Farm is located in Miami County 2 1/2 miles northeast of Louisburg, KS. The farm is owned by brothers, Samuel Leon and David Eugene Harper, who inherited the farm in 2015 from their father, Robert W. Harper Sr.

Robert Harper Sr., early 1940s

Scott and Susie Harper

Samuel Montgomery and Mary Elizabeth Harper moved to Kansas in the 1880s to raise their family, Helen, Irene, Leon, Bruce, Scott and Nathan. They resided on a farm close to what later became the Harper Farm. The four brothers rented and worked the farmland several years before purchasing it in 1907 from J.V. Lee. The purchase price was $6,000. Scott and his wife, Susie, later took sole ownership of the property. Upon Scott's death in 1965, his son, Robert Harper Sr., inherited the land. Throughout the years, the property was used for raising crops, chickens, selling eggs to hatcheries and raising dairy and beef cattle. The current farm encompasses 160 acres and is now used exclusively for raising cattle. The original home (which was on the property at the time it was purchased by the Harpers), still stands. It was initially a 16' x 16' structure. Room additions were done in the early 1920s.

The Harper brothers built a large barn in August 1927, which still stands today. It was used to store hay and grain and served as the "milk barn." Dairy farming was the major source of income for many years. The brothers hand-dug three wells on the property, two of which are still functioning today.

The hand-dug cellar was built in November 1920 and was used to store the abundant garden produce they raised throughout the years. It was also an escape from the tornado that ripped through the property in 1959, which destroyed numerous buildings.

The Harper farm was recognized as a Farm Bureau "Century Farm" in 2015. *Submitted by Samuel Leon and David Eugene Harper.*

HART – Cynthia Hart owns a unique Century Farm in Cloud County. A narrow lane leads from the road up the hill into the pasture that provides a stunning view of the Jamestown Wildlife Area, a historically shallow salt marsh.

Oscar Kaad plowing

Jamestown Wildlife Area

Anne and George Kaad, Sr. purchased the 80 acres in 1896. Both were natives of Denmark who arrived in Jewell County, Kansas in 1878. They married in June 1879 and established a farmstead directly west of the Hart property in 1883.

In the 1930s, the 80 acres was transferred to a daughter – Emma (Kaad) (Behrends) Isaacson and her husband Carl. In 1937, the Isaacsons leased part of the property to the Jamestown Gun Club for 25 years. Organized in the 1920s, the Jamestown Gun Club established a lake in the marshy area northwest of Jamestown, Kansas. The Sportsman's Lake was known for boating, hunting, fishing and picnicking. Powerboats from four states raced in summer racing events in the 1930s and 1940s. Large crowds of 2,000 or more also witnessed sailboat regattas at the lake.

The next owner was Oscar Kaad, Emma Isaacson's brother. However, on December 5, 1960, the State of Kansas condemned 13.7 acres of the Kaad property. Sportsman's Lake was drained and turned back into a marsh for waterfowl hunting. Today, the marsh is a collection of managed wetlands and upland areas.

Margaret (Kaad) and Gerald Hart purchased Oscar Kaad's land from her uncle's estate in 1967, including the tract adjacent to the state's marsh. Gerald Hart expanded the native pasture, seeding five acres of cropland to grass. He also established a small beef feedlot on the southwest corner where the windmill and water tank were located.

Margaret and Gerald Hart's granddaughter, Cynthia Hart, inherited the property in 2010. It remains in cropland and native pasture.

HART – George Kaad, Sr. left Denmark in 1872. Family lore reports that he hid in the straw on a cattle boat to escape conscription by the German Army and that German soldiers boarded the ship and poked their swords into the straw but somehow missed him. Kaad arrived in Kansas in 1878. He purchased a soldier's filing west of Jamestown, married Anne Peterson and lived in a dugout.

Anne and George Kaad, Sr. sold that land and purchased 80 acres of State Normal School land on April 9, 1883. They purchased the adjoining 80 acres from his brother-in-law in 1901. Labeled "Oakdale Farm" in early abstracts, the driveway entrance is still framed by a majestic red oak tree.

The current house was built in 1898 at a cost of $1200. An original stone chicken house also remains on the farmstead.

Following his father's death in 1932, Oscar Kaad inherited the 160 acres, which was the gathering place for many community card parties and family activities. Oscar Kaad's niece Margaret (Kaad) Hart and her husband Gerald purchased the place in 1966. Margaret treasured owning and living in her grandparent's house. Robert Hart, son of Margaret and Gerald Hart, inherited the place in 2010. Robert and Linda Hart continue to own the property. Their daughter, Paula (Hart) Carlson, and her husband Brent Carlson currently live in the home and farm the land. Their son Addison is the sixth generation to live there.

The Hart Century Farm is located in the far northwestern corner of Cloud County. St. Luke's Lutheran Church, which lies just across the county line in Jewell County, was built of native stone in the early 1880s. The church was very important in the life of the Kaad and Hart families, many of whom are buried in the adjoining cemetery.

Please see the Kaad family photo featured on pages 2-3 of this book!

HATHAWAY – In 1878 John L. and Mary Clark Hathaway homesteaded the northwest 160 acres of 34-19-10 in Pioneer Township of Rice County, KS. Originally from Washington, PA they had married in Lee County, IA in 1869 following John's service with the 140th Pennsylvania Volunteer Infantry. He had fought in the Civil War battles at Chancellorsville, Gettysburg and Cold Harbor where he was wounded when two minie balls injured his right leg. Before the move to Kansas, John had traveled by rail to Raymond, KS on one of the railroad company's excursions to encourage prospective pioneers. John bought out a squatter who had built a sod house on a homestead quarter. The Hathaways lived in the sod house for a couple years until their frame house could be built. Their son, Fred Hathaway, bought the farm from his widowed mother in 1907. Fred's son, Ralph Hathaway, inherited the farm in1975. Ralph's children — Marlene Hathaway Carson, Martin Hathaway and Carolyn Hathaway Kern — inherited the farm in 2009. Carolyn and her husband, Ed Kern, live in the homestead house they bought from her grandfather Fred in 1973.

Ralph Hathaway greeting visitors to Ralph's Ruts on the Santa Fe Trail

An important feature of the farm is the existence of very prominent Santa Fe Trail ruts that are easily visible in the 40 acres of native pasture known as Ralph's Ruts. This property is recognized by the Santa Fe Trail Association and the National Park Service as the location of the Plum Buttes Massacre, an altercation between a wagon train and a band of renegades that resulted in the deaths of several travelers. When Ralph Hathaway returned from service in WW II, he built a second house next to his parents where he raised his children. The newer home is now Ralph's Ruts Retreat, a bed and bath on the Santa Fe Trail. *Submitted by Carolyn Hathaway Kern.*

The home of John (seated) and Mary Hathaway about 1898 with sons Frank (with bicycle) and Bert.

HAURY FAMILY FARM – The history of our farm began in 1852 when John Ruth and his brother-in-law, Daniel Haury, migrated from Bavaria to the US. They settled in Summerfield, IL. Their families were growing and they needed bigger farms. In 1874 the Ruth and Haury families moved to Harvey County and settled on adjoining farms near Halstead, KS.

The first building Ruth built in 1874 was the horse barn, which housed the family until 1875 when their house was built. Ruth was a gifted carpenter and he built two additions to the house as his family grew.

Pictured l-r: Hilda, Elizabeth, and Orville, ca. 1918

On March 4, 1908 Daniel Haury's son, Edwin, bought the farm from his uncle and aunt, John and Eliza Ruth and started his career as a farmer. Edwin and his wife, Susie Schmidt Haury, raised their three children: Orville, Elizabeth (Roberts) and Hilda (Goering) on the farm until Edwin was killed in a farm accident in July 1927.

During his lifetime, Edwin built the cow barn (1915) and erected a picket fence around the house. Orville was 5 years old when he rode to town with his dad in a wagon to haul the pickets to the farm. The house, barns and fence are all preserved and used today.

Susan Schmidt Haury in front of the barn, ca. 1920

When Edwin died, the farm was passed to Orville and he and his wife, Lucretia Parker Haury. They raised two children, Kent and Janelle (Karras), on the farm. They remained there until 2003 when Orville died. Kent and Myrla Cole Haury have lived on the farm from 2003 to the present.

In 2010, Scott Person, Myrla's son and Kent's step-son, bought the farm and it will stay in the family for at least another generation.

Wheat, alfalfa hay, milo, atlas, corn, oats, chickens, Short Horn cattle, steers and mules have been raised on the farm over the years.

Today's family includes four children: LeAnn Moran, Grant Davison, Scott Person and Heather Person, nine grandchildren and three great-grandchildren who live in Texas, Oklahoma and Kansas. *Submitted by Kent Haury.*

HAYDEN – Terry Hayden lives at 2131 17000 Road, northwest of Oswego. There he has 80 acres of ground that has been in his family for over 100 years. Terry's great-grandfather, E.A. Black, purchased the acreage in 1904 from James P. Christy and E.R. Haggard. Mr. Black had moved here from Indiana and actually lived on a farm located on what is now the Kansas Army Ammunition Plant land. When the government decided to buy the land for the KAAP, they had to relocate. It is thought that they might have moved a house from the plant area to this farm. The house was later turned into a barn and used to milk cows. On this farm Mr. Black raised poultry and livestock and did gardening.

Daisy Black, grandmother of Terry Hayden

Later Terry's grandfather, Charles A. Black and his wife, Daisy, acquired title to the property. The properly was then passed on to their daughter, Elizabeth Hayden, Terry's mother. Now Terry has title to the farm.

The home that Terry resides in is the original home on the property built sometime between 1900-1904. A room was added to the original three-room house. Later when water and electricity came and a family member moved in, another bedroom and bath was added. The only other early structure remaining is a chicken house that Terry has made into his woodworking shop. Today the land produces milo, wheat, beans, and corn.

Terry and Velma (deceased) had three children: Crystal Kenworthy, Teresa Jones and Katie Hayden. Their grandchildren are Markel and Jacob Kenworthy; Shayanne, Cloey, Brody and Makayla Jones and Niab, Franco and Lazaro Hayden.

HELLER – Andrew Heller emigrated from Wittenberg, Germany. In 1884 he acquired 160 acres of land in Lincoln County, 8-10-9 where he raised alfalfa, cattle and hogs.

In 1905 the farm passed to his son, Charles Heller. In 1971 the farm passed to Emerson Heller, then his son, Loren.

The farm has grown to 320 acres where wheat, milo, feed and beef cattle are raised. At one time, 400 turkeys were herded around the farm to control the grasshopper population. There were also geese, chickens, cattle and work horses.

Loren and Kay still live in the two-story brick house built in 1929. They are still farming with son, Jonathan, assisting.

The Heller's have four children – Marlene Nelson, Lori Back, Joy Smith and Jonathan Heller. Grandchildren are Hannah, Gideon, Noah and Hezekiah Nelson; Kaylee, Colton and Kenna Back and Adilynn and Hudson Smith. *Submitted by Loren and Kay Heller.*

HELWIG – On Jan. 4, 1909 Mathias and Jennie Hook sold 160 acres to John W. and Charlotte Helwig. This farm is located in the southwest quarter of Section 28, Township 33, South of Range 23, east of the 6th PM, Cherokee County, KS.

The Helwigs grew grain, wheat, corn and oats, farming with horses. They milked cows, raised hogs, chickens and had a large garden to provide food for the family.

John and Charlotte had four sons: Harold, Ralph, Howard and Russell. The farm was acquired by their son, Howard W. Helwig, who in turn sold it to his nephew, Lloyd, in July 1987. Lloyd is the son of Russell; therefore, this farm has been in the Helwig family for three generations.

There is no home on the property presently, and the farm produces beef cattle and grass.

Lloyd and Vera Helwig have four children: Teresa, Karen, Dale and Mark. Dale, his wife Jodie and their children – Amber, Dylan and Hannah – currently work with Lloyd and Vera, producing top quality beef and hay. *Submitted by Lloyd and Vera Helwig.*

HENRY – Maxamilan Henry was born in Ulm, Germany, on February 6, 1832. At the age of 23, this farm boy boarded the *Nutemburg*, sailed to New Orleans, steamboated up the Mississippi headed to Kentucky, and backtracked to Leavenworth, Kansas to marry his German be-

The Henry homestead, 1880

Ning, Walter, and William, 1920

William, Pauline, and Walter, 1899

The Henry house, 1899

William and his daughter, Susie, 1923

trothed sweetheart, Laura Kasbohrer, in 1862. A short time later they moved to Northern Riley County, where they homesteaded 160 acres on Otter Creek, built a two story stone home, and had seven children. With the modern equipment of windmills, water tanks, and hay scales the farm grew raising cattle, hogs, mules, horses, and oats.

Their son William began the second generation, marrying a neighbor girl, Pauline Hedke, in 1893 to which three children were born. Working with his father, he expanded the farming operation with more livestock and a binding machine for shocking corn.

It is their son Walter that takes the lead for the third generation. After courting a school teacher, he married Bernice Kunze in 1935. Father and son work together to herd cattle and hogs on foot to the Randolph Stock yards eight miles away. Yet, they were at odds when it came to purchasing a tractor, especially when William's team of horses was needed to pull Walter's tipped over tractor.

Two sons came from the marriage, Larry and Terry, who continued the Northern Riley County farming tradition. Larry married Rita Faulkender in 1955, to which three sons were born. Terry, who lives on the homestead, married Tammie French in 1981. They raise soybeans, wheat, and cattle that drink from the same Otter Creek the cows did more than 150 years ago. Terry and Tammie have one son and three daughters, and although their 5th generation son, Dalton, does not currently live on the farm, he is actively involved owning his own cattle.

HERD – The Rich Rose Ranch Development Co. LLC is a partnership between the descendants of Margaret Rich Herd, who owned the Rich Ranch at the time of her death in 2010. The Rich Ranch, containing 440 acres, is primarily located in Section 22, Township 32 South, Range 19 West, Comanche County, KS.

It was first established by Horace H. Rich and his wife, Nell, in 1885. It has been continuously owned by family members. Margaret's grandfather and father were the owners before her. Her father built a cabin on the property in 1950, which is the primary residence today. Her father operated a rose testing facility on the property from approximately 1955 to 1978, establishing the Rich Rose Ranch. At one time, there were as many as 5,000 rose bushes. The deer are so plentiful today that you could not grow the roses.

There are three ponds on the property and a spring fed creek. Part of the property is cultivated; the rest is pasture.

HESS – In February 1912, Henry Hess bought 120 acres in the northeast 80 of Section 4, Salem Township, Allen County, KS. The farm produced wheat, oats and corn and had a dairy.

Paul Setter, Mike Setter, and Jeremy Setter

Upon Henry's death, the farm passed to his wife, Josie. The farm then passed to Josie's daughter, Mary (Hess) Setter and on to her son, Paul Setter.

The farm now has grown to 1,256 acres and produces corn, wheat, soybeans, milo, alfalfa, prairie hay and beef cattle. The dairy was discontinued in 1997.

Paul and Dorothy Setter have four children: Mike Setter, Melinda Herder, Michelle Setter and Amanda Ames. Grandchildren are David Setter, Jeremy Setter, Anna Setter, Ryan Herder, Janelle Herder Williams, Megan Herder, Briana Ames and Cade Ames.

Mike Setter and his son, Jeremy, farm with Paul.

This farm received a Soil Conservation Award. *Submitted by Jeremy Setter.*

HESS RANCH – John Hess emigrated to the US in 1853. He lived in Hermann, MO and married his wife, Gertrude Schmitz in 1854.

He moved to what would become Wabaunsee County, KS, in April 1858, settling on and later homesteading, 160 acres. The farm was extremely diversified, with the largest acreage in corn at 50 acres. He also owned horses, cattle and swine.

John Hess became a naturalized citizen on Sept. 29, 1864. He helped organize the first school in the area and became the first postmaster of Bismarck, KS.

The original 160 acres has been greatly expanded by each succeeding Hess generation. It is now known as Hess Ranch and consists of a custom-grazing operation along with yearling steers and a cow herd. The original 160 acres is now owned by the fifth generation, Jim and Alan Hess and their wives, Judy and Charla. They live on and manage Hess Ranch. Ownership of Hess Ranch land is shared by Jim and Alan and their cousins: Judy Johnson, Sandy Asendorf, and Robert Shindler. Jerome Hess, sixth generation, lives on the ranch and is assuming all management responsibilities. *Submitted by Jim Hess.*

HIBBARD – In 1869 Edward A. Hibbard and his wife, Mary, came from Pennsylvania to settle at the upper end of Carlisle Creek in eastern Greenwood County. They built a log cabin, which they lived in for two years while building their house. After getting settled, Edward had his father, Hiram Hibbard, come out from Pennsylvania, where he lived with them until his death in 1882. Edward had three children and his son, Charles Hibbard, took over the farm when Edward was injured in a bad accident with a buzz saw in1905. Edward and Charles built the second home in 1903, when Charles and Evie were married.

Charles and his wife, Evie Johnson Hibbard, had five boys and one girl – Claude, Vera, Blake, Andrew, Paul and Olin.

During WW II, Blake Hibbard and his family came back to the farm to help out, as three of the boys had gone into the military service. On July 11,1943 tragedy struck when the Blake Hibbard family lost their three children in a fire that destroyed the first house. After the war, Paul and his wife, Bernita Woodside, purchased the farm and Charles and Evie moved to Toronto, while Charles served in the House of Representatives in the 1950s.

Paul and Bernita had three children – Larry, Mary Alice Hibbard McNitt (deceased) and Gordon and in keeping with family tradition, were very active in the community. Paul served eight years on the Toronto School Board and also served as a county commissioner for Greenwood County. Their oldest son, Larry, came back to the farm after college and on Dec. 7, 1969 married Cynthia Leayman Hibbard and they built their home at that time. All four homes built on the farm were built within a quarter of a mile from each other.

Larry and Cynthia have two children, Clayton, who lives in Fredonia and is a manager for a John Deere agency. Paula Hibbard Blizzard lives in Basehor, KS and is an administrator for Children's Mercy Hospital. The Hibbard's have five grandchildren – Clara, Nicholas and Natalie Hibbard and Peyton and Connor Blizzard.

Larry and Cynthia continue to own and manage the homestead and approximately 5,500 acres of bluestem grass around the surrounding area. They have both been very active in the community and Larry is currently serving his third two-year term in the House of Representatives, representing District 13.

Both of their children and their families come back to help out on the ranch whenever possible and Clayton's son, Nicholas, is Larry's summer help, making him the seventh generation of Hibbard's to work on the farm. *Submitted by Larry Hibbard.*

The Hibbard family, ca. 2017

HIITTER – The Hiitter Century Farm in Lincoln County, KS is currently owned by Mr. and Mrs. Larry Hiitter. In 1892 Larry's great-grandfather, Frederick Hiitter, purchased the 160 acres in Section 3, Township 13 (S), Range 9 (W). He retained ownership until 1908 when it was deeded to his sons, William and Emile Hiitter. It was then passed on to Larry's dad, Walter Hiitter, who owned the land from 1931 to 1998. This is where Larry and his sister, Vera May, were raised. Larry inherited the land in 1998.

Walter Hiitter Family, 1953. Pictured l-r: Ethel, Larry, Vera, and Walter.

The Hitter family raised beef cattle, horses, chickens and swine. Wheat, milo, oats and alfalfa are the crops that have been raised over the years. A large vegetable garden and a yard full of flowers were a must.

Today the Larry and Janet Hiitter family includes two sons, Gerald and wife, Teri and Timothy and wife, Amy and three grandchildren: Kimberly and husband, Mitchell Meyer, Wyatt and Sam. All are living in Lincoln County and are active in the farming operation today.

June is a busy time on the farm with wheat harvest, planting spring crops and putting up alfalfa and you will find all the family helping out, whether it is running a piece of equipment or bringing meals to the field. Readying cattle for pasture or bringing them back home is another time the whole family helps out. On this Century Farm, it is truly a family farm.

HILDEBRAND FAMILY FARM – Many "Little Hildebrand Feet" learned to walk in the sandy soil of this Hildebrand farm east of Bedford. Many hours were spent fishing, wading and skating on the pond dug by Bob Hildebrand. Many excellent meals were prepared, shared and eaten in the little house.

Oct. 18, 1909 Simon Fritzemeier Jr. sold to William Hildebrand the 160+

The Hildebrand home

The Frantz Hildebrand family, front row: Maria and Frantz. Girl's row: Mary, Louise, Annie, Frieda, and Martha. Boy's row: Ernest, Henry, John, Charles, Louis, and William.

acres of ground that is still in the family today. Many William and Frantz Hildebrand children have owned, rented and lived on this farm throughout the generations. Frantz Hildebrand bought it from William, his son and Frantz's children inherited it from Frantz. Henry, one of Frantz's children, inherited some and bought the rest of the quarter which was then inherited by his children: Gladys Fritzemeier, Alice Weigand, Alfred Hildebrand, Lucille Moore and Robert Hildebrand. Jeff and Gerry Hildebrand bought it from Henry's (Jeff's grandfather) heirs. I know that at some time the families of William Hildebrand, John Hildebrand, Gilmore and Opal Hildebrand Priess, the Clifford Hildebrand family and the Robert Hildebrand family called this quarter home. Now it is the home of Jeff and John Hildebrand's Angus cow/calf herd. *Submitted by Mrs. Jeff Hildebrand.*

HISS – The Otto Hiss Century Farm in Barton County is currently owned by Verna (Hiss) Hewitt, Marilyn (Hiss) Button, Loren Hiss and DeAnn (Hiss) Hauser.

Otto Hiss immigrated from Germany and was one of the first farmers in the Barton County area. On March 8, 1899, he purchased, for $10.00 an acre, 160 acres in South Bend Township, Barton County KS. Otto and his wife Wilhemina, "Minnie" farmed, growing crops and raising cattle until they moved to town in 1927.

Otto and Wilhemina Hiss and children in front of their house built ca. 1899

The original house was built sometime before 1899, which has been remodeled. A barn was built around 1900 and both are still original to the farm.

Ownership included: Son, Fred Hiss, 1949-1995; daughters Verna (Hiss) Hewitt, Marilyn (Hiss) Button, son Loren Hiss 1995 — current, son Lawrence and wife Elinor Hiss, 1995-2012, DeAnn (Hiss) Hauser 2012-current.

Fred and Mary Hiss became owners of the land in 1949 and raised cattle and grew corn, wheat and milo. Fred also served on the school board for District 41 school.

The Hiss farm has been family farmed throughout the years. Since 1998 Rick Hiss, (great grandson of Otto), has farmed the ground, raising wheat and milo and occupies the pasture with beef cattle.

The Hiss farm was a favorite place for grandchildren to fish and play while enjoying wonderful meals that the family shared.

When the sun sets, a farmer's job may continue for several more hours. Farming the ground your great grandparents homesteaded is truly an honor and the dedication of farming lies within a farmer's heart. *Submitted by Rick Hiss.*

HOAGLAND – Sailing from Holland in 1624, Levi Martin Decker's ancestors farmed on the island of Manhattan in New York before moving west. After Levi served in the Civil War, he bought his own farm in Kansas in 1866. Between Topeka and Lawrence, his land was in the "Bleeding Kansas" area but Levi set deep roots. He was a founder of the Bethel Presbyterian Church, across Tecumseh Road from the farm. The church has long disappeared, but the cemetery there holds generations of the Decker family. Levi donated land and built the one-room stone Decker School, which all his children and

Pictured l-r: Dirck Dayton Hoagland, Norma Decker Hoagland, and Joe Hoagland

Decker School, built in 1881

Stone marker for Decker Farm

Front row, second from left: Levi Martin Decker. Behind him is Calvin Lewis Decker.

Pictured l-r: Lewis, Mora, Retta, Myrtle, Emma, and Melvin

Calvin Lewis Decker

Decker farmstead, ca. 1940s

Calvin Lewis (on tractor) and Dayton Lewis Decker (on loader)

Dayton Lewis Decker, steering the bike, with Norma behind him

grandchildren attended. The building today has been converted to a private residence.

Levi's son, Calvin Lewis, stayed on the farm. Calvin's son, Dayton Lewis, farmed his whole life alongside his father, except for the years Dayton served in WW II, in the 82nd Airborne Division. Calvin and Dayton began a commercial dairy, farmed their cropland and always kept a meadow of native prairie grass for hay. The horses they farmed with are gone, along with the ice house where they would store blocks of ice from the Kaw River, packed in sawdust.

Dayton and his wife, Jennie (Smith) founded the Decker 4-H Club in Shawnee County. Their daughter, Norma Decker Hoagland is today the sole owner of the Decker Farm. It is now a division of J&N Ranch, which she and her husband, Joe, established. J&N Ranch is the birthplace of Black Hereford cattle. Their son, Dirck Dayton Hoagland, has stepped in to manage their ranches, commercial feedlot and the Decker Farm as the 5th generation and he has two sons, Dayton and Reed. Located at 29th and Tecumseh Road, crops are still raised on the fertile Decker land and the meadow of native prairie grass remains unchanged. *Submitted by Norma Decker Hoagland.*

HOCKERSMITH – In the year 1907, John Mason Hockersmith bought one-half section in Gove County, 36-13-28. Cattle, chickens and pigs were the livestock raised on the farm along with wheat and cane feed.

The farm passed from John to Chris Hockersmith and Charley Hockersmith. Today wheat, milo, cane fed, corn and grass are produced on this land. The Lone Star Schoolhouse and Bank house are on the farm.

Keith and Naomi Hockersmith have three children: JoAnna Trezise, Kate and Amy. Jaylen Trezise is their grandchild. *Submitted by Keith Hockersmith.*

Naomi and Keith Hockersmith

Keith on Easter Sunday

HOFFMAN – Ferdinand Koett came from Germany and became a citizen of the US on Sept. 2, 1876, Pittsburg, PA. He homesteaded 150 acres Aug. 15, 1893 in Barton County in Kansas. Section (12) Township (19) Range (11), eight miles northeast of Ellinwood, KS. $13.44 per acre.

Vic and Evelyn Hoffman

I own the 80 acres on the north. The home he and his wife and family lived in was on the South 80 acres. No buildings remain there.

Great Grandfather Koett's home received notoriety as being haunted. In 1927 an interesting picture and article appeared in the *Great Bend, KS Tribune* headlined "Prankish Ghost" causes Koett to leave his home. This was the son, Fred Koett and family who lived there at the time (My Dad's Uncle Fred).This story was published along with others in "Haunted Kansas Ghost Stories and other Eerie Tales" by Lisa Hefner Heitz. It is titled "The Haunted Farm." My dad, Ferdinand, went to his grandfather's house to check out the story and planned to stay the night. He said he didn't stay very long. There definitely were strange things happening. No one was living there at the time.

Lineage of ownership: 1. Ferdinand Koett (Evelyn's Great Grandfather). 2. Bernard and Catherine (Koett) Johanning (Grandparents). 3. Catherine Johanning and sons: Herman, Leo, George, Bernard Jr.,

Louis (Grandmother and Uncles). 4. Louis Johanning (Uncle). 5. Thelma Dreiling, Arlene Kutina, Evelyn Hoffman (sisters). 6. Evelyn Hoffman (present owner).

Evelyn and husband, Vic Hoffman, have five children, Alan Hoffman, Joan Jerkovich, Wayne Hoffman, Leon Hoffman, Ann Musser, (deceased Aug. 25, 2016); 10 grandchildren, six great-grandchildren.

We have been married 63 years, and farmed & ranched in the Barton County/Hoisington, KS area and are now semi-retired. *Submitted by Evelyn (Johanning) Hoffman.*

HOGAN – The picturesque Century Farm in Lincoln County, KS is currently owned by sisters, Dorothy (Hogan) Hollenbeck and Janet (Hogan) Marshall.

In 1897 the farm was purchased by John W. Wilcox, our great-grandfather. Family ownerships followed: Romeo N. Wilson (grandfather, 1916-1959), Uncles Harold and Eugene Wilcox (1959-1973),

John Wilcox traded wood from the creek area on his property for the stones used to build this fence in the 1870s. At one time the fence surrounded an 80-acre field.

Bernice Wilcox Hogan and Claude C. Hogan (our parents, 1974-1988), Dorothy Hollenbeck and Janet Marshall (1988 to present).

Romeo "R.N." Wilcox and wife, Clara, reared their family on the farm. They produced beef cattle, swine, milk cows and chickens. They raised wheat, milo, oats and alfalfa and added a pond, built to complement the wooded area on Spillman Creek, which runs through the farm. In the early 1900s, they traded wood from the creek area for stone to build a fence, which at one time surrounded an 80-acre field.

Today the family extends down to Dorothy, Janet and great-great-grandchildren: John, Larry and Bradley Hollenbeck and Cathy Gaines and Cindy Knapp, who make their homes in Kansas and/or Texas. *Submitted by Dorothy Hollenbeck.*

HOLMAN – In 1883, Willis Benjamin Warner, the youngest son of Edward and Lillian Warner, left home from Leeds, New York, at the age of 20 to head West. By way of work on a Percheron ranch in Montana, he arrived in Kansas in 1884, traveling to the Sawlog Valley area west of Kinsley in October, 1884, where he purchased a relinquishment quarter. Next he bought some cattle, then added a registered Angus bull in 1891 which was the beginning of the all black herd of commercial Angus cattle which are still produced today. He married fellow homesteader and schoolteacher Charlotte Breakey and had four children, Ruth, Kate, Lucy and John. He and son John raised cattle and grain and forage crops until W.B.'s death in 1943.

Son John married Ruby Johnson and with their children Elaine, Willis, Kathleen, Jack, and Kelvin ran the ranch, calling it Warner Angus Ranch. It became renowned for the all black livestock raised there including black registered Morgan horses, which are also still produced and used to work cattle today.

Son Willis, with his wife Ann and children Monica, Melissa, Marcella, Willis B. III, and Myrinda ran the east side of the operation, becoming Black Diamond Angus Ranch. John passed away in 1996, and following son Willis' passing in 2009 and son Jack in 2011, the land went into a family land LLC managed by daughter Marcella Warner Holman and her husband, Dr. John Holman. Always a family operation, the grandchildren, Mariah, Marissa, and Maleah Walker, Grace Wright, and Emmarie and Aria Freeman and their moms help when they can.

Early photo of the Holman farm

The ranch still runs commercial Angus and grows forage crops for the cattle both harvested and grazed and is still known for its all black livestock, both Angus cattle and parading black Morgan horses!

HOLT – Fifth generation Kansas farmer Michael Holt is particularly proud of the framed Land Patent assigned to great-great grandmother, Elizabeth Hamby Holt, dated March of 1864 and signed by President Abraham Lincoln.

Ancel Holt harvesting wheat with WC tractor and all crop harvester

Ancel Holt loading wheat in a box wagon, ca. late 1930s-early 1940s

Elizabeth, one of 14 children, had moved with her family by covered wagon from the Cumberland region of Tennessee to Missouri about 1840 where she met and married William B. Holt, who was a widower and was moving his grown family to the same area. Together, they had three boys before the father died. The Kansas-Missouri border was a dangerous place prior to Civil War. With her brother, Issac Hamby's urging and help, the now widow, Elizabeth, with young sons, Henry-13, William-10 and James-8, crossed the Missouri river on a flat boat and landed at Iowa Point, KS. They then took up a claim and homesteaded a quarter of land in Doniphan County. Young William would die that same year of typhoid fever.

The sight was chosen because Cedar creek provided a good source of water for livestock and for their small house, which they built nearby. The creek could also flood, as they discovered the morning after a heavy rainfall when their toes dipped in the water as they slipped out of bed.

Next generation landowner James and his wife, Julia Jenkins Holt, would build their house on the top of the hill. They would farm and raise three children before venturing out to Orange, CA about 1906 and would remain there, leaving son Ancel, then about 19, in charge of the home place. Ancel and wife, Ella McCullough Holt, would also try California for a couple years, leaving Ella's brother in charge of the farm. But they apparently didn't "find it there" and returned to the home place in Kansas in 1915, raising three sons and a daughter.

All of these four were farmers at heart but the youngest, Leonard, remained to work the land and raise the livestock. Leonard would spend his 92 years with the land and now his youngest son, Michael, intends to do the same. *Submitted by Michael J. Holt.*

HOPPER – Heather Franklin and George Hopper are proud to own the 160-acre farm settled by their great-great-grandmother, Rosa Burger Tuthill Schiever.

The farm was purchased in 1877 when Charles Tuthill came to Saline County from Randolph County, IL. His family were furniture makers there and he brought furniture with him for himself and other family members. On his land, he built a two-room house, planted an apple and other fruit orchard and grew grains. Hunting pheasant and rabbit to ship to restaurants in Chicago provided needed income.

Charles married Rosa Burger in 1902. Charles died two years later of typhoid fever leaving Rosa with the farm and an infant son, Harry.

Rosa had been a milliner, working in Salina, before her marriage. Her family lived on a farm near New Cambria, Saline County.

Working on a neighboring farm was Fred Schiever, who had ridden the train from Chicago and stopped in Salina. He walked to a farm where they gave him a job. At one time, he was sent to help neighbor Rosa and her brother, John, with building onto Rosa's house. Fred and Rosa were married four years later. She and Fred had two children, Lloyd and Grace.

Harry completed a degree in electrical engineering at Kansas State University and moved to New York state. Lloyd contracted Hodgkin's disease and died around 1940. Grace earned a nursing degree from St. Francis School of Nursing in Wichita, KS. Grace remained in Kansas and married Raymond Devinish. They lived in Salina, part of the time on the farm and had two girls, Rose Marie and Virginia. They purchased the farm and later Virginia and Andy Hopper purchased it from Grace

Virginia moved to Thomas County to work as the Extension Home Economist before marrying Andy, a local farmer. Their family was Heather and George. The farm has now been deeded to them.

Heather and her husband, Howard Franklin, are living on it. George and his wife, Cynthia and their children – Blake, Bradley and Isabella – live in Topeka and help with the manager. All enjoy the land. *Submitted by Virginia Devinish Hopper.*

HÖRLACHER-HERLOCKER – Hans Michael Hörlacher and family emigrated to Philadelphia, PA about 1700. He had married Maria Veronica in Germany. Her maiden name could not be confirmed. Their children were Maria, Daniel, Susanna, Johann Peter and George.

The name on the ship list was written Hörloger, but Hörlacher was written on the oath of allegiance. Hans Michael was invited from Baden, Wurttemberg and Bavaria, Germany area for his farming expertise by Wm. Penn.

Kevan Schupbach, Dr. Mindi Garner and family, ca. 2017

The L.D. Herlocker family

The Herlocker farm

The fifth generation descendant of Hans Michael from Johann Peter was Lorenzo Dow, known as L.D., Herlocker. The name spelling was changed in Crawford County, KS in middle 1800s.

Southeast Kansas Herlockers' are descendants of L.D. He homesteaded a 160-acre farm seven miles northeast of Girard in 1871. L.D. had two brothers, Frank and Tudor. They jointly purchased a threshing machine and for many years they followed this business. L.D. was a politician respected by the community and was a leader with a jovial personality. He was County Commissioner for 12 years, as well as Sheriff of Crawford County for two years. Records reflect his being a Trustee of Sherman Township for several years.

He married Elizabeth Cramer while in Ohio and soon moved to Anamosa, IA, then to Kansas for homesteading. Four children were born to this union: Elmer, John, Edward, and Ida. Their son, John, married Ella Marsh, whose parents, Joseph and Cynthia Marsh, homesteaded 160 acres in 1880 directly south of L.D. and Elizabeth's homestead. Previous to John's marriage to Ella, L.D. and Elizabeth had purchased 80 acres from Joseph and Cynthia. Upon their passing, John and Ella inherited the total 160 acres.

When John and Ella passed, their son, Glenn, purchased the 160 acres, 10 years after sharecropping with his mother, Ella. He purchased the farm from his siblings: Ethel, Lester, Edna and Eileen in 1955. All the five children took care of their mother financially until her death in spring 1960.

Glen and Ruby Herlocker successfully farmed the land until their passing in spring 1993. Their only child, Madrienne, inherited the 160 at this above date. She resided at the homestead from 1993 until 2000. At this time her son, Kevan Schupbach and his wife, Dr. Mindi Garner, are current residents along with their children: Sierra, Skyler, Marrone and Milan.

This is a synopsis of a journey from 1700s, 1800s, to the present time of this writing in 2017, totaling 137 years. As Noah Webster once said, "The least people can do is to have a history of themselves for those who come after." *Submitted by Madrienne Herlocker, great granddaughter of L.D. and Elizabeth Herlocker and Joseph and Cynthia Marsh.*

HORNER FAMILY FARM – John Clough and family sailed from London, England on the ship *Elizabeth* in 1635. Ten and one half weeks later landed in Boston, MA.

Through the years many had migrated to Iowa and further west to Oregon. Naturally they farmed to maintain themselves like everyone else.

Fernando Clough, his wife, Alma and their oldest son, A.E. "Bert" Clough and his wife, Edna, started with other families headed for Texas. They stopped in this area to rest before resuming their journey. Fernando liked the area so much that he bought this farm of 160 acres in 1884. Another 160 acres was bought later on. Their main crops were wheat and corn and they also had cows, chickens and a few pigs and hay.

Bert Clough took over around 1930. After his death the farm was passed to his wife, Edna. She leased the farm out. After her death in 1967, the farm was passed to their daughter, Doris Clough Harrington. My father, William Harrington, was caretaker until 1995 when my husband, Dennis Horner and I, Harriet Harrington Horner, bought the farm to keep it in the family. We have cows of which half are Herefords. We also have a few Hereford pigs. Our main crop is hay.

We have a son and family in Wyoming. A daughter and family here in Kansas. We have seven grandchildren.

I spent as much time as I could here on the farm when I was growing up. I feel it is an honor to walk where my ancestors walked. *Submitted by Harriet Harrington Horner.*

HOWELL – The Howell Century Farm in Marshall County, KS is currently owed by Lyle and Ann Howell. The Howell history reflects 134 years of farming in Kansas. Lewis R. Howell moved to Kansas by covered wagon from Illinois in 1883, looking for a better place to live. Lewis owned a farm in Rice County, KS for two years before he sold this farm and relocated to Marshall County, KS. Lewis purchased the Howell Century Farm in 1908. Ownerships include: Lewis and Millie Howell, 1908-1932; Fred and Mary Howell (son of Lewis), 1932-1974; Lyle and Ann Howell (grandson of Lewis), 1974 to present.

Lyle and Ann Howell

Aerial view of the Howell Century Farm

Thru the history the Howell Century Farm operations have included: Dairy, Swine, Beef Cattle along with crops raised: Wheat, Corn, Soybeans, Milo, Alfalfa and Prairie Grass. Land Stewardship and animal husbandry/welfare have always been key to our operations.

Church, 4-H and school activities were important to the family. Family gatherings for the holidays are always special memories and still celebrated today. Today the Lyle and Ann Howell family includes: Sons - Joe and Pat; five grandchildren - Aaron, Alyse, Matthew, Meredith and Mark; three great-grandchildren - Charlie, Lillian and Henry, located in Kansas, Missouri and New York. *Submitted by Lyle L. Howell.*

HRABE/OCHAMPAUGH – The Hrabe/Ochampaugh Century Farm in Rooks County, KS was homesteaded by Frank and Anna (Rostocil) Hrabe on January 9, 1892. Their first home was a 14 by 16 ft. house on the property. A total of 11 children were born while they lived on Section 26-8-17.

Their first child, Joe J., married Rose (Ruzicka) on Aug. 22, 1916 and set up housekeeping on Section 26, in the former home of Joe's parents. They raised four children. Their second son, Paul J. was born May 29, 1919 and father of Ann Hrabe (Paul's oldest child). She and James Ochampaugh were married July 15, 1962 in Stockton. They later purchased the Hrabe homestead in March 1959.

They built a small dairy barn and the dairy portion of this farm began. No one in either of the families was a dairyman, so it was a matter of building one from scratch. Through the years, many improvements were made. A modern dairy barn was built in 1973 and a large free-stall barn was added in 1979, along with a liquid manure system. Today the diversified farm produces wheat, corn, milo, soybeans and sunflowers.

James and Ann are proud parents of three children: Jeff, Amy and Jason and six grandchildren. All three of their children were active 4-H'ers and FFA'ers. They worked alongside their parents and helped develop the farm as it is today. Jason has returned to work on the farm and carries on a strong tradition of farmers.

Only three couples have lived on Section 26 in 125 years. James and Ann grew up in the Stockton area and did not realize how close they would return to their roots. Ironically the farm on which they've lived for over 50 years is located about a mile from James' grandmother's homestead. *Submitted by Ann Ochampaugh.*

HUCKE – Gary and Lou Hucke live five miles northeast of Mound Valley on the acreage his family purchased in the early 1890s. Gary's grandfather, Oscar E. Hucke, first owned and worked the land. Later it was acquired by his father, Bert Hucke, who farmed it before Gary.

Members of the Hucke family, ca. 2017.

Gary is manager of the Mound Valley branch of the Bartlett Coop.

Gary and Lou still live on the farm and raise American Quarter Horses. Lou raised her first foal in 1958 and has had several hundred since then. Their kids have shown several American Buckskin Horse Assn. World Champions. They are particularly proud of Monte riding their home bred and raised FSF Tennessee Classic winning for the third year in a row the World Champion Halter Mare. She also won the National Snaffle Bit Assn. Walk Trot Class and the same class World Champion.

Gary and Lou have four children: Tracy Hucke, Bryan Huckbuella, Robyn Ankrum and Monte Lanoy. Grandchildren are Tim and Samantha Ankrum, Justin Hermreck and Brent and Tara Hucke. Harlee Ankrum is their great granddaughter. *Submitted by Gary and Lou Hucke.*

HUDSON – Samuel Hudson, my great grandfather, was born in Ohio in 1851. After his mother died at Samuel's young age, he lived with an uncle in Indiana. As a young man, he drifted from community to community, looking for jobs. In Iowa he made friends with four other men and they left for the west in the spring of 1871. Samuel was 20 at the time and he had to wait till he was 2l to file his homestead claim. He lived in a dugout that was built extra sturdy to prevent its collapse should buffalo stampede over it. In 1873 he married Hannah and they lived in that same dugout for several years. Samuel and his four friends built a thriving community named Athol, located in Smith County, KS. They named the town of Athol for the Rock Island Rail Road conductor's daughter.

My grandfather, Jason, was born in that dugout and he became the owner in 1917. Jason and his wife, Pearl, had 14 children, my mother being the oldest. In 1945 my mother and father became the owners of this 1and. They had an orchard that included pears, apples, apricots and mulberries. In 1945, the house had been destroyed by a fire and to this day the only thing that remains is an upright silo.

When my parents' estate was settled in 1996, I became the owner. My daughter, Donna, will take over the farm when I pass away. In 2002 I received the Farm Bureau Century Farms Award and to this day that sign hangs on the upright silo located on the farm. *Submitted by Elzene Kirchhoff.*

HUNTINGTON/COWLEY FAMILY CENTURY FARM – The Huntington family chose to make a memorable trip from England (1633) to Connecticut where they began a new life. Many of them became educated scholars, lawyers, statesmen, doctors, farmers and ministers. A distinguished Samuel Huntington became the First President of the Continental Congress and a Signer of the Declaration of Independence. When Samuel passed away his adopted son, (his brother's son), Samuel H. Huntington, left Connecticut for Ohio and became Governor of that state. A son, S.P. Huntington, was born there in 1824.

Pam and John Cowley of Locust Lawn Farm

The original home built in 1901

S.P. Huntington, his wife and several sons pursued their new home territory from Painesville, OH to Eureka, KS, Greenwood County. He was known as a community dentist and a productive farmer/rancher. The property purchased in 1869 contained 160 acres of river bottom farm ground and native pasture at a purchase price of $11.25 an acre. The two bachelor sons of S.P. Huntington, Colbert Claude "Uncle Bert" and Alfred Phelps "Allie," remained on the property during their lifetime, living in a small home. Uncle Bert, a Civil Engineer, designed the water system of Eureka, building the water plant and Fall River dam located at the west edge of the city. The original dam is still functioning today. Uncle Bert and his brother, Allie, built many of the original buildings on the property, which include the barn, granary twin silos and a home built for Allie's bride, Hattie, in 1901. The brothers also designed the sewer and water system that supplied the stock and house. The outside toilet was connected to the home sewer system and flushed with a bucket of water. A placed water tank in the 1901 home attic also supplied the house with a full bath and latrine.

In 1901 Allie married Hattie Roby and two sons, Colbert and Samuel Alfred "Fred," blessed this union. Fred married Phyllis Denchfield in1932 and one daughter, Pamela, was born. She married John Cowley in 1964 and after several teaching moves, they secured a home in Eureka. With the death of Pam's father, Fred, in 1971, the couple became active in beef and crop production while teaching school youth. The couple raised two sons, Lance and Craig, who were active in the farm management and 4H.

Locust Lawn Farm still supports the usual production of 60 head of cows and calves with grain/hay crops. The south is bordered by Fall River where there is an abundance of wildlife present on the property. The Cowley's have added a watershed structure, workshop, several additional waterers and provided new cattle pens with pasture fencing. The barn and home have been maintained to the original status and Pam and John are living in the 1901 original structure. Their sons are still active in preserving the Flint Hills Bluestem Pasture and Historical Heritage of the property. The property now consists of 1,040 acres located in the productive Tall Grass Prairie of the Flint Hills.

Locust Lawn Farm and the Cowleys have received Farm Bureau awards of the Natural Resource, Century Farm, Farm Family of the Year and John has received the Leader of the Year.

JANZEN – The Janzen Century Farm lies in southern Ellsworth County. E.E. Janzen, who was born in 1872 in the Green Garden Township of Illinois, purchased the first 160 acres in 1907 from D. Noyes Burton who owned the then well-known Burton Ranch. He farmed it between the years 1907-1915 along with raising sheep, cattle and his family.

Thelbert W. and Darlene Janzen with their dog, Tater

Today the 320 acre Century family farm is owned by Thelbert W. Janzen and his wife, Darlene, of 69 years. Thelbert, who is the great nephew of E.E. Janzen and Darlene, bought the farm in 1955 from his Uncle Henry Janzen. Henry had bought the farm from E.E. and when he married his wife, Hannah, the farm was doubled because of the land she owned to the north. They raised a family of two daughters before Hannah passed away. Henry later remarried and when they decided to retire, sold the farm to his nephew, Thelbert.

Gambrel roof barn on the farm place in the process of being repainted

Thelbert and Darlene had a rough start when just after they purchased the farm, one of the worst hailstorms in some time left their wheat fields looking like plowed ground with just straw stubble showing. In April of 1972 a tornado dropped down over the buildings leaving the house alone, slightly shifting the gambrel roofed barn on its foundation and then took half of the roof of a barn sitting to the northeast of the house and completely demolishing a barn further on to the northeast, scattering bits and pieces across the section. Only its foundation was left with a stack of bales, which had once been inside, untouched.

Today Thelbert and Darlene reside in the house they built in 1976 with a good portion of the farmyard looking the same as it did when Henry built on it. Their daughter, Gloria, did move the old garage to her family's lot in Kanopolis where it still stands today. The barn, which lost part of roof, now has a lower roofline and has gone through some interior changes, being converted to a granary at one time and now a workshop with the attached cattle shed still housing calves.

The large round-topped barn is still standing to the east of the house. Henry Janzen built it in 1924 and it is thought to be from a Sears Roebuck mail order barn kit. The reason for this thinking was that a bundle of cut-to-specification boards were tucked in under the rafters. The barn houses stalls used for teams of horses and stanchions for at least seven milk cows with a lean-to on the lower east end still sheltering cattle today. The large hayloft not only held hay, but allowed for roller skating and the playing of basketball as the hoops are still there. At least one Halloween party and one graduation party each were held in it.

The large chicken house was torn down in 2016 to make room for more farm equipment, but the tank house sporting the "Farm Bureau Century Farm" sign still stands beside the windmill in the center of the farmyard today.

Thelbert is mostly retired having turned over the farming to the next generation of Janzens, his son, Michael. He still likes to drive the tractor at times and kept a small cow and calf operation. Cattle sporting the Bar J brand are to be found in the pastures to the north and east of the house and wheat the major crop on the 80 acres on farm ground on the hill.

Today the Thelbert and Darlene Janzen family include four children and their spouses; 10 grandchildren; and 13 great-grandchildren, one who has already gone on ahead to join his Lord. All live in Kansas with the exception of two grandchildren and their families who live in Virginia and Florida. *Submitted by Lonnie Vopat.*

JAY – The ownership of my Century Farm starts in 1875.

The history begins when George and Sarah Jay came west from Marshall County, state of Illinois, in 1874. They came by railroad, not a coach like you would ride, but in a boxcar.

Current home of Susan Kay Weller Corkrey

George and Sarah Jay, my great-grandparents, purchased 87 acres in Saline County, located one-fourth mile east of Kipp, KS from Galveston Taylor on June 18, 1875. Mr. Taylor had acquired the land by homesteading under President Ulysses S. Grant. Here they raised their two daughters, Lilly my grandmother and Rose. At the time of my great grandparents' death, the farm was then owned by my grandparents, Lillie Jay Weller and Clarence Weller. Here they raised six children on the farm. After my grandparents' death, it was owned by their daughter, Beulah I. Weller, my aunt. In the year 1993 my father, Keith J. Weller, was given ownership. He raised wheat, corn, milo and soybeans on the farm, which is now four acres smaller because of a creek and road.

I recently inherited the land. So after 142 years, five owners and four generations, this farm remains in the ownership of a direct descendant, a great granddaughter. My brother, Pat Weller, now is farming the land.

I live in the same house built by my great-grandfather in 1876. This house has survived floods and tornadoes over the years. It has been remodeled and updated, but remains on the same foundation. There have been weddings, births and deaths in the house, many celebrations at holidays, family picnics and gatherings have occurred. My ancestors made the house a home to many in the last four generations. Today I have two children and six grandchildren, so may this continue for many more years. *Submitted by Susan Kay Weller Corkrey.*

JIMERSON – William and Elizabeth Jimerson, and their two daughters, Elizabeth and Julia, came to Kansas in a covered wagon in 1854. They established their home on a squatter's claim in township 6, range 3-9, Rock Township, Marshall County, KS.

The Wanklyn home

It was necessary that they were near a good source of water. They chose a home site by a big spring on the Robidoux Creek, that the Indians told them "never went dry." It still flows today. They lived in their covered wagon for 50 days while trees were felled and sawed into boards, to make a dugout into their temporary home. The dugout is still visible in the bank, along with the remains of the house they later built to the south.

Daughter, Elizabeth Jimerson, married Martin Life. Mr. and Mrs. Life gave a plot of ground for a cemetery on their farm. The Life Cemetery is still a well-kept cemetery to this day. The Life's were instrumental in the founding of the Life School House, District No. 10. The Life School was built in 1859. It still stands at the original location, and is now a private residence.

Martin and Elizabeth Life's daughter, Sarah Elizabeth, married Levi Goin and they were the next owners of the ground. At their death, their son, Martin Goin, became the owner in 1936. After Martin Goin's death, Kenneth E. Wanklyn, a great-great-grandson of William and Elizabeth Life, purchased the land in 1954. He, and his wife, Dolores Harry Wanklyn, still own the farm in 2017.

Kenneth and Dolores have three children living in the area: Tim, Cindy and Jim. They have two grandsons, Ben Koch and Cody Wanklyn. The farm is still an active part of the Wanklyn family for farming, hunting and fishing. Tim Wanklyn farms the cropland. *Submitted by Kenneth and Dolores Wanklyn.*

JOHNSON – The year was 1898 when Sherman S. Johnson acquired 235 acres in Phillips County, KS. The land included 77 acres in S 2 SE 28-2-17 and 158 acres in NE 33-2-17, all in Greenwood Township. Livestock on the farm consisted of beef cattle. Grains were wheat and milo.

This farm passed from Sherm Johnson to his son, Floyd Johnson, then to Neil Johnson (grandson) and to Norma Johnson, widow of Neil.

The farm still retains 235 acres and still produces wheat and milo and supports beef cattle. Craig and Dustin Johnson now farm this land.

In 2003, the Century Farm Award was presented to this farm in Phillips County.

Neil and Norma Johnson had two children, Craig and Blane. Grandchildren are Dustin Johnson, Danielle Landen, Allison Bohall and Arika Luedke. *Submitted by Norma Johnson.*

JOHNSTON CENTURY FARM – This Century Farm was purchased Jan. 16, 1901 by my maternal grandparents, Isaiah and Anna Miller. In 1933 ownership was transferred to my mother, Mabel Miller Johnston and her brother, Fred C. Miller.

The Johnston farm house

In 1952 my brother, Peter S. Johnston and I became Joint owners.

In 1911, I purchased my brother's share of the farm. This Century Farm is just a 1/2 mile from the city limit of Yates Center, KS, yet it is in beautiful, rolling country. It is well-known for its unique spring water source that has never been known to go dry.

Also, it has a sandstone rock quarry where my ancestors quarried rock for the foundation and walls of various buildings.

A two-story frame home was built in 1907 with a rock foundation and rock steps for the two-sided front porch. A rock-walled building for the springhouse and smokehouse still stand after 116 years.

On this farm is a fully equipped railroad caboose from the Union Pacific Railroad. When my brother, Pete and I were young, in the '20s, '30's and '40's, we farmed this land with horse-drawn equipment. Corn, wheat, oats, Kafir corn, alfalfa and prairie hay were grown. We also raised beef cattle, swine, chickens and turkeys.

We milked 10-12 cows, separated milk, bucket-fed calves, sold eggs, churned butter and sold cream to the local creamery.

We had an incubator to hatch baby chickens in an upstairs room, heated by the chimney of the wood cook stove.

This farm is 160 acres with two houses 1/2 mile apart. The land is now share-cropped, with approximately 30 acres under cultivation – 30 acres hay and 100 acres of grazing land. *Submitted by Harold Johnston.*

JORNS FARM – The Jorns' Family is celebrating 125 years of farming this year. The long farming tradition of the Jorns' Family began in March of 1885, when John Jorns and his brother, William Jorns, claimed a stake on a quarter of land north of Preston, KS. It has survived five generations and at one time had eight family farms in operation.

Kc Jorns with his great-grandfather's plow that broke out their land

The Homestead of the Jorns Family farms, now operated by Kc Jorns, was built and lived in by John and William Jorns, both brothers walking to Kansas from Wisconsin in the 1870's and finding work for ranchers in the area. In March of 1885, John claimed a stake for the Homestead north of Preston and lived there off and on for the next 15 years, while brother William married and lived there, rearing three of his four children in the home. In 1899 John returned to claim his farm and William and moved down the road with his family and built his own house in 1900.

The remodeled Jorns home

Today, the Family farming operation is primarily done by Ken, Jan and Burke Jorns. All residing in Turon, KS. Their operation spreads over 5000 acres and entails cattle with both, dryland and irrigated farming.

The original house of the Jorns Family is still intact today. Kc lives on the farm and operates the farm. After Kc's dad, Paul Jorns, passed away in 1997, he rented the farm for many years as he slowly began to make improvements to the farm. In August of 2007, Kc moved from the farmstead he had purchased from his brother, Curt Jorns, back to the Homestead and remodeled the original house, inside and out. He continues to make improvements as he lives there, rebuilding the original barn and other buildings on the property, replacing coral fencing with other fences yet to replace. Kc has been working on the farm that he was raised on - all his life. He has also worked as an auto and truck Technician in the Pratt area for more than 35 years, the last 15 being with Kansas Dept. of Transportation. Kc would like to return to farming his own land but says that he may have to wait until he retires from KDOT. *Submitted by Kc Jorns.*

JOURNOT – On April 10, 1869, Eugene Journet purchased 160 acres located in Labette County (32-31-19), where he raised corn and livestock. Through the years the farm passed to Robert E. Adams Sr., and is now owned by Robert E. Adams, Jr. Robert Sr. planed the first wheat crop using a dump truck to pull the drill. Robert Jr. has helped on the farm his entire life and is still working at 78-years-old. He is married to Carolyn S. Adams; they have a daughter, Dana Jean Adams, and two grandchildren – Justin Carl Willems, Levi, and Eugene Willems.

JOURNOT – On Aug. 3, 1895 Eugene C. and Annettie Journot acquired 200 acres located in Walton (31 South), SW 1/4 and SE 1/4 of NE 1/4 (Section #30), Range 19 E of the 6th P.M. Wheat, oats and corn were crops raised and livestock was cattle, pigs and chickens.

The farm passed from Eugene and Annettie (great-grandparents) to John F. and Annie Journot (grandparents) to George W. and Mary (parents) to William M. and Connie Journot (current owners).

The 200 acres has been retained and now produces cattle, wheat, beans and corn.

William and Connie L. (Baker) Journot have three children: William Aric Journot, Hollie M. Alvested and Andrea L. Dunseth. Grandchildren are Jonathan Beckman, Anastin, Hayden and Mason Journot and Ethan ad Isabel Alvested. *Submitted by Jane Journot Kelsey.*

JUDD FAMILY FARM – James Edwin Judd came to Kansas after the Civil War as a small child with his parents, Uzz and Martha Judd from Springfield, IL. They settled in Crawford County, KS. James married Belle Young in Girard. They remained in the area until their deaths.

Left to right, front row – J.E. and Belle Judd. Middle row – Dora and C.V. Judd. Back row – Willard Judd, Jim and Roy Brooks.

In 1909 J.E. bought 160 acres in Sheridan Township, Cherokee County. He built a house and buildings, farmed the land and raised livestock and chickens. J.E. and Belle had a son, Cornelius V. and a daughter. Life was never really prosperous for this family and I'm sure there were struggles at times.

C.V. married Dora Smith. After their son, Willard, was born, C.V. and Dora moved their house to the Judd family farm where it was set on a foundation and rooms added. A new garage and chicken houses were built. My grandparents were quiet, hardworking and unassuming people. C.V. was one of the early adopters for Farm Bureau insurance in Cherokee County and Kansas.

Both couples, J.E. and Belle and C.V. and Dora, celebrated Golden Wedding Anniversaries on this farm. Both were large gatherings due to proximity of family members.

Willard married Jennetta Brader and farmed with his Dad. The original farm became theirs when C.V. died.

During the years that C.V. and Willard operated the farm, production included wheat, corn, soybeans, milo and lespedeza. These crop rotations continue today. Willard removed all of the buildings, leaving only the pecan trees.

Their daughter, Brenda and her husband, Jim Dooley, received the original quarter from Willard and Jennetta and plan to hand the land on to their son, Scott. The farm never made anyone rich but provided for the hard working families that lived on the land.

JUNG – Grandpa Michael Jung (Young) came to America in 1856 from Wurttemberg, Germany at the age of 15. He used his brother's passport, who was drafted into Kaiser's army. After a 45 day voyage on a sailing ship, he arrived in New York City, then travelled to Ohio where he had another brother living. In 1861, the Civil War broke out and he went to Canada. He worked in lumber camps there for 50 cents a day. After the war, he returned to Ohio.

In 1871, Michael moved to Kansas and in 1872 he built a frame house near Globe, using lumber and sand from Overbrook, KS. In 1874 there

Michael and Katherine Young, ca. 1920s.

Pictured l-r: Edwin, Ivan, and Harold Jehle, ca. 1932.

At left: Anna (Seaman) Fox and Clara (Seeman) Gilbert, and twin boys George Seeman Jr. and Willie Seeman. Also pictured is Alfred Seeman.

was a grasshopper plague where they ate everything to the ground. My mother, Emma Jung, was born in 1891 as the youngest of five children.

Grandpa Jung purchased the home place near Worden in 1898 and farmed 65 of the 95 acres. The farm is1/4 mile south of the Santa Fe Trail and he remembered seeing wagon trains traveling west. A neighbor told me that after Michael retired, they rented the farm and raised broom corn in the field south of the house. In 1921, Emma married Walter Jehle and they all lived in the farmhouse for one year. Michael Jung lived to be 101 years old.

In 1922, Harold Jehle was born to Walter and Emma. My parents continued to farm and raise cattle and hogs. My dad was one of the first people to terrace his land and practice soil conservation. I grew up on the farm with my two brothers, Ivan and Edwin.

After selling my service station business in 1972, we began farming with my son, Charles. In 1974, I bought the home place from my parents, who had moved to Lawrence. We have received two farming awards for conservation practice. This involves maintaining terraces and waterways as well as turning rocky and wet areas into CRP ground. Today, my grandson lives on the farm with his wife and a 2-1/2 year old son. *Submitted by Mark Jehle.*

KATS – On Nov. 10, 1889 William A. Kats bought 160 acres located in Phillips County, Township 8, Section 4, Range 19. Grains crops were corn and wheat and livestock was cattle, hogs and chickens.

The first house on this farm. It is no longer standing.

The farm passed from William A. Kats to his son, Simon, on Dec. 7, 1917. On Dec. 4, 1929 the farm passed from Simon to his wife, Jennie S. Kats. Henry Kats, son of Simon and Jennie, and his wife, Alice, lived on the farm from 1941 to 1947. On Jan. 9, 1948 the farm passed to Martin and Nellie Bayens. Nellie was the daughter of Simon and Jennie. Martin J. Bayens, the son of Martin and Nellie, acquired the farm on Feb.7, 2006.

The center part of the barn is original but all others have been changed. *Submitted by Martin J. Bayens.*

KELTY – The now John and Jane Kelty farm was homesteaded by Jane's grandfather, George Seeman Sr. He was born in Wurttemberg, Germany and emigrated to Pennsylvania with his family at the age of 4 in 1865. In 1874 his family moved to Ash Valley in Pawnee County, KS. He homesteaded the farm on April 24, 1886. It was known as the "White Swan Farm." A Mr. Swan painted the barn red with a White Swan on the front of the barn. The barn still exists but it has been remodeled, however, a White Swan is still on the barn.

The Kelty farm; the barn is in the center with silver roof.

Alfred Seeman was next in succession to own the farm. He was the sons of George Seeman Sr. Alfred's wife was Mary and they had two daughters, Joyce and Jane.

Alfred and Mary were killed in a car accident in 1970. Current owners, John and Jane (Seeman) Kelty inherited the farm.

Their family consists of three daughters and two sons: Jamie Kelty, Jackie and Kirk Holmberg, Jennie and Chad Erway, Jonroy and Michelle Kelty and Joel and Morgan Kelty. They have 15 grandchildren: Kolby and Korey Holmberg; Landon, Madison and Logan Erway; Abbigail, Ben, Clayton, Daniel, Eli, Faith and Grace Kelty and Caden, Brennan and Farah Kelty.

Jonroy and Joel are both involved in the farming.

The farm started with mules doing the work. Now the farm utilizes GPS guidance systems, no-till farming practices and cover crops, etc.

In 1976 John and Jane won the Bankers Award.

In 2009 John and Jane, Jonroy and Michelle and Joel and Morgan won the Bankers Award.

In 2016 John and Jane won the Windbreak Bankers Award. *Submitted by John Jr. and Jane Kelty.*

KENNEDY – The Kennedy farm, consisting of 340 acres, was part of an original 21,000 acres acquired by William Kennedy in Marshall County, KS. The land in Section 34, Township 5 South, Range 9 East of the 6th PM was purchased in

James Kennedy (on horse), 1870-1945.

1866 (20 acres), in 1882 (240 acres) and the rest in 1884 (80 acres).

William Kennedy was the youngest of four siblings (James, Margaret, Bartholomew, and William) who came to America from County Tipperary, Ireland. He arrived in Marshall County, KS in March 1858 at the age of 25 years. He married Catherine Kelley, whose family had come from New York to settle in Marshall County in July 1856.

James Henry Kennedy (standing in front of barns), 1916-1993.

James T. Kennedy (leaning on truck), 1950-1998.

William raised cattle and hogs, grew wheat and other crops and freighted by wagon train across the plains. He was known to be "industrious, enterprising and honest...accumulating a competency of more than the ordinary farmer of the west." (Obituary, November 1897). By 1879 he had a large farm, a splendid stone house, and a fine herd of cattle (*Blue Rapids Times*, July 17, 1879).

William died on Nov. 19, 1897 at the age of 65 years. At the time of his death he had begun to work with pure bred Hereford cattle and his son, James, continued with this, shipping cattle to various parts of the country. William's grandson, James Henry Kennedy, raised Brahma cattle.

Although a Kennedy no longer resided at the farm after 1908, a daily visit was regularly made, driving 10 miles from Frankfort, just to check on how the farm was doing.

After William Kennedy, three generations of James Kennedys were to own and love the Kennedy family farm: James Kennedy (wife, Anna Wendling Kennedy) until his death in 1945, James Henry Kennedy (wife, Mary Clare Bacon Kennedy) until 1993 and James T. Kennedy until his death in 1998.

James Henry Kennedy left the farm to his six children: Mary Patricia Kennedy Solbach, James T. Kennedy, Francis R. Kennedy, Margaret Kennedy Dinsmore, Robert G. Kennedy and Katherine Kennedy Sperfslage. The farm is currently owned by four of the siblings in a family trust. *Submitted by M. Patricia Kennedy Solbach.*

KIMMERER – Our Century Farm that is located at W./2, SW/4 of 23-26-19 of Allen County, was settled by my grandfather, Michael Kemmerer, when he came from Germany in June 1879. He settled in the area and bought this farm in November 1902. Mr. Kemmerer, a stonemason, harvested stone and built a house and a barn from the natural rock of the farm. The barn is still there today. He and his first wife, Mathilda, had one son, Herman, before she died. Michael married Emma Krueger in 1896 and they had four more children: Charles (my dad), Henry, Alvena and George before she died in 1911. Michael and his wife officially purchased this acreage in 1902 where they had cattle, hay and wheat and raised their family.

Charles with the old rock barn in the background

Upon Michael's death in 1940, Charles bought the farm from the other siblings. Charles and his wife, Lucille, raised six children on this farm. They were Mary Kemmerer Stark, Sue Kemmerer Hurst, Carl Kemmerer (deceased), Gene Kemmerer (deceased), Helen Kemmerer Uitts and Carol Kemmerer Tucker.

Shortly before her mother's death, Sue Hurst and her husband, John, purchased the farm from her. We own it today and enjoy spending partial time there in the spring and fall. We have a hay crop every year and now have the pasture rented. The cropland is coming out of CRP and will be planted in the spring.

KING – The King family farm located on US Hwy. 36, four miles east of Fairview, KS was purchased Sept. 10, 1860 by Doodle King's great-grandfather, Ambrose Kesler and his wife, Anna Barbara (Weis). Born in Germany, they settled on the farm located in the Carson School District #1 of Brown County. They had three children. Mary, second born (Doodle's grandmother) married Richard King, in 1883 when they moved onto the farmstead. They had six children of which Frank Sr. (Doodle's father) was the youngest.

Frank Sr. met Irene Ward from Atchison, KS at a church picnic near Leona, KS and two years later married and moved to the family farm. They had three children: Ward, born in1926; Frank Jr. "Doodle" in 1929 and Frances (Wood) in 1931. They all graduated from Carson Grade School and the Hiawatha High School.

Frank Sr. farmed, raised livestock, dairying, custom harvesting, did livestock and grain trucking. It was a family enterprise with everyone involved.

Doodle and Arlene met at church and were married two years later, Dec. 20, 1949. They lived on the family farm. Doodle did livestock trucking, farming and had a real estate brokerage on the farm. Arlene had a beauty shop in the farm home and also was a licensed real estate sales lady. They had four children: Frank III, Richard and twins, Linda and Leisa.

Doodle and Arlene left the family farm to do missionary work, first with the 'Go ye Mission' in Tahlequah, OK, then Foreign Missions with New Tribes Mission, in Bolivia, South America.

Due to Doodle's health, in 1980 they moved back to the family farm to live and do recruitment for the mission until retirement in 2004.

In 1982, the Kings, their tenants, Mr. and Mrs. Elmer Middendorf, received the local Bankers Award for soil conservation practiced for the entire 80 acres.

As Kansas cerebrated 150 years of statehood, Doodle and Arrene, along with their twin daughters, Linda and Leisa, were recognized as one of 17 families in Kansas that had been kept within the ownership of the family for 150 years. The event at the Kansas State Fair Grounds in Hutchison, KS, was sponsored by the Kansas Department of Agriculture and a young alumni group at the Kansas State University. Governor Brownback and Secretary of Agriculture, Dale Rodman, addressed the group.

Counting it a privilege, the Kings continue to live on the family farm sharing God's Word as the opportunities arise.

Their good neighbor, Fred Maschewski, operates the no-till farm for the Kings. *Submitted by Frank Jr. "Doodle" and Arlene P. King.*

KISSEL – Sol Kissel, son of William (born in Ohio) and Kathleen Kissel, was born in Rulo, NE. When Sol got married to Mary, he worked for a lawyer, taking care of a farm near Kelly, KS. Sol decided he wanted his own farm and found a farm for sale and bought it. He bought the 160 acres located in Nemaha County, Granada Township, 16-4-14 in 1909. Cattle and hogs were raised on the farm.

When Sol and Mary died, their son, Theodore Jennings "T.J.", bought the farm from his brother and sister. The farm then passed to T.J.'s son, John. In 2004 John passed the farm to his sons, Jerry has 80 acres and Jon has 80 acres.

The King farmhouse, ca. 1904. Pictured are Uncle Joe (in buggy), Aunt Winnie (in front of pony), Frank Sr. (front rider), Uncle Fred (behind Frank), Aunt Lillian (in front of Ambrose), Grandpa Ambrose, Grandma Mary, Aunt Florence, and an unnamed man to the far right.

T.J. was born Jan. 1, 1899, married Opal Reece on Oct. 10, 1923 and died Jan. 24, 1984. John was born May 7, 1925 married Lucille Kissel April 21, 1952 and died April 2, 2010.

Beans, oats, corn and cattle are now produced on these 160 acres.

While Sol owned the property, he built a four-room house, then a garage and a hog barn.

Jerry Kissel, retired in 2002, has no children and his farm will go to his niece and nephew. Jon and Joyce Kissel had two sons, Jon Russell and Ted (deceased). Jon's grandchildren are Ian, Owen and Evan. *Submitted by Jerry and Jon Kissel.*

KISTLER – The Kistler family name has been known in Thomas County since 1905 when Samuel Allen Kistler purchased the SE 1/4, 5-10-34 from the U.P. Railroad. Brother James Kistler purchased the adjoining NE 1/4, 5-10-34 and later sold to Sam.

The Ernest Kistler home

Children of Sam and Sarah – Elma, Roy and Ruby – all remained in the area as farmers. At 28 years old and with a strong Pennsylvania Dutch heritage, Sam, together with Sarah, provided a comfortable life on this 320 acres. Only 120 acres of the virgin sod was broken, with 200 acres fenced for sheep or cattle. Nearly all crops raised were marketed through livestock, sheep, cattle and eggs. Horses gave way to a tractor in 1932.

Sam and Sarah retired in 1945. The farm was rented to a grandson, Ernest Samuel and his wife, Anita. This recently married couple raised three children – Harvey, Stephen and Janet. Additional and was purchased and in 1964 they purchased the E 1/2, 5-10-34 from Samuel and Sarah's estate.

The farm, all dry land, for the next 67 years has produced cattle, wheat and milo and in later years a rotation of corn with no-till farming.

Sam and Sadi Kistler on the steps of their home, year unknown

The children of Ernest and Anita, all married, have raised their families in the area.

In 2012 Ernest and Anita retired and sold the farm to grandson, Cory Samuel Kistler, son of Harvey and Linda Kistler. With additional owned and leased land, Cory uses the latest technology and equipment in a no-till operation of corn, milo, wheat and cows.

The original 1915 house and 1916 barn (renovated) are still in use as new larger buildings have been added.

Like generations before, Cory remains active in church and community service.

KLOEPPER – The Kloepper Century Farm in Atchison County, KS is a continuation of more than 400 years of farming near Petershagen, Germany. Friedrich "Wilhelm" Louis Kloepper immigrated from Germany in 1883 at the age of 30 years and purchased 80 acres in Lancaster Township, Atchison County, KS. In 1905 he purchased the adjoining 80 acres to the south. He and his wife, Frederike, had added to the original three room house, making it an eight room house, a very large livestock barn, granary, vineyard and orchard.

Standing, l-r: Patricia Snyder, Linda Chartier, Larry Kloepper, Sally Ellerman, Bud Kloepper, Emma Jean Kloepper. Seated: Bill and Dorothy Kloepper

They raised cattle, hogs, horses, chickens, corn, oats wheat, and a very large garden.

Ownerships included: Wilhelm Kloepper - 1883-1913; his son, Louis Kloepper - 1913-1942; our parents, William "Bill" and Dorothy Jane Kloepper - 1942-1992; We the Kloepper children - 1992 to present. The original 160 acres are now owned by Linda (Kloepper) Chartier, Sally (Kloepper) Ellerman, and William "Bud" Kloepper. Bud lives in the home completed by Wilhelm and where three generations of Kloepper babies have been born. Emma (Kloepper) Warren and Bill and Dorothy's grandson, Alan Ellerman, own parts of the remaining 160 acres added by Bill and Dorothy.

The Kloepper family has always celebrated anything to do with family, making certain everyone felt a part of something much larger. We were taught to be proud of the little community of Old Huron where most of our family attended grade school, from 1894 till the school closed in the 1950s due to consolidation. 4-H was a big part of our lives as children, as well as horseback riding and "exploring."

The farm continues to be a diversified operation with cattle, corn, soybeans, oats and hay being raised. The Bill and Dorothy Kloepper family includes six children - Sally, Patty, Larry, Linda, Emma and Bud as well as our children, grandchildren and great-grandchildren. *Submitted by Sally Ellerman.*

KNAPP – The Knapp family farm was founded in 1880 when my great-grandfather, Squire T. Knapp traded a farm near Fredonia for our present farm in Greenwood County, although the family didn't move there until 1884.

My grandfather, Vernor, was 2 years old at the time. Squire and Lida had three sons and a daughter: Clifford, Guy, Vernor and Vera. Vera died at the age of 2, Clifford and Guy moved to California as adults, Vernor stayed on the farm and bought it from Squire in the 1920s.

Dora Knapp in front of their farm house near Severy, Kansas

Vernor married the girl 1-1/2 miles down the road, Dora Templeton. They had five children: Vera, Bill, Lee, Noble and Delbert. They raised the five kids in the two-bedroom house that's still there. A kitchen, bathroom, front porch, and back porch were added in the 1950s.

When Vernor and Dora died, the farm went to the five kids. Noble eventually bought it from them. My Uncle Lee stayed on the farm his entire life. When Noble and Letha, my parents, died, I inherited it from them.

On the farm they raised beef cattle, milk cows, pigs and lots of chickens. The place looked like a small town with buildings everywhere. There was a one-room school directly across the road. My father, Noble ("Nobe") said the kids in school always complained about having to walk so far to school. He complained about being too close. While the other kids were playing at recess, he had to go home and feed chickens! We have the school bell from that school in our front yard in rural Augusta.

I've always heard stories about the pranks Granddad - and later my dad - played and you could never mention the Knapp farm without someone grinning and telling a story about their experiences there. *Submitted by F. Duane Knapp.*

KOCH – The Koch Century Farm in Barton County, KS is currently owned by Richard and Sharon Hiss. John Koch immigrated from Germany to the US in 1857 to Long Grove, IL. In 1879 he and his wife moved to Ellinwood, KS.

William and Henrietta Koch house, built in 1913

In 1905 John and Dora Koch purchased 113 acres of land three miles west of Ellinwood, KS. The land is in Lakin W Township, section 28. William and his wife, Henrietta Koch, became owners of the land in 1913. In 1913 they built their first home on the farm, which was a two story house. Ownership included: daughter, Clara (Koch) Bryant, 1947- 2008; Elinor (Bryant) Hiss, 2008-2012 and 2012 to current, Richard and Sharon Hiss.

William Koch was noted to be an ardent enthusiastic farmer, with a miniature agriculture experiment station located at his farm, where many new varieties of grains were tried out and other experiments were conducted. He also had dairy, poultry and grain crops. William Koch was involved with school, church and many general civic affairs. He had served as an officer in both the church and school district. For a number of years he had served on the Barton County Farm Bureau Board and at the time of his death he was serving as President of the organization.

In 1975, Rick was helping with all aspects of the Koch farm operation. Rick served on the local Coop Board, Barton County Farm Bureau Board and serves the church.

Today the farm operation includes wheat, milo, alfalfa hay, oats and beef cattle.

The original two-story house, granary and large barn still are preserved on the farm. The farm welcomes family and friends to visit and recapture childhood memories.

There truly is nothing more rewarding then farming the land that one's great-great-grandparents owned. *Submitted by Rick Hiss.*

KONOVALSKI LEGACY HOMESTEAD – In 1879, Albert Konovalske (Adalbertus Konowalski) and his wife, Julia Zelinsky (Julianna Zielinska) purchased the Southeast Quarter of Section 12, Township 2, South of

Proud first family in front of their Kansas home, ca. 1900. Sitting in front: Albert and Julia Konovalske. Standing l-r: Rose, Albert, Mary, Joseph, Josephine, Stanley, Frank, and Mike.

Range 4 West, northwest of Belleville in Republic County, KS. The land was obtained from George W. Doles for $550. They and their first two children immigrated from Karniszewo, Poland to the US, in 1872 aboard a sailing vessel. Stopping off in Pennsylvania where they had relatives, Albert worked in the steel mills to earn money to pursue their dream of land in America.

Rosalie and Albert Konovalski (sitting), with Albert and Dorothy, ca. 1953.

Family history has it that soon after Albert arrived in Belleville, he had relatives take him in a spring wagon to look over land available for sale to the settlers. Immediately upon seeing what was to become the "Konovalski Home Place," he said, "This is it! I will build my home and raise my family here! We will be Americans!"

Willie and Dorothy Kennedy smiling amidst the Kansas sunflowers, ca. 1994.

As the land was unimproved, they initially lived in a sod-hut with three children and their livestock. A wood frame home soon followed in which they raised their children: Teofila, John, Stanley, Frank, Rose, Josephine, Joseph, Mary, Albert and Michael. They had cattle, goats, hogs and chickens and grew corn, wheat, oats, cane, fruit trees and vegetables.

Judy Hader and Rosemary Chapin enjoying the cabin's front porch swing, ca. 2014.

In 1882 they donated land for a school to be built on the southeast corner of their land and agreed to house the teacher, who would assist the family in learning English. Victory No. 71 continued to educate children until 1963.

Albert and his half-brother, Joe Dumback, had come to America at the urging of their mother, Anna (Lieskie) Konovalske, to escape the government's cruel draft. In 1889 at the age of 75, Anna joined her sons in Kansas.

In 1922 Albert II married Rosalie Pachta and took over the homeplace. They welcomed two children, Dorothy and Albert. Ingenuity, hard work and the WPA allowed them to survive the Depression with their land intact.

The homeplace passed to Dorothy and her husband, Willie Kennedy, in 1991. They built a small cabin and developed a wildlife habitat, planting over 4,000 trees on the land around where the old home stood.

In 2007 it passed to their daughters, Rosemary Chapin and Judy Hader. It continues as a working farm with the pasture land rented out for grazing. The cropland is also rented out and grows a rotation of wheat, corn and soybean. Markers commemorating the people who have lived on and loved this land remind future generations of our proud heritage.

This information was drawn from *Another Dorothy From Kansas*, written by Dorothy Marie Konovalski Kennedy.

KORTHANKE – Dina Scheer Korthanke and her husband, August Korthanke, purchased the 160 acre farm from J. P. Bowron Nov. 21, 1895.

Lawrence and Evelyn Sabbert, Norris and Eva Loyd, and Earl and Hazel Loyd

Our great-great Aunt Dina sold the farm to our great-grandfather, George W. Scheer, her brother on March 3, 1928 because of some financial problems she was having.

The farm was sold back to Dina April 27, 1932.

Our grandparents, Lawrence and Evelyn (Scheer) Sabbert, then purchased the farm from Evelyn's aunt on Feb. 8, 1955.

Our mother, Eva (Sabbert) Loyd, inherited the farm when our grandmother passed away Sept. 6, 1983.

We inherited the farm with the passing of our mother, Eva (Sabbert) Loyd, May 27, 1999.

Charlott (Loyd) Blanton-Hall has lived on the farm for 48 years and continues to live there to date. She raised three daughters and a son on the farm where her mother was born and raised.

Dina (Scheer) Korthanke, aunt to Evelyn (Scheer) Sabbert, who was the mother of Eve (Sabbert) Loyd, mother to Charlott (Loyd) Blanton-Hall, Virginia (Loyd) Higginbotham and Dorothy Loyd.

KRAFFT – German immigrants Friedrich (Fred) Krafft and wife, Barbara, came to the United States in the early 1870's settling near Mountain Creek in Marshall County, Kansas.

After denouncing Wilhelm, Emperor of Germany, they pursued their dream of owning and farming land made available by the Homestead Act. Arriving in Phillips County in 1877, they acquired 80 acres in each of Mound and Dayton Townships northwest of Stuttgart. Their first "home" was a dugout, but as the family grew to five sons and two daughters, they built a wood-framed house.

Fred and Barbara Krafft

Vernon's sons, Denton and Blaine, grandchildren, Eric and Tina, and great-grandchildren, Maggie, Wyatt and Ellie, with the 1966 combine nicknamed "Grasshopper" on its 50th anniversary.

Stories of life on the prairie have been handed down through generations. One tells of Fred walking about, perhaps seeking a site to lay claim or making the long journey to Kirwin for supplies, when he was forced to climb a tree and spent all night waiting for a herd of buffalo to pass by.

Fred and Barbara were active members of Emmanuel Lutheran Church in Stuttgart where Fred served as one of the first deacons. During their 20+ years of married life, the family farm expanded to 800 acres. Fred's untimely death in 1896 occurred on his 52nd birthday. Around 1900 Barbara had a second wood-framed house built on the Mound Township homestead.

In 1912, their son, Heinrich (Henry) and wife, Bertha built a new house on the Dayton Township homestead where they raised their three sons. The oldest son, Vernon and wife, Martha, lived in the 1900 house raising four children, Vesta, Denton, Blaine, and Ardes. These siblings are now owners of the above-mentioned land.

Blaine and his wife, Candace remain in the Dayton Township house where they raised their family. Their children, Jason and wife, Myndi, and daughters Kinley and Kensi; and Tina and husband, Lane, and children Wyatt and Ellie live on adjacent farms and help farm the original Dayton Township homestead.

Denton and his wife, Sharon, built another house on the Mound Township site where their three children, Jodie, Eric and Jenna, grew up. Eric, along with his wife, Alissa, and daughter, Maggie, are currently in the process of remodeling the 1900 house where they'll soon live and continue farming the original Mound Township homestead. And so the legacy of farming lives on....

KRAMER – The Kramer Century Farm in Elk County, KS is currently owned by Kenneth and June Kramer of Longton. The Kramer family came to the US from Luxembourg and settled near Alva, Oklahoma.

Kenneth and June Kramer and family

The Kremer family at the farm house

In September of 1912, great-grandparents John and Eva Kremer, as the name was spelled at that time, purchased the original farm from M.M. and Sara Fulkerson. The next descendants to own the family farm were Kenneth's grandparents, Jahn and Mary Kramer, who were followed by Vincent and Treva Kramer, Kenneth's parents.

The farmhouse was built about 1910 and is still occupied as the primary dwelling and the early barn and granary are also used. The original farm had 220 acres and was purchased for $68 per acre.

Vincent had a brother, Philip Kramer, married to Myrtle Bothgeb and a sister, Margaret Kramer Kunzman. Vincent took care of the farm and cattle, including the milk cows, and was also a self-taught electrician. Treva helped a lot with the farm and had chickens and a large garden as well as operating a cafe. Kenneth's sister, Karen Cook of Longton, was also raised on the family farm.

The number of acres in the farm has grown to 640 acres at present and the crops raised there are wheat, soybeans and corn. The pastures are currently leased to a local rancher.

Kenneth is a retired heavy equipment operator and June has taught at the Longton schools for 45 years. The Kramer family includes three children: Kristine is a librarian/teacher at an elementary school in Bartlesville, OK; Joni is a CPA for AON in Grand Cayman and is married to Joseph Steffen, a computer network specialist, they will have a baby in June and have a daughter, Seryn. Thomas is an AGCO technical support specialist in Hesston, KS who likes to be involved in the farming operation on occasional weekends. Our family is very honored to be a Century Farm. *Submitted by Kenneth and June Kramer.*

KRUEGER – The Fred Krueger Farm is a picturesque 240 acres lying on gently rolling hills, just north of Yates Center, KS in Woodson County. It was acquired by Fred and Matilda Post Krueger in a trade for their Homestead near Orlando, Oklahoma Indian Territory in 1907.

The Krueger farmhouse

They were German immigrants and they raised six children on this farm. Fred had the farm annexed into the Yates Center School District and all the children graduated from Yates Center High School.

The farm includes farm ground, pasture, hay meadow and timber, with a creek running across the north end. The creek, with its rocky ford and surrounding timber, was a public recreation area, especially during the '20s to '40s, for swimming, boating, fishing, hunting and picnicking--all were welcome. The rocky ford has been used for centuries by ancient Native Americans to today's fishermen. Many Indian relics have been found.

The farm was inherited by Kathy Krueger and Richard Krueger, grandchildren of Fred and is operated by Kathy Krueger today. The picture shows the original 100 year-old eight-room farmhouse, which was destroyed by a tornado in 2003. The only structure left standing was a stone barn. The house has since been replaced by a modern home *Submitted by Richard Krueger.*

KURTH – First generation – My great grandfather, Gottlob Kurth, came out here along on the Santa Fe railroad from Will County, IL and bought this land, S31 T25 R20, a half section in 1884. It was railroad land. He bought it as an investment, because the land was cheap and he wanted some land for his children. He then returned home to Will County, IL. He never did live here.

The Kurth-Lancaster farm at S31 T25 R20

Second generation – In April 1886 my grandparents, Edward and Bertha (Wagner) Kurth came from Will County, IL, with three small children, on the Santa Fe railroad to settle on some of the land that his father had purchased. My father, A.C. was the first child of theirs born in Kansas. Edward and Bertha's oldest son, Herman and wife, Hannah (Galyart), were the first to live on S31T25 R20. Later they left the farm and moved to Hutchinson, KS.

Third generation – My father, A.C. and wife, Lena (Weiss) Kurth, bought this land in 1945. After their deaths, my husband and I purchased the farm in 1973.

Fourth generation – My husband, Duane Lancaster and I, Wilma (Kurth) Lancaster, purchased the farm in 1973. We moved onto this farm in 1981 and have lived here ever since then. My husband died in June 2004. I remain living here.

Fifth generation – Our son, Bret and wife, Debbie, now farm the land. They live a mile north of us on the old Lancaster homestead place. It is also over 100 years old, homesteaded in 1903.

Our religious heritage has meant a great deal to all of us down through the generations of our family. A.C. and Lena had seven children. I was the youngest and the only one still living. Our faith was the most important factor in making it through the hard times the pioneers endured in this area. All of the children of A.C. and Lena Kurth were baptized and confirmed in the Lutheran faith at Zion Lutheran Church in Ford County. My parents were married there, as were the three girls in the family. Duane and I were married there in September of 1947. We had four children. We have seven grandchildren and 12 great-grandchildren. Some of them live close around here, so we have some sixth and seventh generations close in this area. We have been blessed!

LACHMAN – Joseph and Josephine Lachman had 10 children and purchased this farm in 1910 from Anton Trecek. After the death of his wife, Joseph gave land to his sons, who were farmers. In 1925 he deeded this farm to Charles and Marie Shimek Lachman, his youngest son. The railroad ran through the west side of the farm. This acreage was returned to the owners when it no longer was in business.

Charles farmed with horses, which he raised and trained himself. He was partial to polled Hereford cattle, which he raised. When tractors came into the picture, he purchased a Case with steel wheels and lugs. Eventually horses were a thing of the past. A John Deere A was bought, along with a drill which made farming easier for him and his two older sons, Richard and Roger.

The Charles Lachman residence. Pictured from left: Josie, Mike, George, Charles, and Frank.

A large garden was always raised with much canning and meat processing by the fried and in lard in stone jars method. The butchering was always done by all the family members.

Electricity came to the farm in 1948. Three more children were born in addition to the two older boys. An infant daughter did not survive.

Charles served in WWI and his two eldest sons served in WWII. He died on Dec. 31, 1980. Oldest son, Richard, who taught in high school, took over the farming in his spare time. He continued with cattle and the crops.

After both retiring from teaching, Richard and Margaret continued until farming and cattle were no longer feasible for them, so the land and pasture was rented out to a younger farmer.

Today all siblings are deceased except the daughter, who continues to live on the Century Farm. The nephews and nieces come to help her with maintenance once or twice a year. *Submitted by Margaret A. Lachman.*

LALLIER – The Lallier Farm was homesteaded in 1870 by Eugene Lallier, who had grown up in France. He moved first to Fond du Lac, WI before coming to Kansas. He began with gardening, orchards and vineyards. In 1903 he and two of his sons started a dairy. The dairy was continued son, Ernest Lallier. Ernest died while his son, Leon Lallier, was still in law school at the University of Kansas. Leon returned home to the farm and maintained the

Leon Lallier on his tractor, February 1956

Retha Lallier

Grade A Dairy, Retha and Leon Lallier

Grade A Dairy until 1976. Leon and his wife, Retha Lanier, farmed the land until his death in 1988. Retha died in 2014 so the farm passed to their two daughters, Beth Lallier Lucas and Kay Lallier Grant.

Though reduced in size by public utilities, the family is committed to preserving as much of the original homestead as possible. Currently the 212-acre farm raises grain crops and hay. It remains in the Lallier Family as a cherished place for their children and 13 grandchildren to live and gather. *Submitted by Beth Lallier Lucas and Kay Lallier Grant, daughters of Leon and Retha Lallier.*

LAMBORN – Established in 1877, the Horace and Rosemond Lamborn Farmstead as a subsistence farm and later evolved to meet changing agricultural needs. Early in the farm's history, Horace boarded horses for area residents and drove cattle to pasture on the nearby state prison island, transporting them across the Missouri River in flat boats. Beginning in the early 1900s, the family developed a small dairy operation, which continued until 1998. The property has remained in the Lamborn family and continues to function as a working farmstead.

Picture-puzzle of the Lamborn family

The farm was designated a Kansas Century Farm in 2000. In 2011, the Lamborn Homestead was added to the National Register of Historic Places and the Register of Kansas Historic Places. The farm received the 2012 Medallion Award for Excellence in Rehabilitation presented by the Kansas Preservation Alliance for its local significance in the area of agriculture. The barn is more than 100 years old. Restoration was completed in May 2012. *Submitted by Joseph Lamborn.*

LANDRITH – Michael and Brooka Landrith reside on a Century Family Farm in Labette County, located just north of Bartlett. The farm was originally purchased by Michael's grandparents, Claude and Mary Pickering Landrith, in 1914 for $50.00 an acre. The 80 acres was bought from Dee and James Lambert. Since then an adjacent 320 acres has been acquired.

Michael's great-great-grandparents originally came from Jedburg, Scotland and England. The Pickering-Baty ancestors grew up on opposite sides of the Hadrian Wall and came to Kansas through Canada. It is thought they left Scotland due to lack of employment when the soap factory closed.

The first two generations on the family farm raised chicken, beef cattle, sheep, and swine as well as wheat and oats. Michael and his sisters remember Grandmother Mary as always being busy, wearing her apron and sunbonnet, hoeing in the garden, dressing chickens for family meals or making lye soap in black kettles in the backyard. Never was a family meal served without her homemade yeast bread.

The home that Michael and Brooka reside in was built in 1915. An addition was put on in 1928. However, before the house was constructed, a barn was built on the farmstead. That barn is still standing and in use today.

Presently the family farm has a cow-calf operation and backgrounds steers. Much of the crop ground is being turned into pasture although some wheat and alfalfa are grown.

Members of the Landrith family on horseback

Claude and Mary Landrith lived on the family farm from 1914 to 1952 when it was passed down to Michael's parents, Clifford and Arlene Landrith. On this farm they raised their five children - Michael, Melanie, Brenda, Nancy and Lynette - before it was passed on to Michael and his wife, Brooka. He and Brooka raised two children, Darcey and Flint. They have two grandchildren Darla and Daylyn. Their daughter Darcey and her husband, Damon Fesmire, acquired an adjoining 60 acres.

Work ethics has always been an important part of the family upbringing. The children were expected to pull weeds from the feedlot or pick up rocks from a field to earn money for their 4th of July fireworks. Michael remembers that he picked up corn in the field that the corn picker had missed to earn money for his first box supper.

However, the favorite family story was the fact that their dad Clifford would get them up early in the morning before dawn to saddle their horses and haul them to the pasture and then wait until daylight to start the cattle drive.

The Landrith family has always been proud of their Presbyterian lineage and involvement in 4-H and school activities.

Michael will tell you this farmstead has been an important part of all their lives and they all share in the rich heritage of the Landrith Family Farm. *Submitted by Arlene Landrith.*

LARKIN – John Lawrence Larkin immigrated from Ireland, the city of Borrisokane and the county of Tipperary. They sailed to America when he was a small boy. He was born Oct. 27, 1827. The family moved to a farm near Lawrence, KS from Missouri in 1861.

Georg Joseph and Mayme Kennedy Larkin

The town was raided by Quantrill's Raiders in 1863. John was a member of the Kansas State Militia. After barely escaping with their lives, they moved to Ellsworth, KS and later to Deerhead, KS.

George and Mary Elizabeth Kennedy Larkin had five children. My grandmother spent her entire life on the farm. She moved to Kansas from Cleveland, OH to marry my grandfather. I spent many summers with them and had a great time, riding tractors, feeding chickens and pigs, driving farm trucks and enjoying the Kansas sunsets.

The Larkins were surrounded by many brothers and their wives, cousins, aunts and uncles who lived on adjacent farms. They played cards for entertainment almost every night.

All the Larkins attended school in the one room schoolhouse in Deerhead, KS. The men and boys were great baseball players. They would ride horseback to the school for their Sunday games.

Mary Larkin Jones, Ron Jr., Deb, Greg, Melissa, Mike, and Sharon

The Larkin family held a deep and abiding love for the Catholic faith. They traveled 28 miles each Sunday for church in Medicine Lodge, KS. My grandmother prayed every night for her son, Charles D. Larkin, who was in Patton's Third Army in WW II and all the boys from Barber County. She was proud they all came back home safely.

The Larkin Century Farm is currently owned by four cousins: Mary Larkin Jones, Kenneth Larkin, Fred and Ed Stump.

George Larkin worked for the Comanche Pool, the largest cattle ranch in the state's history. The family had many friends and neighbors and were good community members. *Submitted by Mary Larkin Jones.*

LARSON – In the spring of 1903 Ed Larson of Lisbon, IL came to Milan, KS on a train trip sponsored for potential Kansas land buyers. Ed and his wife, JoHanna, were immigrants from Norway who settled in Illinois. They purchased the NE 1/4 of section 13 33S R4W and rented the quarter to the north, SE 1/4 Sec 12 33S R4W. They stayed in Illinois and sent their sons, Lewis and Olen, to farm the land. Ed and JoHanna moved to the Kansas farm in 1927. Later Lewis married a Kansas Gal and the partnership ended. Olen rented the land east of the "Home Place," three quarters all in Sec 18 33S 3W.

Today I, Edward Larson, own my Great-Grandfather's quarter, "The Home Place" he purchased in 1903. It was handed down to my grandfather Olen and his five brothers and one sister. Over the years I have bought out all the other heirs and have purchased adjoining land which today totals over 2,000 acres. 1,960 acres of which are in one field and still rent the SW quarter of Sec 18. I am the third generation to farm this quarter for the fourth generation of owners. Making a total single field of 2,120 acres. It is nice to till over 2,100 acres and never fold up a 52' cultivator'.

My nephew, Austin Botkin, farms Larson land owned by my cousin, Kenneth Larson, on Sec. 23,24, 26 of 33S R4W which has been in the Larson family for almost 100 years.

Since 1903 my great grandfather Ed Larson, my grandfather Olen Larson and I, Edward Larson, have worked hard making land payments to build this farm up. In 1940 Olen Larson purchased where I live now on Sec 7. My father, Richard Larson, never wanted to own land. He said, "all the land you ever end up with is a 6x6 plot." My Great Grandfather, Grandfather, and I never saw it that way. *Submitted by Edward Larson.*

LEWIS – The Lewis Century Farm, located in northwest Lyon County, KS, is owned by Roberta Lewis with son, Jim and girlfriend, Traci and sister, Wiletta running the day to day operation of the farm. The Lewis' have farmed in Lyon County for over 150 years with the first 80 acres bought in 1875 by Isaah Lewis. They have added over 400 acres over the years and rent additional land for the cattle.

They have raised sheep, pigs, chickens, beef cattle and milk cows. Today they currently have a cow-calf operation and background calves. They raise milo, corn, soybeans and hay. Also, Roberta raises a big garden and a lot of flowers. Caring for the animals and taking care of the land is very important and always comes first.

Church, 4-H and FFA were very important to the family, as both son, Jim and Jim's daughter, Sarah, were both Kansas Star Farmers.

Today the Lewis family includes mother Roberta, son Jim and Traci, sister Wiletta, three grandchildren – James, Meranda, Sarah "Cody" located in Osage Beach, MO, Hesston, KS and Kansas City, MO.

LILLEY – Our three Century Farms are all located in Clay County, KS. The first, from 8-10-2 Athelstone Township of 160 acres, was purchased in 1882 by my Great Uncle William Burgess from the Missouri, Kansas and Texas Railway Co. for $480.00. My Aunt Maryann Simpson received this farm in 1931 and I received it in 1984 from her.

Uncle Edward Lewis at Industry, KS in the "Dirty '30s."

The second farm, 31-10-3, Gill Township, of 120 acres belonged to my Great Grandparents Jacob and Marquirette Burst in 1878, then to my Grandparents Jesse and Matilda Lewis, then to my mother, Twila Lewis Simpson and then to me in 2001.

The third farm, 13-10-1, Chapman Township, consisted of 80 acres purchased in 1918 for $6000.00 by my dad, George T. Simpson at the age of 12. I was born and raised on this farm, learning responsibility not only for others, but for caring for the animals and taking care of the land.

Take care of the land and it will take care of you, which I heard many times. These are all still good producing, clean, well-cared for farms.

My ancestors, from many places and countries, were from Scotland, England, Switzerland, Germany, Illinois and Pennsylvania that I know of. In fact, my Grandpa Simpson brought the first Angus cattle to America.

Our plans are to pass this land on to our children through Kenneth and Dorothy Lilly Trust. *Submitted by Dorothy Lilley.*

LIPS – John and Lucy Lips were very early supporters of Kansas Farm Bureau. John came from Missouri and Lucy from West Virginia. Married in early 1900s. Bought present farm in 1910. John was disabled and died in 1945. Lucy took over with hired help, ran the farm for some 30 years. She had a green house and sold tomato plants for 1 cent each, plus raised a large garden. Lucy also raised 1,000 laying hens and sold eggs.

Lucy Lips on the family Ford Jubilee tractor, 1953

Stan Lips (grandson) lived in Coffeyville, KS but spent his summers at his grandmother's house, helping farm. After her death in 1964, Stan and his brother inherited the farm.

Stan and Marty were married in 1959. They continued a farming and livestock operation until retirement in 2010.

The Craftsman home built in 1924 is still occupied by Stan and Marty Lips. There is also a 100-year-old horse barn on the property. *Submitted by Stan Lips.*

LOEFFLER – Shortly after their wedding in February 1909 in Washington County, George and Callie (Bellows) Loeffler came to a 320-acre farm in Rooks County, purchased from Charles Stewart. The farm is located 17 miles west and one mile north of Stockton.

Mr. Loeffler came by "immigrant car," bringing his animals, machinery and household furnishings by train at a special rate. When he arrived at Damar, he unloaded his wagon, placed the rest of his possessions in it, hitched up the mules and drove the nine miles north to his farm. Mrs. Loeffler followed soon after, also coming by train.

Their first home was a two-room sod house in which their oldest daughter was born. In July 1910, they moved into their new four-room frame house.

The farm was a diversified one, growing wheat, corn and sorghums and raising cattle, sheep and hogs. Mrs. Loeffler raised chickens and turkeys and planted a large garden. The farm was enlarged to approximately 940 acres by additional purchases of land.

The barn was built in 1916 and in 1925 two rooms, a bath and screened porches, were added to the house.

George and Callie Loeffler

The Loefflers retired to Stockton in 1958 and about 10 years later the Cecil Roy family purchased the original one-half section. Cecil, Patti, Brandon and Carol Roy continue to make it their home. The Loeffler children, Mildred (1909-2004), Lois (1915-2015) and Betty Mavis (1925-2016), managed the remaining 610 acres of Loeffler Farms, 1973-2016. Blessed with excellent tenants, they benefited from oil and gas leases, pasture rent, and harvests of wheat, corn and sorghum.

Lois' daughters, Sherry and Jan, are the current owners of the land acquired by George and Callie Loeffler from 1915 through 1947. Sherry and Jan's daughters will inherit the land, continuing the family ownership into the fourth generation.

LUDWIG – It was in 1878 that John Martin Ludwig (45) traveled to North America from Germany after being released from the Prussian Service. He settled on a farm five miles southwest of Beloit in Mitchell County with his wife, Gertrude Heller Ludwig (40). They had seven children released to leave Germany with them...John Peter, Martin, Sabilla, Carl, Christine, Catherine and Mary. A daughter, Gertrude, was born in a dugout on the farm that John Martin made for the family only two weeks after arriving to Kansas. Joseph William was born three years later on the farmstead.

The youngest of John Martin's children, Joseph William, married Rose Ann Grennan on Aug. 27, 1907. Their marriage was the first union to be made in the current St. John's Catholic Church built with native stone from the Ludwig pasture and surrounding areas. Joe and Rose lived on the family farm and raised 10 children: Margaret, Martin, Helen, Christine, Mary, Joe, Ernest, Bob, Max and Bill. Rose died Aug. 2, 1952 and Joe died Aug. 27, 1954.

Bob Ludwig, son of Joe and Rose, married Jean Jones on Nov. 23, 1954 at St. John's Catholic Church and moved to the Ludwig Farm.

Joe and Rose Ludwig in the Dirty '30s with the boys ready to work on the farm: Joe W. Ludwig Jr., Ernest Ludwig, Robert L. Ludwig, Max Ludwig, and William L. Ludwig.

The couple raised five children...Ann, Linda, Joe, Mary and Robert Scott. Jean Ludwig continues to live on the family farm at age 91, after losing Bob in 1988 at age 67.

It is known that The Ludwig's were farmers at heart, built with stamina and true grit. They were known across the countryside as fine horsemen and had an eye for good cattle. They were good people with kind hearts, strong faith and each blessed with God given talents. Fun times were plentiful with barn dances, playing cards, throwing horseshoes, reminiscing and eating good food.

Joe W. Ludwig with wife Rose (Grennan). They were the first couple to be married in the current St. John's Catholic Church in Beloit, Kansas, 1910. They are pictured with their girls Margaret, Helen, Christine, and Mary.

The Ludwig Heritage is strong and the Circle of Life continues on the Family Farm 139 years later. *Submitted by Linda Ludwig.*

MADDOX – In 1880, David E. and Martha Ann Maddox were in their late 30s when they moved from Adams County, OH to Miami County, KS. Son McPherson was born in Kansas in 1883, while his older brothers and sisters - Harvey, Frank, Lulu, James, Margaret and David Connor - had made the move to Kansas with their parents. The Maddox Century Farm was established in February 1896 when they bought a 76-acre farm in Richland Township, Miami County. In 1913 David C. and his wife Amelia Maddox bought the 80 acres on the north side of his parents' place and in 1919 they purchased the original 76-acre farm from his father. Their sons, Raymond and Clifford, were raised on the farm they inherited in 1964. The size of the farm has changed over the years as two acres were returned to the family when the rural one-room school closed and again when the government bought the north 80 acres for the creation of Hillsdale Lake. Before her marriage to Clifford, Elizabeth Cavaness taught at the school on the farm. Elizabeth purchased sole interest in the remaining 78 acres in 1983 and left the property to their daughter and current owner, Sara Maddox Walter, in 1991.

David C. Maddox and Sara Maddox Walter

David E. and son, David C., raised cattle and other farm animals. Many years ago the schoolhouse, farmhouse, barn and outbuildings, along with the pond and well, were removed leaving all of the excellent cropland available for the production of soybeans and corn.

The families were active in their community as members of Eden Chapel Methodist Church. David E. Maddox was involved with the Grand Army of the Republic, David C. Maddox was on the school board and Amelia Maddox took part in quilting bees. David E. and Martha Maddox are buried on two acres set aside from the farm at Scott Valley Cemetery.

MADER – John Mader purchased 160 acres located in Anderson County, KS in 1894. The acreage is located in Jackson Township, Section 11,

Range 19E, Township T-20S. Corn, wheat, oats and alfalfa were the grain crops grown here and there was a cow/calf and dairy operation.

The farm retains the original 160 acres and was passed from John down to the fourth generation. The farm passed from John to William Feuerborn, to Cyril Feuerborn, to Rick Feuerborn and finally to Luke Feuerborn about 2000.

Now the farm produces corn and soybeans and still has a cow/calf operation.

Barns built prior to 1920 and the house built in 1920 remain on the farm. The KN&D Railroad ran through here until 1934. An unusual characteristic is the Deep Watering Hole in Cedar Creek.

Rick has three sons: Luke, Levi and Lancer and seven grandchildren: Lily, Wyatt, Kallie, Kamrie, Karlie, Dillon and Delainey. *Submitted by Rick Feuerborn.*

MANETH – In 1875 a group of Moravians left the villages of Unter/Oberlangendorf, Maehr, Truebenz, Schoenwald to settle in Walnut and Grant Townships of Barton County, KS, eventually founding the town of Olmuetz, KS, now spelled Olmitz.

Ferdinand Maneth was born April 11, 1846 in Austria. In 1878 he took up land in Walnut Township and he was one of the men who had much to do with the development of that part of Barton County. Ferdinand Maneth, up until the time of his death in 1906, was one of the best known men in this part of the state. The Maneth's were well known in Barton County on account of the fact that they came here at a time when the county was being developed and just emerged from the frontier stage. This family passed through the hard times of the 1880s and had much to do with the work of making this county one of the best in the state of Kansas and one of the leading agriculture sections of the world.

Land that was settled in 1878 remains in the family today. It passed from Ferdinand to Otto Sr. to Otto Jr. to Eugene. Eugene and wife, Mary Jo, have sons who farm, as well as grandchildren that help. That makes six generations that have farmed and walked on the same land in an unbroken lineage.

For the past 139 years the land has been a general farming operation including cattle, a dairy, hogs , pastures , alfalfa and grains including wheat and milo. Today Gene and Mary Jo have four children: LeRoy, Lisa, Dale and Michael and eight grandchildren. At present all the family, including grandkids and spouses, live within the borders of the state of Kansas, located from Dighton to Wichita and Kansas City. *Submitted by Eugene Maneth.*

MANN – One of the early settlers and representative famers of Summit Township, Cloud County, near Scottsville, KS is Alfred Amos Mann. He is a native of Monroe County, OH where he was born in 1841. He was raised on a farm and at the age of 21 enlisted in the service of the US Army, Company C, 116th Ohio Volunteer Infantry in August 1862. He was mustered out after the battle at Appomattox, April 9, 1865. After the war, Mr. Mann returned to Ohio where he remained until coming to Kansas in 1872.

Alfred (Alf) and Nancy Mann

A.A. Mann married Nancy A. Coen, also a native of Monroe County, in April of 1862. They came as far as Solomon City in 1872 and from there to Summit Township, Cloud County with an ox team and settled on the farm where they now live, their original homestead. They first built a log house with a dirt roof where they lived several years. The first July in Kansas they had no money nor land broke out; a prairie fire came, doing much damage, which added to their hardships.

The Mann homestead, 1946

The Maneth family

Mr. Mann is of German origin. His ancestry was among the early settlers of Pennsylvania. His great-grandfather, Barnet, fought in the Revolutionary War where he was wounded in the thigh.

To A.A. and Nancy eight children were born. Mr. Mann was a trustee of Summit Township, a member of the Christian Church at West Asher, the Ancient Free and Accepted Masons and the Grand Army of the Republic, No. 173, at Scottsville. They were accomplished musicians.

The Mann's homestead consisted of 180 acres of land, which was principally adapted to wheat, with five acres in apple orchard and other small fruits.

A.A. and Nancy's youngest son, Anson, married Bertha "Birdie" Hazelwood in 1909 and became the second generation to reside on the homestead. Anson and Birdie reared four children on the farm. Their youngest child, Hiram Dean, was born in 1920. After returning from the European Theatre in WW II, he returned to continue farming with his brother, James Orwan.

Dean was united in marriage to Helen Spillman in Crown Point, IN, in 1945. Dean and Helen became the third generation to live on the family homestead. Their only child, Juliann Kathryn, was born in 1948. Julie married Alan William Snyder in October of 1968. They combined their current farming operation to include Dean and Helen's after the Mann's retirement in 1985.

Alan and Julie had one child, a daughter, Michelle Rene, born in 1969. Michelle moved to the homestead shortly after her grandparents, Dean and Helen, moved to Beloit. Michelle married Jeffrey Baxa in 2006 and had one child, Brock Michael Snyder Baxa, in 2009. Brock is the sixth direct descendant to call this homestead home. *Submitted by Juliann K. (Mann) Snyder.*

MARCY – My grandparents, Floyd Burton and Ethel (Woodburn) Marcy, descended from sturdy pioneers who made their way from Indiana and Pennsylvania to Kansas as territorial settlers prior to 1860. It was the spirit of adventure and the hope for a brighter future that brought these families west.

Burt and Ethel were schoolteachers. When they married in 1905, she became a housewife and he continued teaching until 1914. With the realization that a teacher's salary could not sustain a growing family, Burt and Ethel purchased a 70-acre farm in Greenwood County in 1914. His brother, J.H. Marcy and mother, Estella H. Marcy, owned two adjoining 20-acre parcels. Estella granted both of the adjoining parcels to Burt and Ethel upon her death in 1920.

Their 110-acre farm produced crops such as wheat, kafir, corn, alfalfa and prairie hay. Livestock and poultry, as well as a large fruit orchard and garden, produced additional food for the family, which by now included 10 children. Burt was very proud of the fact that his farm was able to provide for his family during the Great Depression, a time when so many were hungry and out of work. In addition to farming, Burt was active in his community as a township trustee, school board member, substitute mail carrier and he drove the county "patrol" or road grader.

Clara Marcy Atkinson on her 100th birthday, September 25, 2015

Floyd Burton Marcy and Ethel (Woodburn) Marcy, original owners

My parents, Len and Clara (Marcy) Atkinson purchased this farm from my grandparents' estate in 1950. When my father died in 1964, Mom was a single parent with three children to raise. Well-meaning family members advised her to sell the farm. She rejected that advice and even though she moved away to pursue a career, she leased the farm and continued oversight of the farming operation.

In the past 100 years, this small family farm has provided a modest living for several families and has been a gathering place for generations of relatives. Born in the original farmhouse, my mother (1915-2015) credited her longevity to a lifetime of performing physical labor and a diet of homegrown, nutritious food that was produced on this farm. As current owners, my husband and I are honored and proud to carry the legacy of our family farm into the 21st Century. *Submitted by Keith and Rita (Atkinson) Stapleton.*

MARKEN – Daniel and Nancy Marken purchased the farm in 1879. They came to the farm in two covered wagons with 11 children from Peru, IN. They built the buildings on the farm as the need arose. Daniel planted over 80 rods of Osage Orange seedlings, along what is now Button Road, for a fence soon after he arrived. This hedge has been kept trimmed to a height of about 4 feet for over 125 years. It is trimmed from 10 – 13 times each year.

Osage Orange hedge that has been trimmed for over 125 years

Daniel sold the farm to his son, Roy Marken, in 1911. Roy built the present farmhouse in 1920. Roy and his son, Harland, farmed together, establishing a small Grade A dairy in the early 1930s, when Topeka, KS started Grade A milk. They sold Grade A milk for over 30 years from a small herd of Guernsey dairy cattle.

The Marken barn

The Marken farm

Harland purchased the farm from his father in 1949. Harland continued to farm corn, soybeans, wheat, hay, cattle and hogs until his death in 1990. The farm was purchased by Harland's son, Larry Marken, in 2000 from his mother, Ruth Marken. The land has been rented out since.

Besides building a new house in 1920, Roy, Harland and Larry have built several new buildings. *Submitted by Larry Marken.*

MASON – George Mason was born on a farm near Bowling Green, KY on Dec. 31, 1853.

Shortly after the Civil War, in the fall of 1869, George and his family sold their farm and loaded what they could into covered wagons and started west in search of new land and new opportunities. 15-year-old George and his family settled one mile south of the Bourbon/Crawford County line, near the town later know as Arcadia.

George Mason and Mary James were wed on Dec. 2, 1875 and were the parents of 16 children, 11 of whom lived to adulthood. They remained in the Arcadia area for nearly 40 years, accumulating 320 acres of land. In 1913, they sold the land to the Coal Company and headed north, purchasing 240 acres, which included a house, barn, and several outbuildings near Redfield. On April 19, 1916, a tornado destroyed the house, barn and all of the outbuildings. Miraculously, the family survived the fast approaching cyclone by running out of the house, laying on the ground, with each family gripping a fence post as the tornado passed over them.

After the loss of their home, the family moved to Fort Scott for the next five years, during which time the house and barn were rebuilt. In 1922, the family moved back to the farm, where George and Mary lived for the rest of their lives. After the death of Mary in 1930 and George in 1936, their oldest child, Betty Lockridge, was deeded the west 120 acres, which included the house and barn.

Betty and her husband, John Lockridge, lived on the farm until Betty and John's deaths in 1954 and 1959, respectively. They were survived by their four children: Alice, Ellen, Louise and John. In 1961 the farm was deeded to their daughter, Alice and her husband, P.H. Williams. Their only son, John C. "Jack" Williams, his wife, Doris and their

The house near Redfield when George Mason moved there in 1914.

The George Mason home after a tornado destroyed the house and building, April 19, 1916.

House and barn rebuilt after the tornado – George and Mary are standing in front.

three children - LuAnne, Steve, and Diana - lived there, buying the farm from Jack's parents and receiving the deed in September 1965. Jack and Doris lived together on the farm until Jack's death in 2005.

As of 2017, Jack's wife, Doris and their daughter, Diana, still live on the George Mason farm, which has been in the family for 103 years.

MAXSON – The Maxson Century Farm began in 1898 with the purchase of 160 acres by great-grandmother Flora Stone. The land was purchased from Frank M. and Mattie J. Kine for $3,250.00 (or about $20 per acre).

Willard Maxson farm

In 1937, Grandmother Nellie Stone Shufelt and great-uncles, C.R. Stone, T.W. Stone, E.L. Stone and Arthur Stone became owners. Later that year the title was passed to Grandmother Nellie and her brother, Arthur. In 1941 Grandmother Nellie became the sole owner. Their mother, Lola Shufelt Maxson, acquired the farm in 1953. Between 1953 and 1993, the farm was owned by Lola Shufelt Maxson and her husband, Willard Maxson.

Willard was a dairy farmer as well as raising wheat, corn and soybeans. They lived and farmed on this land until their deaths in 1993.

The three children of Willard and Lola Maxson - Leon Maxson, Jean Paul and Carole Brown - shared ownership of 160 acres and Duane Brown, grandson of Willard and Lola, farmed the land until 2007 when it was sold to Ronald Madron of Coffeyville, KS.

This farm is located in Howard Township, Labette County, about three miles west of the north edge of Edna. None of the original buildings remain. The present house was built in 1951. *Submitted by W. Leon Maxson.*

McCARTHY – The McCarthy Farm, of Edgerton, Kansas, in western Johnson County has been in the McCarthy family for over 150 years. John McCarthy was born in Ireland in 1812 and immigrated to America in 1847. He purchased 160 acres west of Edgerton in 1864. It has been a working farm in the McCarthy name ever since.

The farm was passed down through four generations of sons. John lived in a log cabin after purchasing the farm and later built a large rock home made from native limestone. It is still standing today. The building is listed on the National Register of Historic Places. His son, Florence, purchased the land from his father in 1887. Florence had little education but spoke four languages fluently. Before raising his family of seven children, he was a scout on the Santa Fe Trail and was in charge or freighting wagons between Ft. Leavenworth and New Mexico. Florence moved a four-room home from Lanesfield to the farm. He then added four more rooms and a porch. That farmhouse was home to three generations of McCarthy. Florence's son, Ed, moved his family to the farm in 1920. He farmed the land until his son Neal bought the farm in 1956. Neal tore down the old farmhouse and built the current structure in 1962. He and his wife, Joan, raised 8 children there.

Neal farmed the family land, raising hogs, cows, and for a time, chickens. He grew wheat, corn, soybeans, and alfalfa. There was always a large garden and lots of canning. Neal died in 2007, but Joan still lives there. All generations of McCarthys were very involved in the Edgerton community and especially Assumption Catholic Church. Neal and Joan raised their children in a strong work ethic. Their children, their spouses, 20 grandchildren and 20 great-grandchildren continue to enjoy gathering on the family farm.

McCLASKEY – The McClaskey Century Farm is located in Crawford County near Girard, KS. The land was initially purchased by David E. McClaskey on Aug. 19, 1893. The farm was later listed in the names of David's sons – Clement W. and Charles McClaskey. After the passing of Charles, the owners were

Clement and his wife, Ruth. The three sons of Clement and Ruth - George, Jack, and Dale McClaskey - later became the land owners with Jack being the primary managing owner and caretaker of the land for many years. The farm is now listed as the McClaskey Estate with the majority owner listed as Jack's spouse, Barbara Jean McClaskey and with George's children, George E. McClaskey and Kay McClaskey Martinous, maintaining partial ownership.

Jack McClaskey

The farm has remained in McClaskey ownership since 1893 with an immediate member of the family directly involved with either raising livestock or crops on the land every year since then. Over the decades the farm has provided for the production of Hereford and Shorthorn beef cattle, Jersey and Holstein dairy cattle, Poland China and Yorkshire swine, Shropshire and Suffolk sheep, draft and standard horses and chickens. Crops produced on the land have included corn, wheat, oats, milo, soybeans and alfalfa and prairie hay. Currently the 160-acre farm is producing corn, wheat, hay, pasture land and a flock of crossbred sheep. The flock of sheep that are being raised on the land are managed by a daughter and granddaughter of Jack and Barbara Jean McClaskey, Barb and Katie. Many members of the family have been involved at some point including other children of Jack, David, Dorothy, Carolyn, Mary and Jackie.

The McClaskey family has been blessed, and we thank our Lord for the opportunity to be a part of the farming legacy that is so vital to Kansas.

McFARLAND – Robert McFarland bought the first 453 acres of the McFarland & Son ranch and moved there in 1911. The only building present was a small house. In 1923 he and his son, William, formed the partnership, and McFarland's have lived on the ranch for 105+ years.

Robert passed away in 1947, and grandson, Wiley McFarland became the 'Son.' The ranch still operates with Wiley and daughters Sara and Greta sole owners.

As adjacent property became available, it was purchased. The ranch now comprises 4800 acres and runs 5 miles along the south bank of the Arkansas River. It has about 2000 acres of river bottom with the rest sand hill pasture. It's been used primarily for pasturing and prairie hay, with about 100 acres farmed for forage feed. We settled into a cow-calf operation in 1950.

There have been many changes since 1911 with four river floods that caused damage. We survived the seven year drought of the 1930s (Dust Bowl) and suffered many dry years. The biggest challenge has been the relatively recent loss of the Arkansas River flow, which resulted in a drop of the water table. We acquired water rights from the State to divert water from the river, and at one time sub-irrigated about 1300 acres - until the river went dry. With the loss of this water source, we have had to make drastic changes to our operation.

The last series of dry years brought the most changes and disappointments. Attempting to hang on to our cow herd, we abused our pasture, bought hay, and pastured corn stalks. In the fall of 2011, we sold 1/3 of our cow herd. As dry conditions continued, we sold the remaining herd in February 2012. It was the first time in 101 years that McFarland's didn't have livestock on the ranch.

We'll get past this setback. We're letting our pastures rest. We're "taking in" a light load of cattle, and the pastures do look better.

I used to think cattle were our basic resource, but our basic resource that the Good Lord gives us is the grass. How we treat and utilize it will determine our future. *Courtesy of Sara McFarland*

McCUTCHAN – David Washington McCutchan was born in 1846 in Scottsville, IN and married Fannie Rebecca Bassett in March of 1870 in Leon, IA. Mr. McCutchan was of Scottish descent. D.W. and Fannie moved to Kansas and homesteaded in section 17 of Lulu Township, Mitchell County, Scottsville, KS in March of 1872.

Their first home consisted of a dugout with a sod roof. They remained in the dugout until a snake fell from the ceiling onto the bed, a wood framed house was constructed shortly after. D.W. taught 22 terms of school in the Fairview area of Mitchell County and many terms of vocal lessons. They were faithful members of the Fairview

Edgar, Frank, Cora, Anna, David, and Fanny McCutchan. The frame was made by Maurice Snyder from Walnut taken from the first frame house.

McCutchan homestead, ca. 1952

Christian church located in the northwest corner of section 21-6-6. D.W. held many offices of responsibility. They built one of the first shingled houses in the township.

Five children were born to this union, one dying in infancy shortly after coming to Kansas. Their fifth child, Anna Evelyn, married William Daniel Black in 1904 and became the second generation to live on the homestead. W.D. and Anna Black had three daughters born on the homestead, with the middle child dying at 6 months of age. W.D. and Anna's oldest daughter, Gladys Marie, met her husband at Cotner College in Lincoln, NE in the late 1920s.

Gladys and George Maurice Snyder were married in 1930 in Beloit, KS. After several years of Maurice's managerial position in Safeway Grocery, they returned home to take control of the farm, making them the third generation homesteaders. Maurice and Gladys had four children: the youngest, Alan William, married Juliann Kathryn Mann in 1968. One daughter, Michelle Rene, was born to this union and the fifth generation to occupy the homestead. In 1974, Maurice and Gladys moved to Beloit to allow Alan to continue the farming responsibilities. Alan and Julie are the fourth generation to live on the homestead in 145 years and the longest to reside there. *Submitted by Alan Snyder.*

McGINTY – Dominic (Daniel) and Mary Monaghan McGinty left their family and homeland as young newlyweds. Never would they return to the rugged terrain of County Donegal, Ireland. They sailed to the land of opportunity and journeyed through New York, Pennsylvania and many other states on a path of unknown destination. Their destination and home presented itself as 160 acres of lush green pasture and fertile crop land in the Irish Catholic Community of Emerald, KS (located approximately four miles southeast of Williamsburg, KS.)

The 1890 barn built by Dominic McGinty still stands at the Donegal Ranch

The McGinty house; the front part of the house dates back to the mid-1800s

The barn they built in 1890 still stands to this day and has sheltered cattle, horses and barn cats though more than a century of storms, wind and sunny days. The house, whose walls cradled a family of nine, also still stands and those same walls now whisper the tales of four generations.

Dominic McGinty in front of his barn built in 1890

This cherished land, now owned by the McGinty's great-granddaughter, Jeanne Dagenette and her husband, Raymond Mason, has been proudly and rightly named "Donegal Ranch" and is the base for Donegal Ranch Performance Quarter Horses owned by the Masons. Dominic McGinty always loved fine horses; he would be very pleased. *Submitted by Raymond Mason and Jeanne Dagenette.*

McLAUGHLIN – This Century Farm consisting of 160 acres located in the southeast quarter section 16, township 29 south, range 2 west of the principal meridian in Sedgwick County, KS was acquired by Josiah and Sarah McLaughlin. Back in that era, wheat and beef cattle were raised on the farm.

The McLaughlin-Carter house

The 160 acre farm passed from Josiah and Sarah to their son, Clyde and May (Wilham) McLaughlin. The next owners were Byrl and Mary (McLaughlin) Carter and the farm passed to their son, Larry Duane Carter. Larry and his son, Layne Carter, became owners on Oct. 1, 2015.

Now crops are wheat, soybeans and milo and beef cattle are still raised.

Fortunately, the original home building circa 1874 is still standing.

In 2003 this became a Kansas Farm Bureau Century Farm.

Larry Duane and Carolyn Kay (Popp) (deceased in 1983) Carter had three children: Layne, Sue (Carter) Howell (deceased) and Vicki (Carter) Hoelting. Grandchildren are: Britt Carter, Jordan Carter Shaw, Kramer Howell and Kaitlynn Howell. *Submitted by Layne Carter.*

The McLaughlin homestead and barn

MECHLING – Simon Mechling and his wife, Maruha Ann McGuire Mechling, moved from Lima, Allen County, OH by train bringing machinery horses and livestock with them. They disembarked at Girard in Crawford County, KS and for a short period of time rented a farm two and one half miles east of the present farm. In February 1883 they bought the present farm from a man named Joe Bachelor. I do not know any history of the farm prior to 1883, but Crawford County was part of the Indian Neutral Lands along with Cherokee County, KS. After the Civil War, James Joy, a railroad magnate, purchased 800,000 acres of the Neutral Lands, which consisted of most of the two counties and resold to settlers.

Back row, left to right: Ely Mechling, James Mechling, Edith Mechling. Front: James Webb Mechling.

The farmstead had a rock fence surrounding the barnyard and small rock quarry. The remnants of the fence and quarry are still in evidence today. The original house was a four room two-story structure facing east and Martha had a two story four-room addition added facing south so that it resembled a T.

My great grandparents brought my grandfather, James and his sister, who died in 1883, with them. They had two more children after the move. In 1893 Simon died on the kitchen table while being operated on for a ruptured appendix. After his death my 21-year-old grandfather, his 2-year-old brother, Harry and my great grandmother ran the farm. My great grandmother sold mending patches for pots and pans to help pay the mortgage. In 1941 my father, Ely Mechling, bought the farm from his Uncle Harry and I was raised on the place where my wife, Nancy, and I now live having moved back here after retirement.

My father loved horses. He kept horses after we switched to tractors. One of my most vivid memories of my childhood is his devastation at finding his matched pair of sorrel horses lying head to tail, dead after being struck by lightning. *Submitted by James Mechling.*

MEHARG – Samuel and Margaret's journey to Kansas covered many miles. Samuel was born in Ireland and emigrated to Canada with his family in 1844. There he and Margaret, who was born in Canada, were married in 1865. After a few years, the family moved to Iroquois County, IL where Samuel was a tenant farmer.

Samuel and Margaret Meharg

In 1872 Samuel heard of good prospects and Homestead land in Kansas and in 1873 he and his son, John, age 7, arrived on Sellens Creek in Fairfield Township, Russell County, to make a home before Margaret and two younger children made the journey by train. The first crop planted was corn and later oats, fruit and forest trees. Over the years, an additional 240 acres were added to the 160-acre Homestead.

The family increased to eight children, six living to adulthood. A stone house with four rooms was built and a few years later a wood frame kitchen was added. The stone house is still standing. The family attended church services held in the schoolhouse across the road from their home.

Four of the Meharg children lived on farms near their parents' home.

After the death of Samuel Meharg in 1912, the estate continued to be owned by Samuel and Margaret's descendants. In 1997 a corporation, Meharg Meadows, Inc., was formed by 27 descendants to purchase 320 acres that included the original 160 acres that Samuel homesteaded in 1873 and the adjoining 160 acres that Samuel purchased from the Union Pacific Railroad for $800 in 1897. Meharg Meadows, Inc. is currently owned by 31 family members, grandchildren, great-grandchildren and great-great-grandchildren of Samuel and Margaret Meharg.

Painting of the Meharg Homestead

These 320 acres are grassland, currently rented to a local rancher.

Homestead legal description: S/2 S/2 of 8-15-13. Union Pacific RR legal description: NE/4 of 17-15-13. *Submitted by Rita A. Lingg.*

MERKEL FAMILY CENTURY FARM – The Merkel Family Century Farm in Reno County, KS was established in 1883 by German immigrants, Mr. William Merkel and his wife, Caroline. They purchased 240 acres of land for $1,300 where they raised one son and four daughters. Their youngest daughter, Emma, married George Kleck who became the successive owner of that land. The original farm house was destroyed by fire and was rebuilt in 1914. After his death in 1933, the farm was passed on to their only child, Viola Kleck and her husband, Harvey Farney. There they raised two sons, Robert and James and two daughters, Donna and Shirley.

Front row, l-r: Robert, Harvey and Jack. Top: Garrett Farney. Middle: Chelsi (Farney) Marshall and Logan Farney.

Due to Harvey Farney's retirement from the farm in 1973, their oldest son, Robert and his wife, Lavonne, took over the farm where they raised two sons, Jack and Tyler and one daughter, Gayle.

In 1979 Robert and Jack began farming in partnership raising wheat, milo, alfalfa hay, sheep and cattle. Jack and his wife, Kay, raised two sons, Garet and Logan and one daughter, Chelsi. They played key roles on this farm and they continue their involvement in Agriculture. Garet is in ranching, cattle management and sells feed for WildCat Feeds. His wife, Jaymelynn, is the Southeast Kansas Extension Beef Specialist out of Parsons. Chelsi and her husband, Zack Marshall, farm and ranch in the Flint Hills south of Eureka, KS.

Robert and Lavonne recently retired from farming and moved to an apartment at Mennonite Friendship Communities in South Hutchinson. This allowed Logan and his wife, Megan, to become the present owners of the Merkel family Century Farm and live in the 1914 farmhouse. Along with farming, they have an Angus cow-calf operation, raise Texas longhorn cattle and chickens. In addition to the farm, Megan also owns a graphic design business. *Submitted by Robert and Lavonne Farney.*

MERKLEIN – Michael Merklein and his brother, Leonhard, immigrated from Germany to Marshall County, Kansas in 1870 at the age of eighteen. In 1872 they walked 150 miles to Phillips County seeking land to homestead. The town of Stuttgart was established in 1888 on a 40 acre tract of land obtained from the farms of these two brothers.

Michael's parents, George and Margaretha, and Heinrich and Eva Schneck and their daughter, Magaretha "Maggie", joined them in 1874. Shortly thereafter, Michael married Maggie and over a period of 19 years their family increased by thirteen.

In 1893, after the birth of their last child, Maggie passed away. Their younger children were "farmed out" to grandparents and other relatives. Three years later, Michael married Henriette Hortz helping reunite his family.

In 1881, Michael's father-in-law, Henry, homesteaded 160 acres northwest of Stuttgart. This tract of land was passed down or sold to the next generation of family farmers: Henry's daughter, Barbara and son-in-law, Leonhard Merklein; to their daughter, Margaret; and then to Henry's great grandson, Arthur Merklein (also Michael's grandson). Arthur's children Dennis Merklein and Sharon Krafft now own the land which Sharon's son, Eric with his wife, Alissa and daughter, Maggie (7th generation), farms today.

Michael had also purchased ($10/acre) a farmstead west of Stuttgart. In 1914 his son, Leonard and wife, Mary purchased this tract where they raised their family. After Mary's death, Leonard moved to Stuttgart. Their only son, Arthur was given possession of the land and the original rock/stucco house where he and his wife, Audrey raised their children. Art's love for the land was not only evident from his working the farm, but in his passion for gardening. As someone once said, "His garden's big enough to feed all of Stuttgart." Art built a wood-framed house in 1974 at which time his son, Dennis, who was farming with him part time, moved into the original house. Less than five years later, Art passed away at the age of 53. Audrey later rented the land to neighbors, including Denton Krafft, future husband of her daughter, Sharon. Denton and Sharon had three children, Jodie, Eric and Jenna. Denton continues to farm the land with his son, Eric. Upon Audrey's death in 2016, her children, Dennis and Sharon became owners of the land. Eric and his family currently live in the 1974 house until their remodeling of the 1900 Krafft house is completed. **See related story for the Krafft family.*

MICHAEL – The Michael Century Farm in Crawford County, KS is currently owned by John and Carolyn Michael. The Michael history reflects 125 years of farming by five generations. The original owner was the Kansas City, Fort Scott & Gulf Railroad Co. In 1880, it was purchased from them by William Hasty. It consisted of 160 acres. Land description is the northeast quarter of section 6 in township 30 south of range 22 east of sixth principal meridian, Grant Township in Crawford County.

Michael Century Farm, March 2017. The building at left is the original house.

In May 1863 John Wesley Michael, at the age of 14, enlisted in Company E, 115th, Indiana Infantry and fought in defense of the Union. In February 1864 he received his honorable discharge at Indianapolis, IN. He came to Crawford County in 1892 and purchased the 160-acre farm from Mr. Hasty.

He married and had four children, one of whom was Henry Wesley. Henry lived on this farm into his 90s. He married and had five children, one of whom was Gordon Wesley. Upon the death of Henry in 1991, Gordon and son, John, purchased the farm from Gordon's siblings. John has a son, Jim, which makes the fifth generation of farmers.

Horses, sheep, cattle, pigs, chickens and of course a large garden was once raised on the farm. The farm is currently a cash grain operation consisting of corn, soybeans and wheat. All conservation practices, including terraces and waterways, have always been implemented to prevent the farm from eroding. The original house is still standing but was turned into a shop many years ago. The second home built has tenants. *Submitted by John Michael.*

MILLER – In 1910 Wesley E. Miller purchased a 160-acre farm from Sarah A. Thompson for $6,400. Wesley had three children who inherited the farm just before his death in 1958. They were Martha Lucille Longhmiller, Dorothy May Williams and Ralph Newton Miller.

Ralph Miller

Ralph Miller family

The Wesley E. Miller Farm

On Aug. 16, 1958 Ralph and Irene Miller purchased the property from his sisters for $1. Ralph and Irene had four boys: Quincy, Earl, LeRoy and Raymond.

LeRoy Miller

LeRoy farmed with his dad (as Ralph did) through high school and after. In 1972 LeRoy and his wife, Pat, built a basement on the southeast corner of the land and added house on top in 1979. To this marriage came four children: Kim, Kelly Wayne, Katy and Kathy (stillborn). LeRoy bought the arm from his brothers in 2001 but in 2002 he was killed in a farm accident when heavy portable corral gates pinned him against the inside wall of his stock trailer.

Kelly and his wife, Alyssa, built on the northwest corner in 2006 and operate the farm with his mother. They have two sons, Talgan LeRoy and Cameron, who help on the farm and are active in 4-H showing different livestock but mainly farrowing, selling and showing their Show Pigs.

Through the years the farm has produced milo, corn, soybeans, wheat and alfalfa. Hogs and cattle were raised by all generations.

The children in this neighborhood attended West Union, a one-room schoolhouse until the late '50s. Now they go to Rossville Schools.

MILLER – The Miller Century Farm (SW/4 – Sect. 34 Township 13 South Range (6), Lincoln County and NW/4-Sect. 3 Township 14 South Range (6) Ellsworth County) is currently owned by fourth generation owners, Nancy (Day) Rasmussen and Greg Rasmussen, who inherited it from Nancy's mother, Mary Jane (Meili) J Day in 2014.

William and Mary (Durell) Miller, original owners, in front of the original farm house on the Miller Century Farm.

The original owners were William and Mary (Durell) Miller who purchased 320 acres on May 24, 1899. Mary Miller came to Kansas by train from Indiana to marry William Miller after meeting him while visiting family in Kansas. Mary brought with her the resources to invest in the new frontier and they purchased the land for $15.62/acre. William and Mary lived in the original farmhouse for several years prior to moving to Beverly, KS. They had two children, Nancy Esther and Benjamin Alva.

Nancy Esther inherited the farm in 1955 and owned it with her husband, John Meili, until her death in 1979. At that time, Nancy and John's daughter, Mary Jane, inherited the farm with her husband, Wilborn N. Day. Mary Jane and Wilborn farmed the land until Mary Jane's death in 2014.

The original 320-acre Miller Farm now consists of 170 acres of tilled ground and 145 acres of pasture, after giving up several acres for the construction of I-70. The farm has always been used to raise milo, alfalfa, wheat and corn in addition to grazing cattle.

Current owner, Nancy Rasmussen, is proud to say female siblings have owned this farm for generations and hopes to one day pass the Miller Farm on to her children. Proud owners for 118 years and counting!!! *Submitted by Nancy Rasmussen.*

MILLER – Purchased 80 acres from Albert Schmanke and 80 acres from Anton Schewe in 1878 by great-grandparents Michael and Rosina Muller. Grandparents August Paul and Anna Miller purchased the original 160 in 1900 and added another 160 acres in 1904. Parents Clarence and Lourine Miller inherited 160 acres of the farm and purchased the other 160 acres from Clarence's sister, Hulda Steinmeyer, in 1952. Paul and Nancy Miller inherited 310 acres of the farm in 2007 when Lourine passed away.

Bayling hay on the Miller farm

The Muller's were the first Europeans to reside on the property. They built a log cabin that still is part of the home. The home has been added on around the original log portion. There is a bottom story native stone and top story frame horse barn and a native stone open front cattle shed on the homestead. These were built by Herman Miller, a son of Michael and Rosina Muller.

Through much of the 20th century, three generations of Miller's lived in the house together. When Clarence and Lourine started their family of four children: Doris, Lois, Joy and Paul, they were active in the Lutheran Church, 4-H and school activities. Clarence and Lourine also had a purebred Duroc herd of nationwide renown. They had many sales on the farm in their sale pavilion and shipped breeding stock all over the country by rail from the depot in Alma.

Preservation of the tall grass prairie and land stewardship has always been a driving factor of the Miller family. Clarence and Lourine were awarded the Kansas State Bankers award for pasture management and Paul and Nancy were state winners of the Farm Bureau Natural Resources Award. *Submitted by Paul Michael Miller.*

MILLER – My father, Melvin Bert Miller was born April 1, 1898 on the Century farm I have inherited. I remember hearing stories how Grandfather and Grandmother left DeSota, KS to travel to Madison, KS to a farm they had purchased. Grandmother drove a team and wagon with my daddy in her tummy.

My grandparents were Edward Alexander Miller and Laura Elmina Watson Miller. And to this union were born Melvin, Harry, Olive and Louise.

James and Marilyn Miller

After the grandparents passed, the land went to Uncle Harry and Aunt Gladys. I remember a summer in the '40s when the young men were away to war that my Dad took my sister, Maxine and I out to Uncle Harry's to help hay. We would go out in the morning and wait while Uncle Harry and Daddy would mow and rake with a team. When it was time to bale, my sister and I took turns poking the wire through the bailer. When the day was done, Daddy would load a trailer with hay. After we got home, Daddy would have to unload the hay, milk his cows and other chores. But my sister and I each got $5 for that summer of haying.

At their passing I inherited the original 80 acres along with many fond memories of a little girl being loved by grandparents, a special uncle and aunt. My husband also inherited a century farm from his parents, so I guess farming is certainly our heritage. *Submitted by Marilyn B. Miller.*

MILLER – Reuben James Miller and Almeda Deer Miller purchased the original 80 acres from a Mary L. Miller and husband (no relation) on May 2, 1901 for $1,820 or $22.75 an acre.

Reuben and Almeda came to Kansas from Cass County, IA with three sons: Milow Duncan Miller (my father-in-law who was born June 8 1891 at Massena, IA), Frederick Dwight Miller (born May 1893) and Estle Miller.

Reuben Miller died at age 54. He was born Nov. 4, 1861 and died Sept. 3, 1915. He is buried at DeWitt Cemetery west of Humboldt, KS.

Sometime after Reuben's death Almeda married John Knock. She was born in Iowa on Sept. 10, 1871 and died March 26, 1962 at the age of 90-1/2 years.

Milow Duncan Miller married Ethel Frances Graham Miller at Yates Center on Aug. 6, 1924. They purchased the 80 acres from Almeda Deer Miller Knock on Aug. 11, 1933 for $2,200 or $27.50 an acre. Milow passed away March 27, 1960 and Ethel on Aug. 15, 1987. They are buried at DeWitt Cemetery by Reuben.

My husband and I inherited the land from his mother, Ethel. James Graham Miller was born Feb. 3, 1930 and started farming the land when he was 11. Milow had paid James $1 a day in wages during his childhood. This was in 1940 and they grew wheat, milo and soybeans. No original buildings remain. The house was built by Milow and Ethel in 1957 and is now a rental for a young family.

James was born Feb. 3, 1930 at Iola and died Sept. 13, 2010. He was named James from Grandfather Reuben James Miller and Graham from his mother Ethel's maiden name of Graham.

James married Marilyn Beverly Miller (no relation) Oct. 17, 1953.

James had farmed the land till 2009 and since then we have had excellent renters. The land is owned by Marilyn and their two daughters, Nancy and Cheryl. Marilyn still lives on a farm next to the original 80 acres, which was the core of the farm. *Submitted by Marilyn B. Miller.*

MILLER – William F. and Martha Jane (Clowser) Miller purchased 160 acres in July 1887. The farm is located in Rooks County, Farmington Township, 3-6-18. Grain crops of wheat, corn and milo were raised on the farm.

In 1916 the farm passed to Morgan and Beulah (Cunningham) Miller. In 1945 the farm passed to Orland and Clara (Russ) Miller along with his sibling, Warren and Marie (Wilson) Miller. Brothers Dwight and Marilyn (Schindler) Miller and Kenton and Dixie (Miller) Miller acquired the farm in 1998.

Milo and wheat are still grown on this 160-acre farm.

This farm has been the recipient of a Conservation Award and a Century Farm Award.

Dwight and Marilyn have two children, Theresa Johnston and Jason. Grandchildren are Dalton and Hunter Miller and Braeden Johnston. Step-grandchildren are Shayne and Daegan Hanvey.

Kenton and Dix have two children, Kimberly Pulec and Jerry Miller. Allison and Trevor Miller are their grandchildren. *Submitted by Dwight and Marilyn Miller.*

MOHR – My grandfather, Chris Mohr, moved to Derby (then El Paso) with his parents and several other families from El Paso, IL. He bought this Century Farm after he was married and later built another home on the property when my parents were married – letting them have the farm place. The old rock house was built in 1880. (A large stone above the front door is engraved "H.H. Snyder – 1880") and a recent survey shows it to be the oldest farmhouse in the Derby farming area – pictures are in Derby Museum.

Mohr farmstead – silo, dairy barn, horse barn and house.

Old Avery tractor, water tank, Grandfather on wagon, threshing wheat.

I, Maurine, was born in this house and except for a few years of college, teaching and when first married, have lived in it (about 60 years). My husband, Ray and I raised our family here – moving to a home my parents built when we married. When our children were married – our eldest daughter and son both have homes and live on the property. Ray and I celebrated 62 years of marriage this spring.

The property touches a creek where we have gone as a family and also entertained many groups at Fall Parties!

A KG & E farming award was given to my father. A Century Farm Bureau Award was given to us. *Submitted by Mary Maurine Mohr Holt.*

MOORE – William Pearl Moore, my great-great-great grandfather, and his wife arrived in 1849 in Oshkosh, WI from Ipswich East Anglia, England where they had been farmers. They saved enough money for their own

William Moore farm house, burned in 1926

Cecil Moore disking on the farm, 1937. He was the first one to get rubber tires on his tractor, so the story is told.

land and in 1871 William P. bought 160 acres in Northeast Kansas SE 1/4 Section 34 from George and Ann Morton. William P. died in 1897 and left the farm to his wife and five children: William R. Moore (my great-great grandfather), C.W. Moore, Pearl Moore, Anna Taylor, Emma Dimmick. C.W. sold his share to William R. and Emma. Anna sold to William R. and Pearl.

In 1915 Emma sold 160 acres of SW 1/4 of Section 34 to William R. Moore. Emma got this land from her husband, Oliver Dimmick

In 1926 a fire destroyed the beautiful 10 room brick and wood house.

When William R. died in 1926 he left his share to his wife, Jennie and three sons - Roland, Cecil (my great grandfather) and Urban Moore. Pearl died leaving his share to the three brothers and Emma sold her share to them also.

In 1947 the brothers separated the land as Roland wanted to raise oranges in Arizona. Cecil and Urban split the original 160 acres of SE 1/4 Section 34, each getting 80; Cecil had 80 acres SE 1/4 Section 34 and 160 acres of SW 1/4 Section 34, totaling 240 acres.

My grandfather, Francis, purchased the farm from his parents in 1966. They sold their house and 22 acres when they moved to town.

My father, John, purchased the now 218-acre farm in 2003.

I purchased the farm in 2016.

William P. brought his love of horse and greyhound racing with him from England and enjoyed coursing the dogs with competitors. This love of dogs and horses was passed down with my Grandfather Francis being the last to race the greyhounds.

Each generation has taken pride in protecting the land, practicing good soil conservation and the latest type of farming. Cecil owned the first rubber tired John Deere "A" in the area. Francis said the foot of the owner was the best fertilizer.

I am aware of my place in the history of this land and take pride in caring for it. My siblings - Alana, Brad and Chris - are happy the land will stay in the family. *Submitted by Jeff Moore.*

MORGAN – Willis and Clara Johnson acquired the farm on October 17, 1908, and passed the land to daughter Nona and husband Hugh Wigginton, to daughter Mary (Morgan) and husband Willard, to Myrna, Elsie, Susan, Sam, Judith and Robert (Morgan) and to Solomon Angus Farms, LLC.

When Hugh and Nona Wigginton owned the land, their children Mary and James rode horses to a country school two miles from home. The farm consisted of pasture, grazed by a herd of predominantly Angus and Angus Cross cattle, bottomland cut for prairie hay, and fields of alfalfa and wheat. A small feedlot was built to feed the calves not being saved for replacements in the cattle herd. When Mary married Willard Morgan, they formed a partnership with the Wiggintons.

Willard worked with Hugh and managed the farm when Hugh was no longer able. The partnership was later dissolved and upon James' death, the land was divided. The Morgan children grew up with involvement in 4-H, church and community activities.

The Morgan portion of the land is now owned by Solomon Angus Farms, LLC, which consists of five of Mary and Willard's children; Elsie Rietcheck, Susan Draper, Sam Morgan, Judy Kennedy and Robert Morgan. Son-in-law, Ronnie Rietcheck manages the farm. This farm was honored as a Kansas Farm Bureau Century Farm in 2011.

MORTON – Sherman and Elizabeth Morton moved from Gallia County, OH in 1909 and purchased a quarter section from John and Lucy Coleman on Feb. 12, 1910. Their two children, Earl and Ruth Morton, became school teachers and farmers. Ruth married Milton McFadden and moved to Douglass. Earl bought the 80 acres just across the road to the north of Sherman's farm. Earl married Iva Shoup on Christmas Day in 1923 and went on to raise four children: Florence, Clyde, Oscar and Cora. The original stone barn was struck by the Udall tornado in 1956 just prior to Oscar Morton taking ownership of the farm from Sherman. Oscar farmed successfully, adding additional rental acres to survive the changing times, until his death in 2003.

Mike and Sharon Morton at the Governor's Inaugural Ball, 2003

Oscar married Maxine Able and they had four children: Mark, Thomas, Rita and Don Morton. Don still farms the 70 acres of tillable ground to this day. Maxine was killed in a tractor accident in 1974. Oscar went on to remarry Linda Furgeson and added her three children - Bobby, Kim and Kay - to the Morton family farm. Oscar was a conservation minded farmer and pioneered and taught the construction of terraces with the old fashion mold board plow. Don has become a master of no till farming.

The current family owners are Mike Morton and his wife Sharon Gordon. Mike is the son of Clyde Morton, who went on to teach in San Diego, CA. Clyde was given the original 80 acres that Earl and Iva farmed and upon retirement returned to his home town. Clyde's son, Mike Morton, had a successful Nursery Business in Orange, CA, Country Bloomers Nursery and Sharon is a contract attorney. They purchased the 80 acres east of Clyde in 1995 and moved. Mike and Sharon bought the Century Farm from Linda Morton in 2005, adding it to their 80 acres. Mike has now inherited a share of the original 80 Earl Morton bought in 1923. Mike and Sharon were among the original investors in the first cotton gin in Cowley County and first cousin, Don Morton and his wife, Susan, are successful cotton producers. In addition to the row crops, the farm has a fruit and nut orchard, as well as prairie hay and alfalfa. The original farm house has been restored and turned into a rental. Mike and Sharon received the 100-year farm recognition in 2010. Mike

Harvesting cotton

served four years on the Cowley County Farm Bureau board, two as V.P. Mike, a member of Leadership Cowley County Class of 2002, served eight years on the Legacy Community Foundation, two years as President, in addition to many other community organizations. Sharon served as the first Chair of the Udall/ Cowley County Joint Planning Commission. The Morton family has always valued public service, and has established an endowment for Leadership & Public Policy with the Legacy Foundation of Cowley County.

Oscar, Earl, Mark, and Sherman, four generations of Mortons, ca. 1954.

Although Mike's daughter, Anne Morton Shirck, passed away in 2007, his grandson, Tyson Shirck, loves to come spend time with grandpa and grandma on the farm. He dreams of running those big machines someday, including the John Deere 730 that Oscar bought back in 1954. *Submitted by Mike and Sharon Morton.*

Mike with grandson Tyson Shirck with the dogs

NEELY – Dr. John Kennedy from Erie, KS, guardian for Sarah F. Fox, purchased 170 acres for Sarah with money from her late father's estate on Dec. 26, 1871. Dr. Kennedy then purchased an adjacent 160 acres for Sarah on Jan. 3, 1873. Sarah Fox married Dr. A.F. Neely on Aug. 6, 1873 and they purchased an adjacent 75 acres on April 2, 1877. Dr. Neely and Sarah had two children, James in 1875 and Don in 1878.

Don Neely and wife, Nellie, purchased the 405 acres from his parents in 1913 and raised 11 children on the farm. Don passed away in 1944 and his son, George and his wife, Mary, lived on the farm until 1951. Nellie Neely passed away in 1946. Two other sons, Dick and Glen, purchased the 405 acres from the other nine siblings in 1951.

Dick and wife, Martha, raised three children on the farm before retiring and selling his half interest to Glen and wife, Marjorie, in 1975. Glen's oldest son, Jim Neely and wife, Colleen, moved to the farm in 1975 and they purchased the buildings and 35 acres in 1992. The remaining 370 acres were transferred to the Wm. G. Neely Trust in 2003. Glen Neely passed away in 2007 and his wife, Marjorie, is the surviving trustee to his trust.

Jim and Colleen Neely raised four children on the farm and they continue to live on the farm and manage the original 405 acres. Neosho County has an operating rock quarry on part of the farm and the farm has received awards for Conservation, Century Farm and a Natural Resources Award for timber stand improvement. *Submitted by Marjorie Neely.*

NEIDIGH – Ron and his wife, Ann Neidigh, live at 787 1000 RD southwest of Edna, just a very short distance from the Oklahoma line. About one mile north of their home lays 160 acres that has been in Ron's family since 1909. Ron's great-great grandfather, George Wesley Neidigh, purchased the 160 acres from a widow, Louise E. Read, for $37.50 an acre. It is interesting to note that it was purchased on Sept. 28, 1909.

George Wesley was born in Cumberland County, PA. He came to Kansas from Decatur County, IN, settling in Richland Township in the southeastern part of Labette County in 1884 He moved to the present farm in Howard Township in 1886 prior to purchasing it in 1909.

Other owners were Ron's grandfather, Claude Alva Neidigh and then his father, Dale Elvis Neidigh.

A portion of the original home built sometime before 1886 still exists on the farm. The farm produces wheat, oats, milo, soybeans and cattle. Ron has put together an interesting book including pictures and stories about his family.

Ron and Ann Marie have three children: Samuel John Neidigh, Sara Ann Gibson and Sydney Ruth Neidigh. Solomon Alon Gibson is their grandchild. *Submitted by Ronald Wayne Neidigh.*

NIMZ – Willhelm "Will" Nimz and wife, Celia, bought 320 acres of land in southern Trego County in 1915 and later bought two more quarters close by. They used one of the very few structures that was on the land for living quarters and added on to it as their family grew. They built other structures to house their animals. Two of the buildings, a chicken house built in 1919 and a wash house built in 1932, were made of limestone rock. The too long uninhabited house was demolished a few years ago, but the good old limestone buildings are being maintained and used for different purposes by present owners, Marvin and Frances Nimz, son and daughter-in-law of Will and Celia.

One of two rock buildings on the Nimz farm that have withstood the elements of the last 100 years.

The draw that runs through the original 320 acres and into the Smokey Hill River is known on the map as Rattle Snake Creek.

Marvin and Frances have five children: Daryl, Mark, Jill, Kevin and Larry and 11 grandchildren. *Submitted by Marvin Nimz.*

NINEMIRE – The Ninemire Family Farm or as my wife, Peggy, and I call it, Coffin Rock Ranch, was started in 1898 by my great grandparents, Peter and Catherine Schandler. They were living in Tipton, KS

Mrs. Peter Schandler and baby Leo, with Catherine, Bernard and Joseph. They moved into the rock part of this home when they moved here in 1900, then later added the south room and later the kitchen to the east.

when Peter came to Norton County and bought the first 80 acres from the American Investment Company. In 1900, he bought another 80 acres to make a full quarter. He moved his family out here and lived in the rock house that was on the property when he purchased it that was built in 1887. In 1904 Peter built a wood frame house that was used until it burned to the ground in 1997.

My grandparents, Joseph and Dolorose Schandler, started farming the place in 1924. There were many struggles through the depression and dirty '30s. They made it thru it with an orchard and many chickens to make a living and raise three girls: Imogene, Miriam and Beatrice.

In 1948 Imogene Schandler married my father, Raymond Ninemire and to this union, eight children were born and raised on the Ninemire farm. In 1957, my father and mother moved into the farm home and took over from Joseph and Dolorose Schandler.

Raymond and Imogene worked very hard and added to the farm that they loved until his death in 1978. I took over and ran the farm for my mother until her passing in 2000.

On Oct. 29, 1983, I married the love of my life, Peggy Brooks. We have raised three children: Lacy, Levi and Luke on Coffin Rock Ranch. After our home burned down in 1997, we put a new home on our beloved ranch and are still blessed to have the original rock home that my great grandparents lived in still standing 100 feet from our new home.

Not a day goes by that I don't "Thank" God and all my ancestors that my wife and I are able to live on this beautiful place. *Submitted by James and Peggy Ninemire.*

NIXON – Fertile land and a sense of adventure were combined to bring Perry O.C. and Virginia Nixon to Labette County in 1869. At a young age, Perry was bound out until the age of 18 when he went out on his own. As a young man, he travelled to Pikes Peak to celebrate his 21st birthday. Later, when the Civil War began, he enlisted in the Union Army and served at the front lines for nearly the entire war. After the end of the war, Mr. Nixon returned to Illinois where he farmed and also married Virginia Humphreys. They, along with two other families, travelled to Labette County where he homesteaded 160 acres. He later sold 60 acres to his brother, Isaiah, and continued farming 100 acres.

Children of Clarence and Ruth Nixon

Mr. Nixon's first house was built of native lumber, cut on Big Hill creek. However, tragedy happened in 1892 when a tornado struck the farm, destroying the house, barn, and several outbuildings. Mrs. Nixon was the only one of the family who was unable to reach the safety of the cellar. She was so severely injured that she died in 1898, never having been able to sleep lying down during those years.

During his first year, Mr. Nixon raised sod-corn, which he planted with the aid of an ax and later harvested a very successful crop. Over the years, he raised beef cattle, Poland-China hogs, sheep, and chickens. As a part of the family's self-sufficiency, the Nixon's planted a diverse orchard that provided an abundance of fruit and also raised a large farm garden. In later years, the family put in a dairy, which operated until the 1950's.

Perry was known not only for his farming ability but also his expertise in stone masonry. He hewed all of the stones used in the building foundations, a rock pond dam, and a stepping stone used by ladies to mount their horses. His most outstanding project, however, was a stone cellar built without interior supports. It had a key rock the length of the cellar and the ceiling was oval-shaped.

While the Nixon family throughout the years has been involved in all aspects of farming, their real dedication has to been to the surrounding Mt. Zion community. Mr. Nixon donated the land for the nearby church and was a leader in the church and school. Later generations continued this involvement. Even during a dark episode of the community, Mr. Nixon was a presence. He was in the search party that pursued the infamous Benders after the discovery of the bodies of their victims.

Ownerships of the farm in its 149 year-history have included: Perry O.C. Nixon, 1869-1918; Bert Nixon, son of Perry, 1918-1936; Millie Nixon, wife of Bert, 1936-1943; Clarence Nixon, son of Bert and Millie, 1943-1977; Ruth Nixon, wife of Clarence, 1977-1984; and the children of Clarence and Ruth, Barbara Cook, Brenda Richardson, Gary Nixon, Mike Nixon, and Nancy Kirk, 1984-present. A great-great-grandson of Perry O.C. currently lives on the same land claimed by his ancestors.

NORRIS CENTURY FARM – The story really begins in 1878 when my great grandfather, Steven A. Norris, homesteaded in Mount Ayr Township, Osborne County, KS. The original homestead passed through another line of descent and is no longer owned by the Norris family but in 1901 he purchased an adjoining f-arm and it remains in the family. This was known as the weaning farm. Steven had a family of five boys and two girls. As each of the boys grew to manhood and was ready to earn his own living, they were moved onto this "weaning farm" and it was theirs to use to make a living UNTIL the next oldest son was also ready to take that step, then the older brother had to move on. Hopefully they had earned enough to start another business elsewhere. My grandfather was the net to youngest and his younger brother was able to stay on the original homestead to assist and care for his parents. So there was no one to push granddad off and he got to stay there for the remainder of his life.

The Norris home

As with most farms, products produced have changed to meet market demands. From cattle and work horses, dairy, broom corn and rye to wheat, milo and alfalfa. The latest product added is fee hunting for deer and upland game. An interesting feature of the dairy barn is that it was built into the side of a creek bank with the soft limestone available locally. Over the years many family members carved their names into the soft rock as they lived, worked or passed through there. Of cultural interest, sometime in the 1930s my granddad was able to get the Conservation Corp to plant a shelterbelt for him. One of the members of the work crew that came was a Negro. Many of the people in the neighborhood had never seen a Negro before and they came around to watch the "colored boy" work.

This farm was purchased by Douglas Norris from his grandparents in 1993 to maintain continuous ownership within the Norris family. *Submitted by Douglas Norris.*

O'TOOLE – Patrick William O'Toole bought 320 acres from the US Government on June 1, 1899. The land was located in Trego County, N 1/2 30-14-24. He raised Angus cattle on the farm.

Grandson Patrick Michael O'Toole in the bluff area after deer hunting in 2015.

The farm has always been owned by an O'Toole. The farm passed from "P.W." to his son, Arthur O'Toole. Then it passed to Arthur's nephew, Walter O'Toole. Walter passed the farm to Patrick Walter O'Toole, his son. 100 years after being purchased, in 1999 the farm passed to Patrick's widow, Janice O'Toole.

Although there are no buildings remaining, there are beautiful bluffs from the Smokey Hill River. This makes it a nice place for family hunting. Livestock are still on the farm, but the pasture is rented.

Patrick and Janice had two children, Darcy Pat O'Toole and Shelly Sue Poppelreiter. Grandchildren are Emily Elizabeth O'Toole and Patrick Michael O'Toole. *Submitted by Janice O'Toole.*

OURSLER – Charles Franklin Oursler and wife, Emma Sophia (Elschlager) Oursler with their two children came to Kansas from Adams County, OH. They homesteaded their land in Butler County near Burns in 1901. It was not registered until 1902 in the courthouse at Eldorado. Charles' Great-Great-Grandfather Edward came from England to Baltimore as a young single man. Charles' Grandfather Archibald and wife, Malinda True, moved to Ohio. The family stayed there until several brothers moved to Kansas. Charles' twin brother, Thomas, moved to Newton; John Newton was in the Lincolnville area and W.E.M. Oursler settled by Marion, KS, starting the town of 'Oursler.'

Charles and Emma had six more children born at the farm. They were very active in the Methodist Church. The children all attended the Wilcox one room schoolhouse. The family always had a vegetable garden and fruit trees. They had milking cows, chickens, turkey, pigs and horses. The fields had corn, milo, atlas and prairie hay. People from all over came to buy their famous turkeys, especially around the holidays. Charles' second son, Earl, took care of the horses used for working the fields. He also trained horses for racing on Peabody's Joe Young racetrack on the 4th of July.

Charles died Feb. 15, 1932 and Emma died Dec. 22, 1959. The farm was passed down to their daughters, Bessie and Ruby, for the care they gave their mother during her illnesses. Charles and Emma's youngest child, Kermit Franklin Oursler and wife, Edna Caroline (Butts), along with their three sons - Frank, Harley and David - lived on the same land in a second house. Bessie and Ruby were teachers and lived in Eldorado. They sold the house and land to Kermit and Edna who continued to live on and farm the land. Kermit also drilled water wells and hauled rock from Florence. Edna was a waitress in Eldorado and Newton for 50 years.

Edna remodeled the south side of the original farmhouse in the 1980s. In the 1980s David and Frank bought 80 acres east of the farm.

The Charles and Emma Oursler family in front of their home

In the 1990s Frank and David bought the homestead. They have farmed the land and rented out the house. David's daughter, Trisha, lived at the farm for a while but it is now home to David's' son, Rodney and his family.

Hopefully the farm will stay in the Oursler family for years to come. *Submitted by Janet Oursler.*

OWEN – My great-great grandfather, Joshua D. Owen, bought 159 acres of Railroad land in 1894 for $800.00. One acre given for a school district, 119 acres farm ground and 40 acres of pasture in Marshall County, Bigelow Township, Reserville Community. Joshua purchased the land to give to my great grandfather, Cephas E. Owen and his wife, Louisa. The family had originally settled in Elgin, IL but land was much cheaper in Kansas.

On the farm they rotated crops of wheat, corn, oats and hay. Prior to the move to Kansas, Cephas and Louisa lost their three children to complications of measles or scarlet fever. Louisa would go to the cemetery with her mending and sit with her children. They later had two more children.

My grandfather, Ira M. Owen, participated in a tent revival complete with baptism in the creek. Ira joined the Christian church and quit smoking. Ira and his wife, Blanche, purchased a General Store in 1913 in Vilets, KS 14 miles northeast and moved their family. They tended to the store and rented the farm out. They would visit the farm and renters on Sunday afternoon. Sometimes I could ride along, it was "such an adventure". It was in 1941 that my father and mother, Earl A. and Ida Owen, began overseeing the farm from the store in Vilets. In 1964 my husband, Luin D. Larson and I (Eunice E. Owen Larson) began farming the land. Through the years we completed extensive conser-

Aerial photo of the farm taken in 1955

The Owen farmhouse, pictured l-r: neighbor, hired man, Cephas Owen, Ira Owen, and Louisa Owen

vation projects. Luin died in 1998, since then our son, Kurt M. Larson and grandson, Kurtis D. Larson have farmed the land. There are still indentations of the buffalo wallows by the creek. *Submitted by Eunice E. Owen Larson.*

PASTINE – Heinrich Schroder was originally from Logumkloster, Denmark and Anna Maria Jensen was from Hostrup, Denmark. They were married in the Lutheran Church, May 10, 1885 in Tonder, Denmark.

Heinrich Schroder had lived in Chicago, IL prior to returning to Denmark and meeting and marrying Anna Jensen. They went to the US and purchased rand on the original Homestead Act of 1862. They had two living children, Sophia and Henry. Prior to moving to Kansas they lived in New York City. Heinrich had also lived in Chicago, IL. They had also lived in Omaha, NE and in Otoe, NE.

Harvey, Oscar, and Gerhard Hillman

Otto Schroder concrete dugout on the farm.

The land they purchased was in Trego County, KS 1/4 11-22-11 north of Ogallah, KS and east and north of Wakeeney, KS. The property is pastureland, about 160 acres – grassland, wildflowers, wildlife, water for cattle (via underground springs) and shale banks, and limestone rocks.

There is a concrete dugout on the property built on the property by Maureen Hillman Pastine's great grandfather, Heinrich Schroder, more than 100 years ago.

Maureen Pastine received a 2011 Farm Bureau 2011 Century Award for this pastureland.

Maureen and Jerry Pastine now own this pastureland and it is to go to her brother, (Gordon Hillman and Susan Hillman, Brian Hillman, Clint Hillman, and Dion Hillman upon her death). *Submitted by Maureen Hillman Pastine.*

PATON/MUELLER – In the early 1860s, George Paton Jr. migrated from his native Scotland to New Zealand to join cousins who had previously gone to that country. After about six years, George left New Zealand for the US. In 1871 he brought his parents, George Sr. and Ellen Paton and his five siblings to the US and settled them all on a farm he had purchased for them southwest of Oxford in Sumner County, KS.

In 1884 he married Mary Massey and they purchased and moved to their own farm in the immediate area. However, in January 1898 he purchased a farm southeast of Winfield in Cowley County and moved there with his wife and six children. Unfortunately, George died in March 1902, leaving his widow and five young sons to maintain the farm. They did so with the help of a compassionate neighbor who provided advice and assistance as needed.

In April 1918, the second son, Wm H. Paton, my grandfather, purchased a nearby farm (which will become a Century Farm in 2018) and moved there with his wife, Viola. Their two children, Willa Jean and Maurice were born on that farm. Mary Paton continued to live on the original farm until her death in 1944. In 1940, Will Paton purchased the interests of his mother and siblings in the farm and in 1961 deeded it to his children, retaining a life estate. He continued active involvement in the operation with Maurice until his death in September 1980.

Bill Mueller, Willa Jean's son and his wife, Belinda, acquired Maurice's interest in the farm prior to his death in 2002 and continue operating it, raising wheat, milo, soybeans and cattle.

Willa Jean Mueller died in 2007, leaving her share to Bill and Belinda and Bill's sister, Deanne Stevens. *Submitted by Bill Mueller.*

PATTERSON – Hickory Creek Farm has been home to many throughout the years. In its early days the Osage called it home, living on the banks of Hickory Creek and farming the fertile valleys.

The Hickory Creek Farm house and barn

In February of 1869 this land was given to James Joy with the intent to be sold off in 40-acre plots to farmers. In January of 1869 the northwest quarter of section 6 was sold to George Mosteller for $165. In February of 1880 he sold it to Luther Wooley for $800. On Feb. 10, 1888 he sold this 160 for $4,000 to Sarah A. Patterson of Mt. Vernon, OH, who was my great-great-grandmother. In that same year, her son, Abishia Clark McGown, moved from Ohio and took his turn farming the land we call Hickory Creek Farm. A 23-year-old civil war veteran who had just lost his wife, he set out to make a new life. He met and married Sarah Frances Roper. They had 11 children, one of which was John Logan McGown, who was born in the home that still stands on the farm today. In 1924 he married Florence Burch. They had six girls, one of which was Dorothy Jean. She married Luther Zimmerman in 1949. They had four children: Jeanne, Susan, Steven and Betty Anne. Like my ancestors before me, I took the tools of the trade and started farming. I have learned to love the land, not just as a job, but as part of my very being.

In 1990 I married Ruth Toliver. We have five children: William, Andrew, Jonathan, Mary and Sarah. When farming alone couldn't provide the income to raise a family, we turned to Agritourism to help supplement it. My wife and I now share Hickory Creek Farm with others through our pumpkin farm. *Submitted by Steve Zimmerman.*

PENNER – The Penner Century Farm was originally purchased by Frank and Elise Penner on Oct. 9, 1911. They purchased 80 acres and produced wheat and milo and used the pasture for cattle. In 1945

The Penner barn

Frank, a widower, sold the farm to his son, Will C. Penner and his wife, Vinnie L Penner. Will and Vinnie constructed a barn for cattle and mules. They moved in a building from nearby Halstead to use as a house after adding some additional rooms. They also grew wheat and milo and raised cattle and mules. They were also known for their large flock of chickens and sold eggs to many of their neighbors and townspeople of Halstead. They also had a daughter, Norma Lee, who grew up on the farm. She has many fond memories of growing up on the farm.

The Penner homestead

Norma married Thomas Wendling of Halstead. After Will passed away, Vinnie moved to town and the farm was sold to Norma. Several years later, Norma sold the farm to her son, Patrick L. Wendling. Patrick is now the fourth generation to own the farm and lives there presently. The farm still produces wheat, milo and beans. Patrick and his wife, Bonnie, also raise Hereford cattle as well as chickens. It is still the original 80 acres that Frank and Elise purchased and the pasture is still original prairie that has never been broken. *Submitted by Pat and Bonnie Wendling.*

PETER – My great-grandfather, Jonas Peter, came from Switzerland. He came to the USA in 1834 to Jackson County, IN. he was a local preacher in the German Methodist Church. He was naturalized Feb. 10, 1840 in Jackson County, IN. His occupation was then a farmer. He came to Kansas in 1859 and purchased 500 acres of land from the US Government by US Grant, President Seal in Riley County, KS.

The Peter farm house, built in 1900

The land was later divided between his two sons. His son, William F. Peter, came to Kansas and located on the property. His share consisted of 200 acres of pasture and 70 acres of farm ground.

William was a prospecting farmer and stock raiser. He served in township offices and later was elected on the Republican ticket and a member of the board of county commissioners. In 1904 he was elected to the State Legislature from Riley County, he was re-elected in 1906.

After the death of his father in 1916 my dad, Arthur J. Peter, farmed the land in Fancy Creek area west of Randolph, KS. On the death of his mother in 1939 he bought the farm from his brothers.

In 1959 the Tuttle Creek Dam was built and they were forced to move to Clay Center. His son-in-law, Dale Wohler, farmed the land for many years. On the death of Arthur, Dale and his wife, Margie Peter, bought the land in 1987. When the dam came in, all the buildings had to be removed. Some of the land is now easement land. The crops are now corn and soybeans. The pasture has a spring that has never failed and has good grass. The farm is located on some of the best fertile ground in Riley County. *Submitted by Dale and Margie (Peter) Wohler.*

PETERS – Seeking land to farm, thirteen pioneers arrived in southeast Kansas on April 10, 1866. Among this group was Sophronia Peters (Figure 1), a widow, and her sons, Aurelius V Peters and William "Had" Peters both veterans of the Civil War from Ohio. The Peter's family staked three claims in the Cherokee Neutral Lands later to become a part of Cherokee County, Kansas. President Grant granted their patents for land on January 1, 1870.

Aurelius, suffering from consumption acquired in the war, is credited with naming the county seat Columbus after Columbus, Ohio. Governor Crawford appointed him as one of the first Cherokee County Commissioners. He died without marrying in 1872. Sophronia died in 1879 and William added some of her land to his farm.

William brought with him varieties of trees planted in sardine cans from Thornville, Ohio, which took root in the prairie land; hence, he named his land "Greenwood Farm". He married Ellen Gates from a

Sophronia Martin Peters

Myrtle, Adelia and Ralph Bradford, 1921

Pete and Adelia Clark, 1945

Adelia Clark, 2017

Art, Myrtle, Ellen, "Had" and Will Peters in buggy about 1898

Pete and Brad Clark, 1986

neighboring pioneer family and eventually replaced the cabin initially erected with a two-story home. Had and Ellen raised three children, Art, Myrtle and Will (Figure 2). Following his brother Aurelius' example, William "Had" became a county commissioner and maintained his farm.

Anita Beth, Brad, and Cathy, 2017

Upon William's death in 1905, Ellen and her children continued to farm with the help of hired men from the community. Myrtle married Ralph Bradford in 1906 and they lived in the Peter's home. Ralph and Myrtle Bradford had one daughter, Adelia, born in 1920 (Figure 3). Will married Ralph's sister and eventually moved away to work as a surveyor. Art served in the military and retired back to the farm in the 1940s. He did not marry and died in the mid-1950s leaving his interest to Myrtle and Will. Will sold his inheritance to Myrtle. The original two-story home burned in 1937 and was replaced with a one story home.

In 1943, Adelia married a neighbor, Alvie E. "Pete" Clark (Figure 4). Upon his return from the European theater of WWII, they built a house on a portion of the land and farmed with the Bradford's.

Pete eventually took over the operation with the death of Ralph in 1957. Adelia and Pete Clark had three children, Anita Beth, Brad and Cathy (Figure 6). They all worked on the farm in various roles. Brad Clark, born in 1949, returned from the Marines in 1970 to farm with his father. Anita Beth and Cathy married and moved from the farm. Adelia took an active role in farming through the years and was easily recognized hauling wheat to the elevators.

In 1991 Pete died suddenly leaving the farming with wife Adelia and son Brad. Adelia Clark will be 97 in July 2017 and transferred the farming operation to Brad Clark. The land is in the names of her children, continuing the ownership from 1866 until the present through five generations of family history in Cherokee County, Kansas.

PETERSON – I am the fourth generation living on the farm. I am having my 83rd birthday on March 26. My wife Charlotte and I still live here and have been married 55 years. We have one daughter, Darlene and two granddaughters, Andrea and Amanda and one great-granddaughter, Kinsley.

Due to heart health problems, I had to give up livestock but I still farm the 152-acre farm, which is mostly soybeans and wheat. I till the ground and hire combining. Then I rent pasture out and also waterways for hay. I still maintain the fence

In 1971 a tornado took down our big barn and damaged the second one, so no big barns are left. We live in Walnut Township, section 30-3-6. We are in Marshall County.

The Peterson farm in 1970

It was 1899 when our Danish ancestors (last name Grosebeck) moved into this area, mixing with German cultures. They all had a common thread, that being Lutheran faith.

In years past, my parents raised hogs, cattle and chickens and had milk cows. A big orchard also helped sustain an income. We remember the party lines and how life was in years past. In fact, I wanted my granddaughters to know things of their past so I recently wrote a book, *From the Coal Oil Lantern to the Internet*. It is informative on many things like country schools, putting up hay, hand crank cream separator.

I am an only child as well as my wife and then Darlene is our only child. So the farm holds special meaning to us all. It has a view for almost 20 miles in the distance. We hold every day a joy when we can see the sunsets, changing seasons and view the wildlife and listen to their calls. We also note that every year, the swallows return close to April 19. Every day has been a journey, some hard, other times easy. *Submitted by Arlyn Roy Peterson.*

PETERSON – The Peterson Century Farm in Marion County, KS is currently owned by Robert Sellers, the great-great grandson of the original owner, Maria Peterson. The Peterson family consisting of Maria, who was twice widowed, along with her four children, emigrated from Denmark in 1871. They first settled in Wisconsin, but made Kansas their home when Maria homesteaded the 80-acre farm. Ownerships include: great-great grandmother Maria Peterson (1874 - 1884): great-grandfather Mogens Peterson (Feb. 20 1884 – April 9, 1884), great-grandmother Caroline Peterson Olsen (1884 - 1912); grandfather Eli Peterson (1912 - 1964); uncle Willard Peterson (1964 - 1976); Robert Sellers (1976 to present).

Wheat harvest on the Eli Peterson farm. Eli Peterson is on the far right side of the picture.

Determined and strong farm women were a mainstay on the Peterson farm. From a widowed single mother homesteading the land to her future daughter-in-law Caroline, also left a widow with two small sons, each worked hard to care for the land. During the 1920s oil boom in the surrounding area, Eli's wife, Lydia, opened her farmhouse kitchen to oil field workers as a place to purchase a hot meal. The extra income helped provide for the farm and her family of nine children.

The Peterson family raised wheat and corn on their land along with chickens, milk cows and horses for farming. They also kept pasture for their livestock. Large vegetable gardens provided fresh produce as well as for canning. The nearby Doyle Creek provided fish and the hillside wild plums and gooseberries. Despite difficult times, large gatherings of family and neighbors were commonplace on Sunday afternoons.

PIPER – In Labette County, KS William Pfaff, from Kentucky, homesteaded 160 acres of land in 1868. This acreage is located in Fairview Township, SW 1/4 3-33-20 and produced wheat, corn, grain sorghum and Mr. Pfaff sold mules.

The farm passed to his daughter, Alma Pfaff Piper to her son, J.C. Piper and then to Mark Piper in 1961. Now wheat, corn, soybeans, fescue and prairie hay are grown and calves are raised.

The brick home built in 1885 remains and is now rented to a family.

Mark and Charlene Piper farm

Before Mark had to deal with poor health, he loved to experiment with different crops, along with what the farm is growing today. He experimented with growing cotton and sunflowers. He sold many 25# bags, as well as 50# bags of sunflowers to feed birds.

In 1970 Mark was the "Outstanding Young Farmer" and received second place at the state level. He was a charter member of the National Cattlemen's Beef Association.

In 1998 Mark and Charlene won an award of merit for Outstanding Accomplishments in resource conservation.

Mark served on the Kansas Livestock Association board of directors.

At the farm he hosted a Kansas forestry meeting about the care of trees.

Since Mark's death on Dec. 20, 2014 Charlene has been managing the farm with the help of others.

They had two children, Letitia Petitjean and Julianna Talkington. Grandchildren are Laneah Knuth and husband, Cody; J.C. Petitjean, Bailey Talkington and Sydney Talkington. *Submitted by Charlene Piper.*

PLAMANN – August Plamann was born in Tehevelein, Germany in 1829. Mary Magdalena (Rusch) Plamann was born in Labes, Germany in 1838. In 1868, August and Magdalena, their five surviving children and Magdalena's father and niece immigrated to the US, originally settling in Wisconsin. Another child was born in Wisconsin and four children in Kansas.

Jackson Meyer holding the Century Farm Award, March 2015

August and Magdalena disliked the cold weather in Wisconsin, so in 1874 they purchased land and moved to Brown County, KS. They built a two-story log cabin in Walnut Township (Section 20 Range 2 Township 16) northwest of Hiawatha. Due to flooding, the house was placed on logs and oxen were used to pull it to higher ground. The land consisted of 31 acres purchased and added to the original Plamann land. Remnants of the cabin still exist in 2017.

The Plamann-Meyer Farm, 1950

The Plamann-Meyer Farm on Goldfinch Road, 2015

August and Magdalena owned 420 acres and grew cane for molasses. The first known community 4th of July celebration in Brown County was held in 1857 in the timber, at the site of a saw mill, where Mulberry and Walnut Creeks meet on the current family farm.

In 1903, August and Magdalena moved to 503 Iowa Street in Hiawatha. Their daughter, Rosa, married Edward Meyer and they were the next to live on present Goldfinch Road. Rosa loved African violets and grew many of them in the bay window of their house. When Edward and Rosa owned the land, there was a gravel pit on the east side of the road about 200 yards south of Mulberry Creek. They donated the gravel for the church road that ran from Highway 35 to the Immanuel Evangelical Lutheran Church northeast of Fairview. They had nine children that made it to adulthood. After Edward passed, Rosa moved to Oregon Street in Hiawatha.

The Meyer boys devised a wagon with a dump floor made out of 2 x 5s that could be turned on its side to dump the gravel. A mammoth tusk was found in the gravel pit. The youngest son, Stephen, took over the farm. He and his wife, Helen and their family lived there. Their son, Doug and his wife, Brenda, currently reside at the farm. Their son, Clint, his wife, Kelly and their son, Jackson, live in the house where Thomas Mellenbruch and Lena (Plamann) Mellenbruch lived. Lena was a sister of Rosa. This quarter section is also part of the 665 acres presently family owned.

Doug and Clint are the 4th and 5th generation to farm the Plamann/Meyer land. Through the years, the farm has raised Persian Horses, dairy cattle, beef cattle, swine, corn, wheat, grain sorghum, soybeans, walnut trees and honey. Stephen, Doug and Clint have also sold for a major seed company. The Meyer family has held a deep abiding faith in God through the generations, along with family, friendships, and community pride. *Submitted by Doug Meyer.*

PLOGER – At the persuasion of Henry Fromme, brother of Fredricke (Fromme) Ploger, Wilhelm and Fredricke Ploger brought their family from Heidelbeck, Germany to the US in 1891. Along with their children - William, Henry, Fredricke and August - they completed their journey to the Fromme farmstead (the site of the National Historical Society "Fromme Round Barn") at Mullinville, KS.

August and Elizabeth Ploger, 1905

After working at the Fromme farm, the Ploger family moved to Edwards County, in 1900, north of Kinsley. All the children eventually married and branched out by buying land and building up farmsteads of their own. August, the youngest of the siblings, purchased the "Century Farm" in 1907. He and his wife, Elizabeth, raised seven children on the family's 1,120-acre farm. Wheat, grain sorghum, forage sorghum, cows, steers, horses and chickens (plus eggs) were among commodities raised to support the family.

The farm was eventually split up among the siblings to branch out and expand on their own. Donald Ploger, the youngest of the seven children, took over the site of the farmstead upon returning from K-State in 1950. This was the same time when his father, August, retired and moved into Kinsley. Donald, and wife, Mary, raised three children on the original farmstead from 1950 until 1974. At that time it was passed on to son, Mark. He remained actively involved in the operation from that time until 2006, when the land was rented to a neighbor.

During the century of ownership, all the generations remained stewards of the land. Sod was broken out, terraces and waterways were constructed and windbreaks were planted on the farm. Presently, this Farm Bureau Century Farm consists of 480 acres of dry land farm ground and pasture. It has been in the Ploger family continuously, ready for the fifth generation to carry on. *Submitted by Mark Ploger.*

POLCYN – On June 19, 1900, Ignac Polcyn, a Polish immigrant living in Topeka, KS purchased 160 acres of land in west central Russell County. The ground was purchased for $15.62 an acre from John and Rose Collins of Warren County, IL. Four years later in 1904, he purchased an adjoining 160 acres for $21.25 an acre from Ed and Mary Van Ingen of New York City.

After Ignac's death in the late 1930s, the farm was handled by his sons, Frank, William and Florian "Mike." Ignac's widow, Catherine and their daughter, Helen, moved from the farm into the town of Russell. After his marriage in 1940 Florian and his family occupied the home place until 1958 when they also moved into Russell. From 1975 to 1981, Florian's son, Richard and family lived in the original farm home until they moved in to Russell. The original house caught fire in 1990 and was not rebuilt.

The land was divided by Ignac's four children in the early 1950s, with Florian retaining the original home place. Richard returned to the Russell area in 1975 and resumed farming until the present time. Wheat and milo are the main crops, but alfalfa and soybeans have also been grown.

For 117 years, the farm has helped support three generations of the Polcyn family. As Richard is the last male in the lineage of Ignac Polcyn, he will likely be the last of the family to farm the land. He hopes to keep ownership of the original 320 acres in the family.

POORE – The Poore Farm in Rooks County started in the spring of 1905. R.D. (Fred) Poore rode his horse west to find the highest, driest hill he could find. His house north of Concordia had flooded once again as the Republican River came out of its banks.

He purchased 80 acres in the SE quarter of 10-8-16. After planting corn he rode the train back to Concordia. His wife Julia and oldest son Delbert loaded in a covered wagon for the move. They arrived late in the day and decided to sleep in the wagon one more night. A thunderstorm hit the wagon and homestead cabin. A large part of the cabin roof blew off. The storms followed them through two more moves until they settled in the SE quarter of 11-8-16. This farmstead is still resided in today by grandson Larry.

R.D. (Fred) Poore is holding the horses. Delbert and Earl are holding the other team. Roy and Alvin are seated.

Over the years, R.D. and Julia were blessed with 4 boys and 5 girls. The first 80 acres passed on to Delbert. Delbert's brother Alvin bought the ground from him. The ground is still in the family. The ground is partially pasture used for cow/calf pairs. The other portion is in a wheat, sorghum, fallow rotation.

The Poore farm has been passed down to the third generation with the fifth generation (great, great grandsons) prepared and training to continue the family tradition.

PORTER – My grandfather, Daniel Albert Crist, immigrated from Virden, IL to Gove County, KS in 1886 where he staked out a homestead claim on the NE 1/4, 20-12-26.

When he got off the train at the Melota Crossing (now Quinter) he had a satchel and $40.00. There was only one building and no official name had yet been designated. He purchased a horse and spring wagon and drove to Wakeeney to file the claim on 160 acres of prairie sod. By "breaking out" and cultivating 40 acres and living on the homestead, one could then "prove up" as the legal owner and receive a deed to the land.

The Crist family on the farm, 1916

Dan was attracted to a beautiful young lady who worked at her folks' boarding house, where he stayed – and soon they were married. A sod house was built on the new homestead and there they established their home and started a family where their first child of nine was born. As the family increased and with savings from their sale of butter and eggs, they were able to build a larger home in 1906, where we now live (see background in picture).

My grandparents' fifth child was my mother, so as it turned out, I was a "depression baby," born in 1930.

Early memories include the dirt storms ("black blizzards") of the mid-'30s when I would help my mother spread flour paste on rag strips to seal up the windows and doors in a feeble attempt to try and keep the ever-present dust from coming into the house. Then the grasshopper invasion came and they ate most everything, including the upholstery on the Model T Ford roadster.

When the rains finally came in 1936, the prairie once again blossomed, the wheat "filled" and the corn tasseled. Many folks left, but our family was glad they stayed here on the "Western Frontier."

This legacy has continued for five generations to become a part of the Century Farm program. *Submitted by Irwin Porter.*

POVENMIRE – John A. Povenmire was a Civil War veteran who took advantage of the opportunity to homestead in Kansas following the passage of the Homestead Act of 1862. He came from Allentown, OH in a covered wagon with a train of seven wagons. They settled in Linn County for several years before coming to Coffey County, KS. John had a homestead certificate, #2438, for the North Half of the SW 1/4 of Section 14, Township 22, Range 13 for 80 acres in Liberty Township in Coffey County signed by President Rutherford B. Hayes and dated April 30, 1880.

Across the road to the west, is the Century Farm 80-acre plot in Greenwood County. It was State of Kansas school land, then went to W. O. Claycomb, Sheriff, then went to Ottawa Land Company and then was purchased by John A. Povenmire for $550, free and clear of mortgage with 1894 taxes due. Forty acres of this land was farmed, but the other 40 acres was never broken up. There was a sorghum mill to the northeast on this section.

After John A. Povenmire retired from farming, he and his wife, Lucinda, made a trip by covered wagon, with a driver to help them, so J.A. could see the places where he fought during the Civil War. The picture shows a campsite.

Farm products have been hay and cattle.

No one lived on this piece of land in the past four generations of the Povenmire family.

John Povenmire's daughter, Alnora, remembered Indians in the area who came to trade for food. John gave them grain and they went on their way.

One story told by Harry Frank Povenmire, John's son, was that Jesse and Frank James, the bank and train robbers, stopped at the Povenmire place around 1880. The James brothers fed and watered their horses, never letting the Povenmires out of their sight. He remembered that their horses were very fine ones. A cave south of the Povenmire place is known as Jess James Cave. *Submitted by Frances Force.*

PRAY – Migration from the East dominated Kansas settlement after the passage of the 1862 Homestead Act. The Pray family farm in Dickinson County embodies that agricultural and pioneer heritage. Seeking a better life, two brothers, Alfred and Joseph Pray, left their home in Rhode Island and migrated to Kansas. In 1872 Alfred Pray (1840-1916) filed a Homestead claim on 160 acres in southern Dickinson County and purchased an adjacent 160 acres. That same year Joseph Pray (1848-1924) purchased 160 acres on the same section of land. Alfred and Joseph married sisters Anna and Alberta Ableson, whose family had immigrated to Kansas from Canada. They created two farmsteads—Alfred, Anna, and their six children living near the west side of the section; Joseph, Alberta, and their five children living near the east side of the section.

Charles and Georgina Pray in front of the Pray farmhouse, 1960

Although engaged in regular farming, Alfred became well known for his Percheron horse breeding operation. But as mechanization developed, the need for horses waned, and in 1895 Alfred and his family left Kansas, eventually settling in Nevada. His Kansas land was purchased by his brother, Joseph, who continued the family farm.

It is Joseph Pray's descendants who retain the Pray farm today. Joseph's second oldest son, Charles Arthur Pray (1880-1974), later known as C.A. Pray, married Augusta Dillon (1882-1975) in 1911. While his parents continued to live on the Joseph Pray farmstead, C.A. and Augusta took up residence on the former Alfred Pray farmstead. In 1911 C.A. built a new house for his bride, and added outbuildings around 1912. Those buildings still stand today. C.A. and Augusta had three children: Mildred (1913-2000), Doris (1915-1999), and Charles Albert (1919-2001).

Following the death of his parents (in 1924 and 1939), C.A. Pray purchased his siblings' shares of the Pray land and became sole owner of the farm. To those 480 acres he added another 400 acres that he purchased on adjoining sections from 1921 to 1946. In the following years his oldest daughter Mildred added to the farming operation by purchasing another 160 acres.

Apart from Alfred Pray's Percheron operation during the earlier years, the Pray farm always functioned as a combined beef cattle and varied crop farm. Wheat, corn, and oats were grown, and in the 1920s the farm changed its cattle operation to raise only Herefords. Until about 1946 additional farm animals were also kept for the family's subsistence including a few dairy cows, pigs, and chickens, and the farm utilized mules and both draft and riding horses.

While stationed in California during World War II, Charles Albert Pray met and married Georgina Richmond (1917-1997), and in 1946 Charles brought his new wife to live on the Kansas farm. Forming a partnership with his father, Charles continued the farming operation, raising Hereford cattle and primarily growing wheat, with corn, prairie hay, alfalfa, and milo grown for cattle feed. For his farming improvements, which included building ponds, waterways, and terraces, Charles won several conservation awards. The long lane that winds into the farmstead was built on one of the terraces, and in 1962 Georgina Pray appropriately named the farm Terrace Lane Farm, which it is still called today.

Charles and Georgina had two daughters, Christine (born 1947) and Bobbie (born 1949), who both grew up on the farm and for many years raised Arabian horses. The Pray family was always involved in the community, working with neighbors and supporting the school, church, and many local activities. Today Christine Pray Smith and Bobbie Pray retain ownership of the farm.

PRIDDY – The founder of the Priddy Century Farm was John Wilson Priddy, who was born in 1813 in Butler County, OH. He married Jane Scott of Fayette County, IN, in 1843, and after their marriage they moved to the farm that he owned near Warren, IN.

Not long after the Civil War had ended, John bought 640 acres of land in Menoken Township in Shawnee County, KS. In 1869 his three oldest sons - Joseph, James and Robert - drove across the country from Indiana in a prairie schooner. The 26-day journey brought them to the property that would be their new home.

The four Priddy brothers, l-r, from youngest to eldest: Henry Clay, Robert Scott, James Mitchell, and Joseph Wilson.

This was land purchased from the US government by the Atchison, Topeka and Santa Fe Railroad, which had originally been acquired through a treaty with the Pottawatomie Indian tribe. In 1871 the youngest Priddy son, Henry, joined his brothers on the Menoken Township property.

At some point the brothers had two loads of lumber cut and shipped to Kansas from the Indiana farm, to build a large house, which they shared for a while. In 1884 John and Jane Priddy came to Kansas to be near their sons, selling their Indiana farm in 1889'

Descendants of the Priddy family still own and live on parts of the original farm property. Although much of the native prairie grassland in Kansas has been lost over the years to cropland production, there are still 126 acres of it on the land that belonged to Henry Priddy, the fourth son. Today it remains just as it was over 100 years ago. His descendants who have inherited this property consider it their duty and privilege to preserve it now and for the future. *Submitted by Patricia A. (Priddy) Tabbert.*

PRUTER – In Round Mound Township of Osborne County, KS along Township Road 290th Ave., there are eight family farms over 100 years old. One of these is the George Pruter family farm. In 1879 George journeyed to Kansas from Waverly, IA with his parents, Fredrick Carl and Christina (Rohde) Pruter and his six siblings. They settled in Round Mound Township, 3-1/2 miles north of Natoma.

The Pruter farm ca. 1906-18 taken before the tornado. After the tornado, the barn had no cupola and the house was one-story.

In 1905 George bought 240 acres in Round Mound Township in sections 30, 29, 31 and that acreage is still in the Pruter family. In 2005 was honored by becoming a Farm Bureau Century Farm.

George built the farmstead including a 40' by 80' by 16' barn. The foundation and floor used one railroad car of cement. The barn had a "no sag roof" which had been designed by Louis Beisner, a native of Natoma, KS.

The barn housed six teams of horses, a milking part for 10 cows and grain bins. He also raised sheep, pigs, chickens and geese. They butchered their own meat. In the winter the meat hung from the end of the tall barn. They had an orchard and raised a big garden. Wheat, sorghum and they harvested the native grass.

On May 20, 1918 the barn was destroyed and the house severely damaged by a tornado. May 18, 19 and 20 are called "Cyclone" Day because the area from Codell, KS to the northeast in to Osborne County – a 50 mile stretch – had a cyclone on the same day for three years, destroying many farms.

George rebuilt the farmstead, the house into a one story on the same foundation and the barn the same but two feet decrease in height.

The Pruter Farm has had four generations living and owning the farm.

From 1905-1942, George and Goldie (Howe) Pruter were owners. A son, Alvin and Yvonne (Goad) Pruter owned it from 1942-1988. Orville and Betty Pruter were owners from 1988-2013. Four great-grandsons are the current owners: Gale, Daryl, Douglas and Kevin. *Submitted by Orville and Betty Pruter.*

QUATHAMER – The Quathamer Century Farm is owed by Thad and Karen (Quathamer) Osgood. It is located in Labette County, KS on Highway 59, six miles south of Parsons.

Karen's grandfather, Gerad Quathamer emigrated from Germany in 1877. He married Augusta and lived in Nebraska and Missouri for several years. Gerad wanted farmland more like he was used to in Nebraska. He found an agent who had a client willing to trade equity in a farm in southeast Kansas for his Missouri Land. Gerad made the trade and moved to a 160-acre farm, which is the current Century Farm. Gerad farmed the land and raised Holstein dairy cows. Upon his death, ownership was turned over the Augusta and all their children. Her son, Henry, Karen's father, continued to farm the property and took care of his mother. Later ownership of the farm was turned over to Henry. He raised draft horses, sheep and maintained a herd of Holstein dairy cows along with his crop acres. After his death in 1968, Thad and Karen Osgood, along with her mother, acquired ownership of the farm. Thad and Karen now own the property and have raised grain, hay and livestock (cattle, sheep and hogs). Now the farm is all pasture land and rented to a local cattleman.

Working the farm was a family affair.

The home that Gerad and Augusta moved into in 1914 burned in 1929. A house in Parsons was bought, torn down and hauled to the farm on hay racks and rebuilt. In 1980 Thad and Karen built the home they now reside in. A part of the barn is the only original building still standing. In 1961 it was damaged by a tornado. The barn was rebuilt with a lower gable roof and without the loafing shed. It is still in use today.

The Osgood's have three children, nine grandchildren and three great-grandchildren. *Submitted by Karen Osgood.*

RAKESTRAW/STILLWAGON – William Kimble (1861-1942) and Louise Rakestraw, originally of Ipava, Illinois, moved from Nelson, Nebraska in 1908 by train to purchase the farm near Buckeye, Kansas, in Dickinson County, for $58.75 per acre.

Their son, John Rakestraw (1890-1928) married Eula Derrick in 1914, living on the farm, and had two daughters, Mary Evelyn (1916-2010) and Ethel (1919-2008). Due to the death of their parents, the girls were raised on the farm by Grandfather Kimble. Mary Evelyn married Charles (Buck) Stillwagon in 1931, living on the farm and struggling during the Dirty '30s. Buck worked for $15 a month for another farmer using horses. Cattle, hogs, sheep and chickens were raised, and they sold eggs and cream at Buckeye and Chapman.

W.K. Rakestraw, with Lousia and son, John, Abiline, Kansas

Stillwagon family, l-r: Linda Melton, Thelma Lexow, Dale Stillwagon, Mary E. Stillwagon, Nadine Iseli, and Chuck Stillwagon. Sitting: Norma Sharp

Buck later had a custom silage and hay business, and worked as a mechanic. Mary told of Ft. Riley Army soldiers camping across the road with German prisoners, and watering their horses at the farm. The WPA built a fine outdoor toilet, and the farm had a drive through granary with unloading pit and a concrete dipping vat for hogs and sheep. Water was hauled from a well.

It was not until 1963 that a newer home with modern conveniences was moved to the farm. Mary and Buck raised six children, the first three born at home. Buck died in a tractor accident in 1973. Sons Dale and Chuck continued farming. Chuck died in 1996 in a truck accident. Mary spent her entire life on the farm, receiving the Century Farm Award in 2008. She passed away at age 93 in 2010, with eleven grandchildren, twenty-three great-grandchildren, and nine great-great-grandchildren.

The family continues to grow with son Dale Stillwagon, daughters Nadine Iseli, Norma Sharp, Thelma Lexow, and Linda Melton all living in Kansas. The farm was sold in 2012. *Courtesy of Linda Melton*

RAYL – In 1882, brothers Charles and Willard Rayl left their home in Kokomo, IN to search for farmland to homestead or purchase in Kansas. Charles found the quarter section he wanted two miles northeast of Kingman. On June 22, 1882, he purchased it from J.O. Davidson, a former mayor of Wichita. Davidson purchased the land from the US Government just three weeks before selling it. According to family history, Charles Rayl broke the sod on the land while walking barefoot behind a horse and plow. He was barefooted because he wanted to save his only shoes for church on Sunday! Two years later Charles returned to Indiana and brought his wife, Emma, to Kansas. Willard Rayl purchased land nearby and the brothers farmed and raised their families together.

Looking out over the fields at the Rayl farm.

In addition to farming, Charles Rayl was a brick mason who built many Kingman buildings, including the current United Methodist Church. He was also a founder and former president of The State Bank of Kingman. After Charles Rayl's retirement, operation of the farm was taken over by his son, P.W. Rayl, a lifelong farmer who added many acres to the farm. P.W. Rayl's daughter, Margaret Rayl Pfalser, inherited the land and currently her daughter, Martha Pfalser Stuart, is the owner. Since P.W. Rayl's retirement in the early 1970s, the land has been leased by neighbors, the Schwartz family. Three generations of the Schwartz family have now farmed the land, so it is truly a multi-generational farm.

It is an honor and a privilege to own such a valuable piece of ground. The monetary value pales in comparison to the sense of stability, the pride and the sentiment involved in owning a property which has been and continues to be, after 135 years, such a huge part of our family and its history. *Submitted by Martha Stuart.*

REDGER-SCHMIDT CENTURY FARM – The location of this farm is the NW 1/4 of Section 16, Township 30, Range in Kiowa County. It was established by David Redger who was born in Karolswalde, Russia (Prussia). He and his wife, Lena, migrated to America in 1874, settling in the Pawnee Rock area. As land became higher in price, David looked elsewhere to find cheaper land. On Aug. 25, 1911 David purchased this quarter, paying the sum of $9,000.00. However he did not homestead this piece, rather he bought it for his son, Samuel, who became the first family member to live on this farm.

Sam and Elizabeth, his wife, raised wheat and a few heard of cattle. Wheat was the first crop harvested on this farm. Sam farmed with a team of work horses and his father built a barn to house them. Sam's father also built the couple a house and "sold" them the property in December 1929 for the sum of "one dollars and natural love and affection."

Sam's nephew, Allen Schmidt, began farming the quarter in 1944 while he was still in high school. His first crop was barley. After serving in WW II, he married and returned to the farm. Allen's father and brothers did the farming in his absence. Allen and his wife, Juanita, lived in the house that David built until 1968. They built a new home and moved the old house to Greensburg. The couple purchased the farm 80 acres at a time. Their principal crops were wheat, hogs, some feed and cattle. Allen won several Soil Conservation District Awards and a Windbreak Award in 1986 for his conservation efforts.

On Dec. 26, 1987 Rex Schmidt, Allen's nephew, moved to the farm, purchasing the quarter several years later. Allen had enrolled much of the quarter in CRP but for a 30-acre and a 40-acre parcel. Rex and his wife, Janie, now live in the house that Allen built. They have raised wheat, milo, alfalfa, soybeans, feed, sheep and cattle. The farm is and has always been non-irrigated. Rex won a Soil Conservation District Grassland Award. Today, the entire quarter is newly fenced and has all been returned to grassland. Cattle graze there, the old barn still stands and both farm and family endure the unpredictable test of time. *Submitted by Janie Schmidt.*

RHINE – John and Dorothy Rhine now own the 160 acres that his great grandfather, Thomas H. Rhine purchased on Nov. 30, 1883. Thomas purchased the farm in Republic County, Farmington Township, 36-2-1 W. 6th P.M. Grain crops raised here were wheat, corn and alfalfa. Livestock consisted of hogs, cattle and work horses.

The farm passed from original owner Thomas to his son, Lester Rhine. On Jan. 17, 1981 the farm then passed to Lester's nephew, John Rhine.

The farm now consists of the original 160 acres plus 120 more acres in the same section. Corn, beans and wheat are the grain crops now grown and cattle are on the farm.

Windmills and holding tanks were located on high ground and provided running water for the house and livestock. There is also a blacksmith shop and feed mill.

John and Dorothy have two sons, Monty and Heath. Their grandchildren are Sara, Chandler, Lyndzee, Abigail and Gracie. John is still farming and is a real estate broker. *Submitted by John and Dorothy Rhine.*

RIEDL – Located in Hodgeman County, KS, this 800-acre farm was purchased on April 9, 1910 by I.C. Ridel. Cattle, wheat, sweet sorghum and grass were grown on this farm located in the SE 1/4, Section 7 and Section 8-23-25.

This father passed from I.C. to his daughter, Helena Elliot. Helena passed the farm to her daughter, Marie Hays. Virginia Stramel-Strong, I.C.'s great-granddaughter, now owns 240 acres of the original farm. Wheat and grass are now produced here.

Virginia, a widow, has four children: Patricia, Dan, John L. and Joe and 10 grandchildren. She retired from ASCS-USDA in October 1993. *Submitted by Virginia L. Stramel-Strong.*

RINGLER – Glenn Jr. and Sharon Ringler own two Century Farms, both located in Pleasant Township, Lincoln County, KS.

The first farm, (Glenn's) consists of 80 acres acquired by Fred Meyer Sr. in 1871. This farm passed from Fred Sr. to his son, Fred Jr., then to Fred's niece, Nora Ringler and in turn to her son, Glenn Ringler Jr.

The second farm (Sharon's) is 400 acres acquired in 1875 by Frank Urban Sr. It passed from Frank Sr. to his son, Frank Jr., to Glenn Sr. and Glenn Jr. Ringler and great niece Sharon (Urban) Ringler.

Both farms originally produced wheat, cattle and hors-

Glenn and Sharon Ringler

es and now produce wheat, milo and cattle. The full acreage of both farms has been retained.

The home stone on the first farm still stands. The current home on the second farm was built in 1912. The barn built on the second farm is still used each day.

The Ringler barn

Glenn Jr. and Sharon have three children: Justin Ringler, Sasha Ringler Fallon and Shantel Ringler Grace. Jack, Gabe, Nicki, Leo, Sonny, Paige and Liberty are their grandchildren. *Submitted by Glenn Jr. and Sharon Ringler.*

The Ringler home

ROBERDS – Sometime in the 1800s Charles Roberds purchased 160 acres located in 22-31-25 of Cherokee County, KS. Corn, wheat, soybeans and cattle were raised on this farm.

The farm has been owned by a Roberds since that time. The farm passed from Charles to Lester W. Roberds and then on Dec. 1, 2002 to Dale Roberds. Still retaining 160 acres, the farm still produces corn, wheat, soybeans and cattle.

The original house on the property was replaced in 1961.

Dale and Phyllis Roberds have three children: Keith Roberds, Diana Turnbull and Glenna Jenkins. Grandchildren are Greg Mowdy, Lacie Waterman, Todd Lock, Stacey Lock and Rachel Roberds. *Submitted by Dale Roberds.*

ROBINSON – Our century farm was homesteaded by Samuel Robinson in 1880, a Civil War veteran and our great grandfather. The farm was inherited by a son, Francis Simon Robinson, who sold it to Edmund Jr. and Martha A. Brant in the year 1954. Our plan for a Grade A Dairy farm began.

The concrete block barn

Our three sons were very young and adjusted to a home with electricity. Their fun time was radios, horses, hunting, fishing and swimming.

Salvaged lumber from a large barn on the farm provided lumber for the roof of the newly constructed concrete block barn. There was repair construction on buildings and tear down on others.

The original barn

Cows consisted of Jersey calves ordered from Wisconsin. These helped though we learned Holsteins were better producers.

Each of our sons had an interest of his own and all too soon, left for school. Rick became an Electrical Engineer, Kyle a Farrier and Wes a Mechanical Engineer.

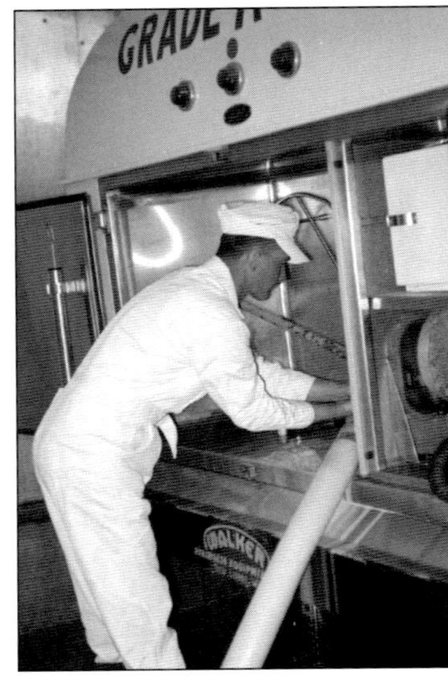

Working with the dairy equipment

Ed and Martha milked between 30 to 50 cows twice a day, cleaned the equipment and scrubbed the barn. The boys helped during vacations and weekends. By August 1992 Ed thought it was time to close our dairy. Our cows were sold to a dairy in Nebraska. We continued to live on our farm. Ed gardened and shared cucumbers and tomatoes with neighbors. Also, he won prizes on his produce at the county fair.

Five grandchildren have played on the farm in its isolated location. Very important, it was close to God and remains the same. *Submitted by Edward Jr and Martha A. Brant.*

ROHRBACHER – Henry and Sarah Rohrbacher purchased 160 acres in 1903 at the cost of $3,000. It is not known who owned it prior to the purchase. The farm is located in South Bend Township, Barton County, KS, 30, 20, 13. Wheat, milo and alfalfa were grain crops grown as well as cattle were on the farm. Henry Rohrbacher was born in Ohio. He worked in a dye factory, dying fabric.

The land then passed to Howard and Fannie Sharp. More land was purchased over the years and 80 acres was left to each of the five children of Howard and Fannie Sharp. The Sharp Family Farm Partnership was formed. Eighty acres were purchased by James and Judi Sharp form their cousins, Deborah Salbreg and Sarah Hall, in 2008.

There were never any homes or buildings on this 80 acres. The farmstead where James grew up is located in the south half of the original 160-acre purchase. James' mother still lives on the farmstead.

The farm currently produces wheat, soybeans and corn. *Submitted by James and Judi Sharp.*

ROSS – On March 10, 1886, the family of John G. Ross left the homestead in Catawba, MO and arrived by train at Collyer, KS on March 12, 1886. The family then traveled with relatives to Shields, KS by covered wagon where John had staked a homestead.

Leslie and Elsie Ross

Charles, the oldest son, 21, staked his claim on 32 16S 28W in Shields. In 1887 the Denver Memphis and Atlantic Railroad bought a strip 100 feet wide for a railroad. Charles received $1,480.82 for the condemned land. That money enabled Charles to withstand the hardships of the area and remained on the land until his death in 1955. On Sept. 10, 1890, Charles married Mary Elizabeth Boshell in Great Bend. Charles moved his wife to Shields and raised their children: Wilbur, Lee, Lester and Leslie (twins), Eula and Hazel in a soddy on the southeast part. In the early 1900s Charles built a three-bedroom frame house, barn, washhouse and an outhouse. Charles worked hard to break all but 30 acres, which he left for horses and cattle.

Charles was the only Ross to remain in Lane County, as the Missourians discovered that 160 acres in Kansas could not support a family and moved on.

In 1935 Charles developed a cataract. He traveled to Dodge City to have it removed. During surgery the doctor blinded Charles in his right eye. After surgery it was unsafe for Charles to live alone. His son, Leslie, married Elsie Sanders in Arkansas, returned to continue farming. Two children, Dora and Lola, grew up on the farm where they were involved in church, school and 4-H. Lola and Dora attended a two-room grade school in Shields and Dighton High.

To supplement the income, Leslie and Elsie raised cattle and sold cream and eggs to pay for groceries. Leslie also worked on the WPA, the railroad and hired out as a farm hand. Their faith carried them through difficult times.

Charles died in 1955. His heirs shared the profits until 1968 when a granddaughter and husband, Dora and Jim Nelson, bought the homestead. They farmed it until 2011 when it was leased to a neighbor.

Jim and Dora were teachers in WaKeeney and had two children, Penny (Lane) and Paul. They have three grandchildren: Erika Lane, Jordan and Jake Nelson. *Submitted by Dora and Jim Nelson.*

ROWE – Thirty-six years after Kansas became a state, the Rowe homestead was established. Robert Rowe and his wife, Emma Williams, along with their children moved from Newton, IA to Scranton, KS. Robert purchased a quarter section south of Scranton where a house and out buildings were built; crops, orchard, gardens and animals were raised.

Robert and Emma Rowe and family in front of their home

September 1916 Robert died from a freak accident on the farm as a team of horses became spooked and threw him from his wagon. His son, Clarence, took over the daily operation of the farm. Clarence married Emma Nester June 1937. Oats, corn, wheat and hay were grown as well as chickens, dairy cattle and Poland China hogs. Clarence became a prominent hog farmer in the community, winning many ribbons and awards at fairs. Their children, Howard and Esther, were very active on the farm as well.

May 1965 Clarence was killed in a farming accident while cultivating. His son, Howard, helped on the farm until Esther and husband, Clarence Hollar, moved to the homestead and eventually bought the 160 acres from Emma in 1970.

Clarence, Esther and their children - Rex, Sherri and Gary - continued to farm the land by growing various row crops and raising beef cattle, sheep and chickens. January 2001 Clarence passed away, making three generations of men (to Rex, Sherri and Gary) to have died on the farm.

Esther still lives on the homestead raising beef cattle and brome hay. She has eight grandchildren and 14 great-grandchildren.

The only items left from the original structures are the hand dug well and one building. Tornados and age have taken care of the rest. *Submitted by Sherri Dean.*

RYAN – The property was homesteaded on Sept. 27, 1879 by Sallie Bidwell. The United States patent was signed by Rutherford B. Hayes.

Benjamin F. Enloe and his wife Harriet L. (Dinwiddie) Enloe purchased the 160 acres from Ebeneezer C. Jones, Oct. 10, 1907 for $37.50 per acre.

Mayme and Kermit E. Ryan

Nellie and Wilmer Ryan

Benjamin died Aug. 7, 1921 and the property was willed to Harriet. Harriet died Aug. 20, 1930. Having no children, but having raised her niece, Nellie Dinwiddie, she willed the farm to Nellie.

The property was occupied by Nellie and her husband, Wilmer Ryan, and their children. The family actively farmed. Financial problems of the Great Depression prevented Nellie from obtaining clear title and facing foreclosure, the property was purchased by Nellie and Wilmer's son, Kermit E. Ryan and his wife, Mayme C. (Chadwick) Ryan. Kermit had been born in the farmhouse and lived on the farm his entire life.

K.E. Ryan farmhouse, Anthony, Kansas

Kermit and his wife worked the farm and raised their family of three children: Harold, Larry and Linda. While the main crop was wheat, other entities played a role, aiding the survival of a working farm. A dairy herd provided grade A milk for sale and, at one time, a small peach and cherry orchard was harvested. The family was involved in 4-H, raising livestock, home economics and leadership roles.

Kermit died Nov. 24, 1981, wherein the farm was owned by Mayme and the children. Mayme died Oct. 5, 2003.

The 160 acres are currently actively farmed and are owned by the family as K.E. Ryan Farm, LLC. *Submitted by Harold D. Ryan.*

SCHEPMANN FARM – Herman Friedrich and Anna Maria Louisa (Dubber) Schepmann were born in Arenshorst, Germany and had moved, with their families, to Holland, IN where the family had a sawmill. In 1884 they acquired land in Ellsworth County, Valley Township, NW 1/4 and NE 1/4 of 22-17-10. Cattle, wheat, oats and hay were originally produced on this farm.

The Schepmann family homestead, 2017

When they moved to Kansas they brought poplar wood, which was used in the four original rooms of the house in 1884. The first addition was completed in 1890 and new kitchen and attic were added in 1918. The granary, with walkout basement for livestock stalls, is still used today for cattle and for feed storage. It is said that the family lived in the upstairs of this building during construction of the house. This upstairs loft is quite cozy with hay stored in it! The house was restored by Vern and Karen Schepmann, who have lived there since 2007.

This farm passed from Herman and Louisa (1st generation) to Friederich H. (son) and Amanda (2nd generation) to sons Nathan F. and Jo Ann (3rd generation) and Elroy F. and Rose Anna (3rd generation). Ronald E. (son) and Damaris along with Vern F. (son) and Karen (both 4th generation) then passed the farm to Matthew B. (grandson) and

Leah Schepmann (son of Paul and Karin, 5th generation).

Polled Hereford and commercial cattle, wheat, milo, alfalfa, native prairie hay are produced today.

Soil and conservation has been practiced since the mid-'50s. Elroy constructed several miles of terraces in the '50s, using a three-bottom plow so they could be easily farmed immediately. He said, "You don't have to be crazy to do it this way, but it sure helps."

SCHEPMANN – Donald F. Schepmann, son of Henry and Louise Wachs Schepmann, was given a Century Farm Award on Sept. 16, 2005. He was a fourth generation farmer in Ellsworth County. His great grandfather, Herman F. Schepmann, was born in Arendorf, Germany in 1833 and came to the US Aug. 1, 1850 at the age of 17. Herman purchased 160 acres June 13, 1889 and 80 acres Oct. 31, 1890. That property was passed to George Gerhardt Schepmann, his son, in 1893. George deeded the property to his son, Henry Schepmann, in 1918. The property was originally part of a Land Grant from the US Government to the Union Pacific Railway Company for the purpose of aiding in the construction of a railroad and telegraph line from the Missouri River to the Pacific Ocean.

Don and Marie Schepmann

Don and his wife, Marie, made their home in Claflin, KS. Together with their children they ran a family farm, feeder operation and trucking business. Don was known as a cattleman and cattle buyer. In 1992 he won the Outstanding Volume and Performance Award, presented by Caprock Industries. Son, Don Jr., owned and operated Schepmann Trucking. Son, Jerry, continues to run the farming operation. Daughter, Michelle, kept books for the farm and trucking business.

Don and Marie received a 2003 District Cooperator of the Year award in Ellsworth County for their conservation work with terraces, waterways and shelterbelts. For Don, conservation meant conserving the land for future generations.

Don's legacy will continue to live on through his three children, nine grandchildren and nine great-grandchildren. His children have each built homes on the acreage he designated for them.

SCHLENDER – Frederick W. Schlender came to America from Germany as a stowaway on a steamer in 1852 at the age of 19 and settled in a German settlement at Ebenezer, NY. In 1883, Frederick W. moved his family to Christian, KS in McPherson County.

The fourth of his eight children, Frederick August Schlender, was the original owner of the Schlender Century Farm in Harvey County, near Moundridge, KS. In August 1900 while inspecting his newly purchased 160 acre farm, Frederick A. came in contact with bad water, contracted typhoid fever and died, leaving his wife, Lydia, with three young children. They went back to Maryland but after a short time returned to the farm where Lydia raised their family.

A small house came with the land. In the 1900s a major addition was added. In 1927 a kitchen was added. The barns were built in the 1910s and 1920s.

One of the children, Milton Schlender, and his wife, Edna, raised their eight children on the farm.

Crops raised were corn, kafir corn, wheat, alfalfa, native grass. Animals raised were chickens, cows, hogs, sheep. The sheep were Milton's favorite. During lambing season, he would stay in the barn at night in a maternity ward, an area covered with a tarp with a small heater. North Garden School was moved to the farm and converted into a chicken house. In the late 1940s, terraces were built on the farm, these were some of the first terraces built in Harvey County. Milton passed away in 1963.

Milton's son, Fredrick Allen Schlender Sr. "Fred," farmed the land for 40 years. Fred passed away in 2006.

Today Fred's son, Fredrick Allen Schlender Jr. "Rick," farms the ground of native grass pasture, brome waterways, alfalfa, corn, milo, soybeans, wheat.

The Milton Schlender family, ca. 1940-41

Ownerships included: Frederick August Schlender; widow, Lydia; son, Milton Schlender; widow, Edna; Milton's children; Milton's son, Fredrick Allen Schlender Sr.; widow, Gwen. *Submitted by Joyce (Schlender) McDowell.*

SCHLICKAU – At the age of 18, George Schlickau came to America from Hanover, Germany, settling in Wisconsin with his older brother, Henry and Henry's family. From there George went to Illinois briefly before coming to Kansas in 1872, where he was one of the first settlers in the area. Descendants still live on the land he homesteaded near Haven.

James Schlickau family: Caitlin, Cooper, James, Charlene, and Ashton

George married Clara Rathke and they had six children. Six years before the town was established, the Schlickaus were one of nine families to organize a Lutheran Church. The eldest child, Albert, was the first church baptism. In 1904, George purchased additional land just east of Haven, which Albert then purchased in 1905 and 1912. Albert and his wife, Florence, raised a family of four children. Albert recognized the importance of implementing diversified farming and ranching practices. Seeing the benefits of using a registered bull, in 1913 he turned his entire cattle herd into a registered Hereford operation.

Albert's two sons continued with the pure bred cattle. Harry moved to Argonia while George Hans, the youngest, stayed on the home place expanding both the farming and cattle operation. For 55 years, the firm hosted an annual production sale in addition to exhibiting cattle at shows across the nation. This Schlickau farm became eligible as a century farm in 2004.

Picturea l-r: Harry, Albert, and George Schlickau

George, who passed away in 2005 and his wife, Lois, have four children: Bruce, Susan, James and Nancy. The younger son, James, now owns and lives on the property. He is the fourth generation operating this land, utilizing innovative and conservation farming practices. James and his wife, Charlene, have three children: Ashton, Caitlin and Cooper'

The roots of the family run deep in the Haven soil and the Schlickau descendants feel they are blessed with a special and priceless family heritage. *Submitted by Lois Schlickau.*

SCHON – Robert Neil (1941) and Karen Kay Coffman (1942) are the current owners of the original farm and another 320 acres in Harper County, KS.

The original owner was John Hie Schon (1849-1919) born Bagband, Germany. Emigrated to Kansas in 1881.

John bought the NW 1/4 Section 24/32/5 Harper County, KS on Sept 18, 1882 for $1900.00.

John Hie Schon, Hiram Jansen Schon, Hiram Everett Schon, and Robert Neil Coffman, the men who farmed the Schon farm for four generations.

Hiram Jansen Schon (1880-1956), his eldest, also farmed the original land. Later Hiram owned a threshing crew, harvesting for the neighborhood and farming with mules.

Hiram "Bud" Everett (1910-1975) married in 1934 and they lived with Hiram and farmed rented ground. In 1941 Bud moved to Harper on a rented farm, returning to the family farm in 1944. He raised cattle, sheep, wheat, alfalfa and feed. They had one daughter, Karen Kay.

Over the years Bud acquired more land and at the time of his death he owned the N 1/2 of sections 24/32/5 and the NW 1/4 of section 24/3/15 Harper County, KS. The farm totaled 480 acres.

Karen Kay married Robert Neil Coffman. They had three sons: Steven Kent (1963), Shawn Alan (1966) and Jason Ron (1973). We moved to the farm in 1969 as Dad's health was failing. Though Neil was not raised on the farm, he was an excellent farmer. Neil raised wheat, alfalfa and feed. He kept a small cow herd. Neil farmed this land longer than any of the Schon's. Neil held jobs off the farm. In 1999 he became a rural mail carrier and in 2000 he rented all but 65 acres of our farm to a neighbor. We raised feed for the cattle and had a small acreage of wheat.

On May 29, 2004 a tornado destroyed our home, eight outbuildings and our machinery. A decision to no longer farm was made. We rebuilt the farmstead and are currently living there. We kept our cattle until 2008. The crops for 2017 will include wheat, soybeans and canola.

This family has held a strong love for this land, remembering how important it was to former generations. *Submitted by Karen Coffman.*

SCHROEDER – William Schroeder arrived in Smith County, KS July 7, 1871 and applied for his homestead July 13, 1871 at Concordia. A tailor by trade, he knew nothing of farming except having worked on one for a month. He purchased 160 acres in Smith County, Oak Township, Section 34, Range 11, and raised hogs, cattle, chickens, horses, wheat, and milo.

Great-grandparents William and Malissa Schroeder, 1871-1928; grandmother and great uncles Bonnie Bloomer, Paul, France, Erwin and Elbie Schroeder, 1928-1934; grandparents Oscar and Bonnie Bloomer, 1934-1974; parents L. Dale and Phyllis Bell, 1974-2012; children Rise" Bell, Vicky Collins, Gloria Snow, 2012 to present.

The Schroeder family and homeplace

Today there are 160 original acres still in the family farm; no livestock, but wheat, milo, and soybeans.

William lived on the open prairie until he made a small dugout in the creek bank. Later, he built three near the creek; one for himself, his horses and his chickens. In 1875, there was a flood that flooded the dugouts. Fritz Dressler, a neighbor who was a carpenter, helped him build a 14'x 20' log house. The excavations and log house are still there.

A cottonwood north of the house was planted in 1891 when William stuck a cutting into the ground. It has been possibly the tallest and oldest tree in the county. A Mr. and Mrs. Faulk lived with William in the log house. She died in 1878 and is buried on the farm.

Religion, education, self-sufficiency, community service, helping others and family were important. A large pond was built. On Sundays, many folks fished, swam and took boat rides. Bonfires were built in the winter and people ice skated. Holes were dug into the bank building a fire and a large kettle sat on top, cooking eggs. Rabbits, prairie chicken or quail were roasted.

SCHWARTZ – Andrew Schwartz, a Swiss-German Mennonite farmer from Horodish, Russia, came to America in 1874. He first settled in South Dakota before moving to Reno County, KS in 1884. Here he purchased a farm where he raised wheat and many varieties of trees. He had a Concord grape vineyard, which yielded wine for medicinal purposes and for church communion services. Andrew donated a tract of land for a church. He was an avid hunter and a carpenter, building furniture and caskets from his trees. He was skilled in leatherwork, making shoes for family and neighbors in addition for harnesses. He also was skilled in setting broken bones. In 1903 he purchased a large house for $250 and had the house moved 18 miles, also costing $250. It took 35 teams of horses to accomplish this feat.

Andrew and Elizabeth (Graber) Schwartz, 1923

Church has been important for all generations and music and education were and are priorities.

The farm passed from Andrew and Elizabeth Schwartz to his son, Ben and Emma Schwartz, then to Ben's son, Milton and Sadie Schwartz. Fourth owners were Paul and Edna Ruth Schwartz Kemp.

The farm retains 160 acres and now produces wheat and hay, plus beef cattle.

Paul E. and Edna Ruth Kemp have three sons: Todd E., Timothy P. and Christopher S. as well as seven grandchildren. Edna Ruth Kemp was a teacher for 50 years. *Submitted by Edna Ruth (Schwartz) Kemp.*

SCHWARTZ – John Freidereich Schwartz was born March 7, 1834 in Wuertenberg, Germany. Frederick and his brother, Charles, came to Kansas in 1855. In 1860 the 160 acres near Worden in Douglas County was officially assigned to Frederick. As was the usual practice, he chose a home site near an everlasting spring. The first home was a dugout back into the bank to shelter his animals, with a cabin on top for living quarters. Part of the dugout, lined by Frederick with chunks of native limestone, stands today.

Pictured at the Schwartz original dug out and cabin site are Don Schwartz's six grandchildren. Pictured from back to front are: Charley Jo Schwartz (Pittenger), Jacob Enick, Camdon Schwartz, Rachael Enick, Hayley Schwartz, and Callie Enick.

Very soon Frederick volunteered and left for active duty in the Union Army. Soon after he returned, he and his wife left the dugout and built a new house about 1/4 mile west, still near the same spring. Here they raised eight children. They were very active in helping to establish the church and school nearby. Ironically, Frederick gave the acre of land and was a board member when Union School began. During its tenure son, Charles and grandsons, Lester and Virgil, also served and when the school closed in 1963 great grandson, Don, was a board member. Fred's eight children, three grandchildren and three great-grandchildren attended the school. As was the custom the acre of land was given back to the landowner.

Son Charles chose to remain on the family farm. In 1910 he hired a local carpenter to build a barn that became well known for its four-sided manger to feed stock. In 1962 the barn was destroyed by a tornado. The house was badly damaged and had to be tom down.

Charles and his sons, Virgil and Lester, maintained a successful cattle operation for many years. Later, Lester's son, Don, became part of the operation. He still lives near the original home site. More adjoining acres have been added over the years and with his son, Greg, they own and maintain 320 acres including the original homestead. *Submitted by Don Schwartz.*

SCOTT – William Scott was born in Williams County, OH in 1845. In 1862 he enlisted in Williams County, OH in Company 111th Reg. Ohio Volunteer army and served in the Civil War until 1865. In 1868 William married Marion Sanford and began his family.

William served three years apprenticeship in the plastering trade and followed that trade until 1873 when he set out to locate land to begin farming. His travels brought him to Parsons, KS where he purchased the North 1/2 of Section 24-31-18 in Labette County from G. Everhart in 1874.

William's family included his wife, Marion and five children. Three children passed away in 1878 from diphtheria, leaving daughter, Lovisa and son, Howard. Prior to William's death, the east l/2 of the farm was deeded to daughter, Lovisa and the west 1/2 to son, Howard.

Howard married Adda Craft and they had five living children: Charles, Lon, John, Irene and William "Bill" G. Howard passed away in 1908.

After the death of Howard, Charles, Lon and Bill farmed the home farm.

William Scott "Bill" began farming on his own at the age of 13 after his brothers, Charles and Lon, were drafted into the army during WW I. During his lifetime he developed excellent farming practices which improved the land quality and in the 1950s received awards for his farming practices. Bill purchased the farm holdings of his siblings and became the sole owner of the farm. He also purchased additional land to increase the farm to 350 acres.

William "Bill" married Ethel Ruth Ligon in 1928. They had three children: William G. Jr., Jack Robert and Barbara Ellen.

After the passing of William Sr. in 1980, the farming land was divided between the children and Bill Scott, Jr. (after purchasing 40 acres from sister Barbara) became sole owner of the N.W. 1/4 of section 24-31-18, which is the current Scott Century Farm.

Bill Jr's son, Thomas "Tom" is currently living on the farm and is in line to inherit it, making the fifth generation of Scotts to own this farm. *Submitted by Jack R. "Bob" Scott.*

SCOTT FARM – On Dec. 27, 1870 Adam Scott purchased 240 acres located in Pottawatomie County. The land is located in Section 3, SW 1/4, Twp. 8S, Range 9E and E 1/4, SE 1/4, Section 4, Twp. 8S, Range 9E. This was a sheep farm that produced crops for feed and horses for power. This farm grew to a total of 1,400 acres.

The Scott home and outbuildings

The farm passed from Adam to Alex Scott, then to Erwin Scott and on Jan. 24, 1995, 360 acres passed to Marian (Scott) Bradley.

The pasture land is rented out and brome/corn crops are either rented or shared.

The native limestone house that was built in 1889 remains. The stone was quarried in the pasture. Unusual characteristics are the Oregon Trail Park and swales/ruts in the farm pasture.

Marian and husband, James E. Bradley, have two children, Scott and Beth. Dean Murray is their grandchild.

Jim Bradley retired in 1980 from USAF and from Boeing in 1995. *Submitted by Marian E. (Scott) Bradley.*

SCOTT-BROERS CENTURY FARM – The Scott-Broers Century Farm is currently owned by Harley Broers and Janet Scott Broers, the great-granddaughter of the original owner, Cornelius P. Scott. Cornelius came to Franklin County in 1855. He earned the money to buy 160 acres of farmland as a teamster on wagons trains taking supplies to Santa Fe, NM and Pike's Peak, CO. He received the deed in 1860. Located in Franklin County just one mile south of the Douglas/ Franklin County line, the property was in the middle of pre-Civil War action. Cornelius was a Free-Stater and joined the Union army as part of the Missouri State Militia.

Nancy Van Meter Scott and grandson, Winfield Scott

After being wounded in Civil War action in Missouri, he came home to the farm and married Nancy Van Meter on May 1, 1863. In August of that year, they ran to the woods when Quantrill came through on his way to burn Lawrence.

Upon Cornelius' death in 1879, Nancy was left a young widow with six children, 13 and younger. She managed to keep the farm until the children became adults. The youngest son, John "Jack" Scott stayed on the farm. He and his son, Jay J. Scott purchased the rest of the farm from the other siblings (Arthur, Todd, Maud Broyles, Amber Eggleston and Agnes Scott) in the 1930s and '40s. The land was passed on to Jay J.'s daughters: Jean Radcliff, Janet Broers and Joan Smith. Janet and husband, Harley, became sole owners in 2002.

The original crops raised on the farm were corn, wheat, oats and sorghum. The native prairie was put up for hay. Shorthorn cattle and hogs were raised for home and market. Jack and wife, Lillie, had an orchard of apple, peach and cherry trees, a large garden and beehives. Mules and horses furnished the power for farm equipment until the 1940s when a tractor was purchased. During WW II, German prisoners worked on the farm. During this period, Jay J. Scott began implementing soil conservation techniques.

Harley and Janet Scott Broers moved to the farm in 2005, after retiring from the FAA (Harley) and as a school principal (Janet). They planted new fruit and hardwood trees. Livestock consists of six Paso Fino horses, three cats and a dog. Currently crops raised are wheat, corn, soybeans, brome and prairie hay.

In 2007 the Broers received the Century Farm award. They have received the county Natural Resources Award and in 2015, they were recognized as county Farm Bureau Family of the Year. They are active in many community organizations and activities. They have two daughters, four grandchildren and two great-grandchildren. They plan to pass the family farm to the next generation. *Submitted by Harley and Janet (Scott) Broers.*

SCRIPTER – The Scripter Century Farm in Dickinson County, KS, was settled in 1902 by Fred J. Scripter. His father, Samuel David Scripter, first came to Kansas from Indiana in 1882 and settled in Sherman Township.

Fred J. Scripter

Fred Scripter's son, Stanley David Scripter, inherited the farm when Fred passed away. In 1971, Stanley Scripter farmed until his passing in 2004, at which time his son, Ron E. Scripter, took over and currently farms with his son, John Wayne Scripter, which is the fifth generation.

During the history of this century farm, many different crops and livestock were raised, including wheat, corn, alfalfa, milo, soybeans, oats, switch grass, and forage sorghum. Apples were also raised in an orchard on Chapman Creek bottom ground, north of the building site. In 1929, the orchard died and later became cultivated ground. To this day, this field is still known as "The Orchard".

In 1952, Fred applied for and received the irrigation rights for this farm. Over the years, many crops were saved from drought.

Fred Scripter raised cattle, sheep, chickens, and mules, which he farmed with for many years. Currently, the farm is a Hereford cow operation.

SELLERS – The family of David Sellers came from Illinois in 1877. The first 100 acres of the property was purchased for $520.00 from Samuel and Elizabeth Moore. In 1883 the Sellers bought an additional 20 acres from Samuel and Elizabeth Moore. This consisted of a one-room house upstairs and one room downstairs for S225.00.

David and Elizabeth raised three children at this site. More land was bought through the years. David's son, Howard and wife, Ida, bought the property in 1927. Howard's son, Eugene Sellers, along with his wife Marjorie, purchased the property in 1964. Currently owning the property is Steven McNutt, the grandson of Eugene and Marjorie Sellers and the great-great-grandson of David Sellers. Steven and his wife, Sherri, purchased the property in 2013.

Eugene and Marjorie Sellers were presented the Century Farm award in 2007

The Sellers farm currently contains around 250 acres. During David Sellers' ownership he added rooms to the two-story house along with two barns. In 1964 the barns unfortunately burned. After Eugene purchased the farm he built one barn back along with a machine shop, woodworking shop and a machine shed. Eugene and Marjorie also modernized the home in 1964, which included putting steel siding on the house. Mr. and Mrs. Sellers worked hard planting trees, shrubs and flowers throughout the property. Current owner, Steven McNutt, has since started building back the orchard that was once present. Many animals have called this farm home throughout the years. Cattle, horses, mules, sheep, swine, chickens, llamas and goats have all been a part of the Sellers farm. In addition many fields of hay, wheat, sorghum, corn, oats and soybeans have been harvested from the farmland.

East of the main house is a quarter section pasture. Snake Creek runs on the eastern border. This property contains a spring fed pond and has brought many hours of enjoyment to all generations through fishing, hunting, camping and picnicking. It is currently a pasture for a cow/calf operation.

The Sellers century farm is located on US Highway 160, which was just a dirt road when David Sellers bought the property. Through the years the Sellers watched as sand was added to the road then eventually asphalt. Today, Marjorie Sellers enjoys many hours sitting on the porch watching traffic as they speed by headed to their destination.

Tisdale United Methodist Church is west of the farm, where the Sellers attended services. To the north of the Sellers farm is the Tisdale Cemetery. This cemetery is the resting place of many generations of Sellers family members.

We are thankful that so many of the Sellers family worked hard to care for and improve this property in order to make it what it is today. This farm is loved by many, and I am thankful it will be enjoyed for many years to come by family members. *Submitted by Marjorie Sellers.*

Original owner David Sellers, Lydi, Marvin Miller, and their dog, Watch, ca. 1917-18

SELLERS – The Sellers Century Farm in Marion County, KS is currently owned by Robert Sellers, the grandson of the original owner William Henry (W.H.) Sellers. The Sellers family lived near Versailles, KY where they farmed and operated a mill. In 1886, they decided to move to Kansas as it was a prohibition state and W.H. and his wife, Mariah, thought it would provide a better environment in which to raise their seven surviving children. They came to Kansas by train and purchased the 160 acre farm in

1886, adding additional acreage in 1888 and 1893.

Ownerships include grandparents William Henry and Mariah Sellers (1886-1946); parents August and Helen Sellers (1946-2003); Robert and Leah Sellers (2003-present).

Although the family left Kentucky, they brought a piece of it with them — thoroughbred racehorses. In 1887 W.H. had a large native limestone barn constructed and it remains a central feature of the farmstead.

William Henry Sellers, original owner of the Sellers Farm.

The discovery of oil in the area brought a rush of excitement in 1919-1920 and the opening of the Covert-Sellers field where numerous wells were drilled. In 1920 the family rented their Kansas farm and moved to Hope Hull, AL near Montgomery where they purchased a plantation and continued farming. August Sellers returned to Kansas and rented the farm himself in 1933, eventually purchasing it from his father's estate in 1946.

Through the years, the family has raised wheat, corn, milo, soybeans, alfalfa-brome and native prairie hays as well as raising cattle, hogs, sheep and chickens.

Robert began farming the land in 1957 for his parents and then purchased the farm in 2003 from his mother's estate. He continues to make the 130-year family farm his home today.

SEUFERLING – Francis M. Seuferling now owns 160 acres located in Johnson County, KS – Aubry Township, 17-15-25. On July 10, 1866 W. Holdren bought two 80-acre parcels of land for a total of 160 acres. Corn, wheat, oats, alfalfa were grown and dairy and beef cattle were raised.

Aerial view of the Seuferling farm

This land was sold by Mr. Holdren to Mr. Drake and in turn to Mr. Schell and then to William Schwartz, great-grandfather of Francis. The farm then passed to maternal grandmother, Dora Kelly; to mother and aunt, Dorothy and Agnes Kelly; to mother, Dorothy Kelly and then to Francis M.

Corn, soybeans, wheat, hay and pasture are now raised along with beef cattle. The house and dairy barn built in 1927 remain on the property. Francis still lives in the house.

The first soil conservation plan was implemented in 1947 by Francis' dad. In 1982, the Kansas Bankers Soil Conservation Award was received. The farm was named a Kansas Farm Bureau Century Farm in 2012.

Francis M. and Linda L. (deceased) had one daughter, Susan Jacobs. Grandchildren are Taylor, Liz and Thomas. Francis is still actively farming 400 acres. *Submitted by Francis M. Seuferling.*

SEYFERT – The Ron and Karen Ziegler Seyfert farm is located in the northwest part of Ottawa County, KS. The farm was purchased Aug. 26, 1903 by Joseph Ziegler. At his death, Nov. 26, 1967, the owners became his widow, Mary and his three children: Gladys Ziegler Blaha, Helen Ziegler Jindra and Frank Ziegler. Upon the death of Mary

Joe and Mary Ziegler and three of their children in front of the original home.

Ziegler, Dec. 30, 1972, Gladys, Helen and Frank became the owners. Frank was Karen's dad. Ron and Karen agreed to purchase the farm on June 11, 1973.

Joseph Ziegler purchased 160 aces, and it is still part of the farm. It is located on the W 1/2 and SE 1/4 of the SE 1/4 of 3L-9-5 and the SW 1/4 of the SW 1/4 of 32-9-5 in Ottawa County.

The Zieglers produced wheat, sorghum, hogs, poultry and cattle. The farm now produces wheat, soybeans, hay and has some CRP grass.

The original limestone home still stands on the farm. In 1994 the farm was given the Kansas Bankers Association Soil Conservation Award for work done on the farm. The Century Farm Award was given to the Seyferts in 2003. They have been recognized as the Ottawa County Farm Bureau Farm Family of the Year and received the Kansas Bankers Association Award for Water Quality in 2015.

Ron and Karen have two sons, Mike (Christy) and Mark (Teresa) and three granddaughters: Sydney, Madison and Lila. *Submitted by Ron and Karen Seyfert.*

SHAGLEY – On Feb. 20, 1883 James Wilson Shagley purchased 160 acres for $1,050.00. The farm is located in Jewell County, Highland Township, NW 1/4 of Section 23, T-1S R-10W. Original crops were corn, wheat, hay, alfalfa and grass. Livestock consisted of hogs, chickens, dairy cows, beef cattle, horses and mules.

In 1901, James' brother, William Alfred Shagley purchased the farm, paying $3,500. Nellie G. and David Clayton Gillett, daughter and son-in-law, paid $6,400 for the farm in 1945. Their son, Samuel D. and Alyce Gillett purchased the farm in 1977.

The original 160 acres are still retained by the family, along with the remaining acres in the section.

Today crops grown are corn, milo, soybeans, wheat, silage, alfalfa and grass. Livestock is a cow calf operation of Black Angus.

The original farmhouse that was built in 1884 is still standing. It has had additions put on. Nellie Shagley Gillett lived in the house from 1901 until 1988. The house is now home to the fifth generation. Son, Jim and Kim, Dereck and Jerrod Gillett lived in the house but now live in a new home in the same section. The house is now home to daughter, Beth and Brett Walker, Lily, Claire and Wyatt. Sam and Alyce Gillett live in a house they built on this section in 1976.

One acre of this farm was deeded to United Brethren Church in 1898. The church building built that year is still standing. It was a very important part of the life of the Shagley and Gillett families. It contin-

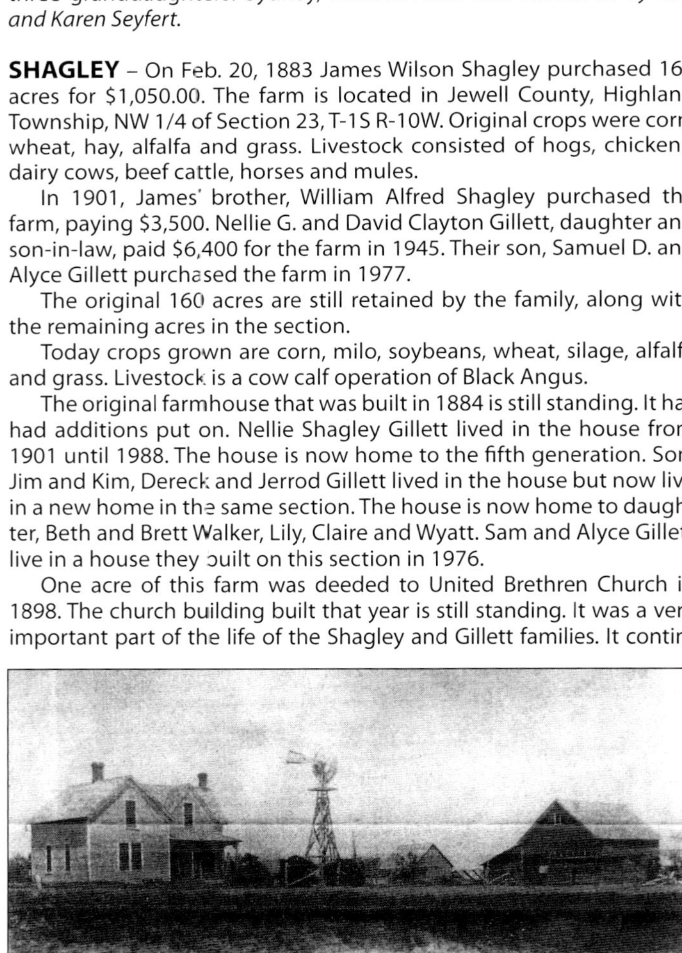

William Shagley purchased this farm from his brother, James, in 1901.

ued to be used until 2008 when the remaining members decided it was time to close. The Gillett family bought it from the United Methodist Church and it continues to be used for weddings and birthday parties.

Today the Sam and Alyce Gillett family consists of nine children – John, Jim, Susan, Marcia, Carrie, Janell, David, Victoria and Beth plus 22 grandchildren who are located in Kansas, Arizona and Minnesota.

The farming operation is now run by son, Jim Gillett and son-in-law, Brett Walker. Sam is semi-retired but still helps when he sees a need. *Submitted by Samuel D. Gillett.*

THE SHEEP HEAD RANCH – In the early spring of 1878 Benjamin Cheever Decker and Rosalia Bates Decker completed the sale of their combined properties in the Great Smoky Valley of Nevada. All of their belongings and supplies were loaded on two covered wagons, each pulled by three teams. They stepped aboard the lead wagon with their two young daughters. Then called to Doc Winders, who drove the second wagon to loose the herd of 75 horses that they had gathered for resale. And set out towards the Wyoming Pass over the Rockies to make their way to a new home in what was to become Sheridan County, KS.

Decker and Lela Mosier and family in the 1940s

They eventually purchased land on a bend in the South Fork of the Solomon River that they called the Sheep Head Ranch, in recognition of the single most important product that was to sustain them through the early years.

Their two daughters, Alice Sarepta and Mary Ann, herded the family owned flock of sheep on owned land and free range, grew to adulthood, married and started their own families.

Alice Sarepta married Jacob Henry Mosier, a young homesteader from upriver. They lived in the sod headquarters and assumed a partnership and the management of the Sheep Head when Benjamin and Rosalia retired and moved to Hoxie. Three children were born to them: Rosa, Decker and John.

By the time Decker married Lela Mowry in 1920, his parents had purchased an "alfalfa farm" downriver. The newlyweds took over the partnership and management of the Sheep Head with his Grandfather Decker, eventually purchasing the entire holding. They spent their entire working life managing, improving and expanding the ranch.

Decker and Lela were never able to complete schooling beyond high school, but they established an early goal to see that their children would secure as much education as they could be brought to desire. Eleven children blessed the union. All but the eldest sister received at least one college degree and she attended college long enough to receive a two-year teaching certificate. Upon the graduation of their youngest son from the School of Veterinary Medicine at Kansas State University, they were able to count 18 college degrees including five Doctors of Veterinary Medicine, one Doctor of Education, one Medical Doctorate and an assortment of bachelors and graduate degrees among their children.

Today the Sheep Head Ranch has grown to 1,760 acres and is owned by the families of six of Decker's and Lela's children. We grow corn and wheat and lease the pastureland to a cattleman.

Hard work, good times and determination have been the hallmarks of the Sheep Head Ranch for more than 130 years. The greeting to the family Gathering Place on the Sheep Head reads, "Welcome to the Sheep Head where men and women worked to build a future and gave us ours." *Submitted by Richard H. Mosier, Manager.*

SIEMSEN CENTURY FARM – Johann Friedrick Siemsen, who was married to Anna Margaretha Johansen, immigrated to the US in 1882. He purchased the SE 1/4 of 3-17-10 in Valley Township, Ellsworth County, KS in 1895 from the Union Pacific Railroad. The railroad had received it from the government as part of the program to develop the railroad system. His son, Marcus, immigrated to the US in 1880 when he was 15 and lived with an uncle, Henry Johannsen. He purchased the land in 1901 from his father, Johann.

Marcus and Ernestine Frees married in L892. They had eight children, four boys and four girls; the last three, all boys, were born on this farm. The original house on the farm was two rooms. Later a granary was moved from Holyrood, adding two additional rooms. A two-story house with eight rooms was built in 1908. Additional buildings on the farm were a large barn, a granary, the original barn made into a granary, a granary with a lean-to for cars and machinery, a machine shed, chicken houses and a wash house.

Plum Creek ran through the farm, starting near the southwest corner running mostly north to near the north boundary, then exiting on the east. There was a natural horseshoe shaped area, which became a favorite place for picnics, carnivals, ballgames, etc. This was referred to as the Siemsen Grove. The younger boys made money by hauling water for the steam engine when there was a carnival. Gypsies would also park along the road by the creek when they came through.

Harvest on the Siemsen farm, July 1909. In the buggy is Anna Siemsen; the two small boys are Max Siemsen (age 7) and Albert Siemsen (age 5).

Siemsen farmstead in SE ¼ 3-17-10, after 1908 when the house was built, probably ca. 1910-12.

Max, Marcus' son, inherited the land in 1949. Albert, the seventh son, purchased it in 1951. The land is now owned by Marilyn Siemsen Smith, daughter of Albert and her husband, Dennis Smith, purchasing it in 2006. *Submitted by Donald Siemsen.*

SLOAN – Rachael Fulton Sloan was the daughter of William Fulton and Emma Mae Stall who were married in Cass Co. Missouri in 1893. The Fultons knew about Kiowa County through Emma's sister Belle who was married to early Kiowa Co. settler BA Sidner. In 1899, The Wm. Fulton family including Rachael traveled to Kiowa Co. on the train. In 1903 William purchased 320 acres south of Mullinville (S1/2 of S30 T28 R19) for $3,500.00. By this time, William and Emma had four daughters: Rachael, Vesta, Cassie and Hattie who were later each deeded a quarter of land. Rachael Fulton married Chalmers S. Sloan in 1915. When William retired, he deeded the land to Rachael and Chal for the sum of "One dollar and natural love and affection". The original farm house was a wooden structure joined by a second with a dog trot. In 1926 Chal received an inheritance from a distant relative and they built the current brick house. By the 1940's the Sloan family included Beatrice, Harry, Jack, Royce, Faye, Don, Clark and Carole. All five sons were in the service. The Sloans were all active in the Mullinville community through the United Methodist Church and 4-H. Many happy times were spent on the farm working together, enjoying neighbors, and making homemade ice cream. Chal and Rachael were blessed with over 50 years of marriage and 19 grandchildren when Chal died in 1970. Clark and Don then farmed together until Don's death in 1972. Crops included wheat, sorghum as well as cattle. Rachel lived on the farm for over 80 years until her death in 1986. Clark farmed until his retirement in 1991. The Sloan Century Farm is currently owned by Nat and Marcia Sloan Bascom.

Emma Mae Sloan and her daughters: Beatrice, Faye, and Carole

The William and Emma Mae Sloan family

William Sloan and his sons: Harry, Jack, Royce, Don, and Clark

SMALL – After moving here from Grant County, OK, Jacob B. Small found and purchased a tract of 880 acres of limestone pasture and some wheat ground. He paid $15,000 on Dec. 15, 1915.

One son, Wright, moved here and the next son, Emery, soon followed, with his wife, Addie and two sons born in Nash, OK, Melvin in 1913 and Adrian in 1915. This was a start for the family farm as they

Arthur and Sidna Small (left), with son Randall and Nicole and grandchildren Dexter and Maddix.

moved into a two-story home with a barn that was built in about 1880, as the "old timers" tell me. The beams are pegged together and many still have the tree bark attached. The barn has been added onto three times to make it one of the largest well-kept old barns in the county.

In 1941 Melvin and his wife, Alice Johnson (married in 1934), had a son, Arthur. They moved to the old two-story home in 1945, when the families traded homes in one day.

In 1965, Arthur married Sidna Trimmell. The Trimmell family had lived around Neodesha and Coyville since Anslum and his father, John, came after the Civil War from Illinois.

To that union, two sons were born, Russell in 1970 and Randall in 1971. Both boys graduated from K-State and Randall returned to farm with the family. Arthur and Sidna were honored as Master Farmer/ Master Farm Homemaker in 1995.

In 1999, Randall married Nicole Kidd from the Fredonia area. Nicole was a graduate from OSU.

To this union, two sons were born, Dexter in 2003 and Maddix in 2006. This is six continuous generations to live and work on this Century Farm.

Randall and Nicole were awarded the Aldo Leopold Conservation Award for Kansas in 2017.

Good conservation practices are used on this 6,000 acres in Wilson and Montgomery Counties. *Submitted by Arthur and Sidna Small.*

SMITH – William Brian Smith of Bucks County, PA, a direct descendant of William Smith, a Quaker who came with William Penn Fleet to Bucks County in 1683, was one of the first of the family to venture West. William was born April 12, 1839 in Bucks County, PA and first traveled to Kansas in 1870. After he visited his Uncle Ross Smith, who had settled in Richland Township, Harvey County near Annelly. William purchased 160 acres, Richland Township, Section 26 SE 1/4 T24R2E, March 1872. Today this farm is located south of the Whitewater Cemetery. He quickly purchased the adjoining 80 to the east from the AT&SF RR. Section 25 SW1/4 T24R2E. At his death in February 1917, he owned 1,228 acres.

William Brian Smith, son of Ezra and Deborah Smith, ca. 1870

William was a farmer and stockman. His main livestock were cattle and hogs. He sold his livestock on the Kansas City Market, making many trips by train. He had an orchard of apple, cherry, apricot, peach, pear trees. Two of the original pear trees still stand today and produce pears. He grew oats, potatoes and maize, corn. Leaving much of the land in pasture for his stock, these pastures survive today.

William married late in life to Hattie Parks. They had two children, Ben and Letitia. The homestead farm passed to his son, Benjamin Ross Smith, who continued farming. Ben married and had four children. The homestead passed to his daughter, Lillian Smith. She and husband, Ray Cave, purchased the east 80 from Lillian's mother, Pearl Smith. Upon Lillian's death, the two 80s passed to the two oldest daughters. Carolyn Cave Fisher (Monte) has the West 80 and Marilyn Cave Ghere (Raymond) has the East 80. Marilyn and Raymond's son, Jeff Ghere, wife, Jana, and family live on the farm today. *Submitted by Sue Ellen Cave Ash.*

SNYDER – The Snyder family roots were planted in Ford County, KS in the early 1880s, when Sarah Jane Snyder and two sons, John and Robert, arrived. According to oral history, they headed west from Eldon, MO, following the death of her husband, Daniel Sylvester, at age 45, after contracting Measles.

They dug into a rise upland from Sawlog Creek in Royal Township, section 20-25-25W, approximately 15 miles northwest of Dodge City

The Snyder "Soddy", ca. 1908. Pictured are Daniel and Josephine Webster, with children Grace, Ruth, and Russell Sylvester.

and filed a Homestead claim, which today remains the family home of fourth generation descendant, Raymond, son of Elbert, son of Daniel Webster — third son of Sarah Jane — who with wife, Sarah Josephine Russell, arrived in 1886 from Missouri. Their shelter that winter was an abandoned homestead. They filed claims in section 18-25-25W and later built a "soddy".

Son Elbert's family — Ervin, Esther, Myrtle and Raymond — grew up on Sarah Jane's original claim.

In 1885 John homesteaded in section 22-25-25W and was appointed Postmaster for "Snyder," KS, still noted on some older maps. He made periodic trips by horseback to the Post Office in Dodge City, returning with mail for neighboring homesteaders to collect from his home.

Relying on faith and providence, beef cattle, a milk cow, chickens and a large garden, Daniel Webster and Josephine built a legacy with five children - Otis, Grace, Elbert, Russell and Ruth - and, with their neighbors, built a community. Josephine was the neighborhood schoolteacher. Today, their descendants continue their examples of leadership and service to others, though many are scattered into new communities.

Youngest son, Russell Sylvester, raised his three daughters: Floris Jean, Carolyn and Rowena, on acquired, Section 29-25-25W. In their family-owned Section 17-25-25W, still in grassland, are barely visible wagon ruts headed to Dodge City for supplies, doctor, or treasured visits to friends in town. *Submitted by Floris Jean Snyder Hampton.*

SOELTER – In 1898, Henry Soelter purchased 160 acres in Wabaunsee County. When Ed Soelter farmed the land he raised a lot of wheat. Bud and Irene raised a lot of wheat, corn, and put up a LOT of hay - prairie hay as well as alfalfa. Ed as well as Bud and Irene were known for feeding hundreds of feeder lambs and sheep. Ed was about the first one in the area to purchase a new machine called a combine. His brother Adolph made the comment that "he had bought himself a white elephant".

Ed, Adolph, Ernest, and Emma Soelter

Ed Soelter's portion of the original farmstead was 40 acres and an additional 140 that he purchased in 1912 made his total land owned 180 acres. Irene and her husband Elwin "Bud" Thoes farmed all of Irene's Dad's (Ed) farm as well as the portions given to other family members of the original Henry Soelter land.

The original 40 acres had a large stone barn that was three stories with entrances on two levels. It held livestock, grain, and hay. It

The original Soelter house, with Ernest, Elza, Eric, Henry, Emma, Ed, and Adolph

was gutted by fire in 1983. But the shell still stands today. The original house was enlarged and is still standing and occupied today. Corn, soybeans and alfalfa are still raised.

Although not farming anymore, Irene keeps up on the markets and share crops the land with a neighbor. She fills her time by working at the Community Thrift Store and helping at her church with funeral dinners and any other meals that are served. *Submitted by Diane Hoobler for Irene (Soelter) Thoes*

SOLBACH – John Charles Solbach (born in Germany in 1859) acquired the farm in 1881 when he was about 22 years old. Six years later he married Magdalena Diehl. They had five children: a son born in 1891 (William), a daughter born in 1890s who died as a child, a son (John Mathias Solbach, my grandfather) born in 1898, a daughter (Della) born in 1900 and a son (Leo) born in 1908. John Charles acquired many other farms after this original one. He sold everything he had in three or four counties (except for this farm) around 1919 when he retired. Son John and his new wife (Martha L'Ecuyer) rented the farm after they were married in November of 1919.

John Charles and Magdalena Diehl Solbach, with their youngest son, Leo, ca. late 19-teens.

John Charles and "Lena" retired to Clyde, KS with their then 12-year-old son, Leo. John Charles had done very well, he had what in today's dollars would be close to seven figures in the German bank in Beloit, KS.

John Charles Solbach and Magdalena Diehl Solbach on their wedding day, 1887, in Washington County, Kansas

While John Charles' son, John Mathias and his wife, Martha Solbach (my grandparents), lived on the farm beginning in 1919, Martha gave birth to a still-born child in 1920 (was to be named Loretta). Daughter Catherine was born in December of 1921. My father, Melvin John, was born in January of 1923; my uncle Ivan was born in December of 1923. These were the first of 14 children born to them.

The German bank in Beloit failed in 1923. Depositors lost all of their money. So in 1924 John Charles called up John Mathias and said they would have to move because John Charles and Lena were going to start over, having lost everything in the bank failure in Beloit. At age 65, John Charles moved back to his original farm and began farming again in 1924. He borrowed $1,000.00 from his son-in-law, Joe Caron, to get started. John Mathias and Martha moved half a mile south to the Bowmaker place.

In the next decade, John C. and Lena were able to make enough money to acquire two more farms and the money to buy a fourth farm. In 1936, before his death, he and Lena deeded the three farms they owned to their three sons: Leo, John and Bill. After his death, Lena bought the then "upside down" mortgage on Joe and Della (Solbach) Caron's farm and burned it.

They reputedly made two fortunes on this century farm, the first one they lost and the second they transferred to their children.

John Mathias Solbach, ca. 1920.

John Charles Solbach's father, Frank Solbach, brought the family to the US in 1868 from Germany. They arrived in New York City on the

Harvesting oats in the early summer, mid-1920s. John C. Solbach is at the far left and Della Solbach Caron (his daughter) is at the far right.

John Charles Solbach, Magdalena Diehl Solbach, and their four children and spouses: William and Ethel Solbach, Joe and Della Caron, Leo and Agnes Solbach, John Mathias, and Martha Solbach, ca. 1935

12th of March 1869. They first settled in Tipton, MO, then moved to Barton County, KS in 1873, where in 1879 Frank's wife, Elizabeth, died along with two teenage daughters, Mary and Carolyn. John Charles had gone in 1873 to live with an older sister, Ann Felder, in the Strawberry Community of Washington County.

Family lore says that Frank brought his family to the US because he did not want his sons to be drafted into the military for European Wars. None of his sons, nor his grandsons, ever had to serve in the military but a great grandson of Frank Solbach landed on the beaches of Normandy on June 6, 1944 as part of a medical unit, another great grandson served in the Korean War, another great grandson in the Vietnam War. Two great-great grandsons served in Vietnam and a great-great grandson served in Iraq and several great-great-great grandsons and a great-great-great granddaughter served in Iraq or Afghanistan. There are currently more than 200 living descendants of John Charles and Magdalena Diehl Solbach.

I acquired the farm from my great uncle Leo (who has no living descendants) but I also worked for him for more than 30 years. *Submitted by John M. and Patricia Solbach III.*

SOWERS-CRAWFORD – In 1879 Mathias Sowers II and his wife, Rebecca (Bowman) left Indiana to join his father and mother who were already in Douglas County. Mathias II purchased the original 80 acres in a location two miles from the homestead of his parents. Soon after their daughter, Martha Ellen Sowers and husband, Lanty Crawford, purchased the adjoining 80 acres. Ownerships have included: grandson, Homer Crawford and Homer's daughter, Geraldine E. Coffman (recipient of Century Farm Award). The two 80 acre tracts are currently owned by Geraldine A. and Martha J. Coffman, daughters of Geraldine E. and Floyd H. Coffman. In 1976 Geraldine and Floyd were recipients of the Kansas Bankers Association Soil Conservation Award in Douglas County.

The Mathias II and Rebecca (Bowman) Sowers farm located in southwest Douglas County.

The Sowers-Crawford Farm is located on a high point in southwest Douglas County in Marion Township. The original barn (see photo) remains a historic landmark and even today is used in the raising of Hereford cattle for the Coffman Cattle Company owned by Geraldine A. and her brother, Hurst Coffman. Original features of the barn still remain intact including the hay fork and stables for the working horses. The original chicken house has been renovated as a storage shed. The farmhouse was replaced in 1926 when Homer and Hazel Crawford built a modern house still in use today. The crops of corn and soybeans are rotated each year.

In September, the descendants of Mathias II and Rebecca gather in nearby Centropolis, KS for a family reunion. For all ages, childhood memories of family gatherings at "the farm" are favorites to share. Adjacent land is owned by Bruce and Hurst Coffman (sons of Geraldine and Floyd). Levi Terrill, a great-great-great grandson, takes a special interest in the care of the original land. *Submitted by Geraldine "Gerry" Coffman.*

SPLITTER – In 1847 Wilhelm Splitter Jr. was 14 years old when he and his parents and younger sisters left the farm in Mecklenberg, Germany. The family sailed from Hamburg on the Panama with their possessions in trunks and arrived in New York in June. They sailed the St. Lawrence River and the Great Lakes to Milwaukee. The family farmed 80 acres north of Milwaukee. In 1853 Wilhelm purchased 160 acres in LaCrescent, Minnesota, west of LaCrosse, Wisconsin.

The 1892 house was built for $3,000 by Wilhelm Splitter Jr., standing on the porch. On the left are his sons, Louis B. and Simon Splitter. Today the remodeled and updated home provides shelter for the current generation.

In 1878 Wilhelm, responding to the promotion of a German Baptist colony, sold the hilly and rocky Minnesota farm and purchased level prairie land in Kansas from the Union Pacific for $3.60 per acre. The railroad boxcar was loaded with their household goods, some farm implements and one span of horses arriving in Ellsworth March 31st. The next day they drove in a wagon across the prairie the purchased land near Lorraine. The house was built first and was ready to move into at the end of May. Then the barn was built. Wilhelm planted spring wheat and the whole family went to the field to cut it with a scythe, bind and shock by hand.

Horses, cattle and sheep grazed the prairie grasses. Corn, wheat and oats were grown. Fences were constructed using limestone posts. In 1892 a two-story square house was built with the basement and foundation made of limestone rock.

Irrigation started in 1977. Melvin won the state soybean yield contest in 1981. In 1997 the farm was awarded the Ellsworth Windbreak award. Today's crops are irrigated and dryland wheat, soybeans, grain sorghum and corn.

The land has passed to Louis, 1919-1964; Vernon Splitter, son of Louis, 1964-2003; Melvin, son of Vernon, 2003-2010, Janet Splitter, wife of Melvin, 2010–present. Matthew, son of Melvin, began farming in 2012.

SPOHN – Nicolaus Spohn was born in Norka, Russia. He and his family immigrated to Superior, NE. They later moved to the Tampa, KS area. On Feb. 19, 1887 Nicolaus acquired 160 acres located in Marion County, KS, 28-17-3, NE 1/4 in Colfax Township. Wheat, oats and corn were the crops raised.

Anna and Adam Spohn

This farm has been owned by four generations of the Spohn family. The farm was passed from father to son – from Nicolaus to Adam, from Adam to Edward and from Edward to Gary.

After the present home was built in 1902 by Adam Spohn, Nicolaus and Katerina lived with their son, his wife, Anna and their 12 children. Gary and Carole Spohn with their two sons, Steven and Ryan moved into the house. When Gary's grandchildren – Adam, Lauren and Luke Spohn – come to visit, they are the sixth generation to sleep in the home. The barn built in 1920 also remains.

The 160 acres of this farm have been retained and now produces wheat, soybeans, milo and corn.

STEGMAN – George and Elizabeth Stegman purchased 160 acres from Herman Schwarz, with the legal description of NE 1/4 11-25-21, in Ford County, which is located one mile west of Offerle, KS. They paid $3,300 on May 28, 1906 (or $20.63/acre).

George Stegman married Elizabeth Urban on Nov. 15, 1898 at Holy Cross Church in Pfeifer, Kansas

George was born March 2, 1880 and was the sixth child, but first to be born in the US at Pfeifer, KS. The five older children were born in Pfeifer, Russia, of German descent. George had 13 living brothers and sisters.

George married Elizabeth Urban, Nov. 15, 1898. They had 14 children and lived on this homestead near Offerle about 40 years. They raised broom corn, sewing and selling brooms, watermelon, cantaloupe, turkeys, hogs, cattle and wheat. With a large family, children learned at an early age to work hard to help make ends meet. Their Catholic faith was very important to them. George's favorite hobby was hunting coyotes.

Julius Stegman and wife, Alice, inherited this ground upon his parents' death.

George's grandson, Rean Stegman and wife, Jerri, purchased the ground after Julius and Alice's passing in 2001. Wheat and milo have been raised for many years. There is a Santa Fe Trail marker posted to document the trail passing across this ground.

Marker says: Pre 1859 Dry Route Santa Fe Trail crossed here.

Colby, Rean's son, moved the St. Joseph's parsonage to this homestead in 2007 and lives there with wife, Kim and sons, Grayson and Langston. These little boys are the fifth generation of Stegmans.

STIGENWALT – The Stigenwalt Century Farm is located in Labette County, rural Cherryvale. It has been in the Stigenwalt family since 1867 when Jim Stigenwalt and family, including two sons and two daughters, came to Kansas from Indiana and homesteaded the 40 acre farm along Big Hill Creek where small encampments of Native Americans still lived but never seemed to bother the homesteaders.

Earl and Johanna Stigenwalt

One son died and Jim was killed when kicked by a horse in the barn they had built of lumber hauled in from Arkansas by horse and wagon. Jim's son, Grant, married and lived on the home place and they had one son, Clyde. The house burned to the ground on a February night in 1919 and Grant's wife perished in the blaze. One year later, a new house was built in the same location.

Clyde married and had two sons, Ernest and Earl and one daughter, Louise (Line). Earl and his father continued farming. After serving in the army in WW II, Earl returned to the now 360 acre farm and resided there with his wife of 47 years, Johanna (Miller), until his death in 1996. They had two daughters, Linda and Patti.

Earl took great pride in raising Polled Hereford cattle and Duroc hogs while raising crops of wheat, corn, soybeans and alfalfa hay in the rich black so l along Big Hill.

Linda and Patti are now the owners of the Stigenwalt Century Farm. Linda married Gale Logson and has two sons, Aaron and Kris. Aaron has three daughters: Kayla, Lexie and Sienna. Patti married Clide Shaw and has two sons, Travis and Jason. Travis has two sons, Drake and Cooper. Jason has one daughter, Brylee and one son, Grayson.

Although the Kansas weather, over the past 100 years plus, has destroyed the barns, the Stigenwalt generations continue to embrace their heritage and ancestors and continue to make their own memories in the 98 year old home on the original Stigenwalt Homestead.

STOSKOPF – Traveling west by covered wagon, Samuel Spencer and Mary Catherine Elizabeth (Beall) Shoemaker arrived in Jewell County in December 1873. According to family lore, Samuel Shoemaker was a traveling blacksmith, the only member of his large Quaker family to leave the Goose Creek community in Loudon County, Virginia. Historical records show Samuel and Mary C. E. Shoemaker moved their growing family at least eight times before settling southeast of Jewell, Kansas. The family later moved northeast of Randall, where family members homesteaded several pieces of land.

The patent for the "Shoemaker Place" was recorded on February 17, 1885. Samuel Shoemaker died in November 1885. Following his mother's death in 1892, David William and Margaret (Houdesheldt) Shoemaker inherited and lived on the homestead. Their daughters Ina Kaad and Jessie Shoemaker owned the land after their parents passed away in 1921.

In 1951, Ina Kaad's daughter Margaret and her husband Gerald Hart began farming – just as a major drought hit Kansas. They moved to the Shoemaker Place, which had no water well or indoor plumbing. Their children attended Hillgrove School, located on the southwest corner of the homestead (1872 – 1956). In 1959, the family moved to

David W. And Margaret Shoemaker

David W. Shoemaker

a farmstead near Glasco but continued to keep cattle and grow crops at Shoemaker Place.

In 1969, Gerald and Margaret Hart, then living at Jamestown, purchased the Shoemaker homestead. Daughter Mary Anne (Hart) Stoskopf and her husband Dean inherited the Shoemaker Place in 2010. Today, the sixth generation, Mary Anne's niece Paula (Hart) Carlson and her husband Brent, farm the land.

The stained glass window from the Shoemaker Place hangs in the Stoskopf dining room. The piece serves as a reminder that, despite hardships, the Shoemaker family and their descendants found beauty and joy in their life on the homestead in Jewell County, Kansas. *Courtesy of Mary Anne and Dean Stoskopf.*

STOSKOPF – "We are," said the little girl in the John Deere t-shirt. "We are," she insisted when asked again "Who is the farmer in the family?" Only four years old, she understands the importance of family working together on the farm. Emaline represents the sixth generation to explore the Century Farm owned by Stoskopf Farms, Inc., Dean and Mary Anne Stoskopf.

Henry C. & Rebecca Hull purchased the 160 acres north of Hoisington on October 16, 1900. Following the death of her parents, Frank B. & Mary (Hull) Stoskopf acquired the land in 1915. They moved into the house when the youngest of their two sets of twins were just six weeks old (1917). The family operated a small dairy, delivering milk in glass bottles to homes in Hoisington and local creameries. Lawrence "Wayne" Stoskopf and his new wife Mildred continued the dairy when they moved to the farm in 1938. When Dean Stoskopf joined the farm in 1977, the focus was on crops and a small beef herd. Dean and Mary Anne Stoskopf moved to the farmstead in 1984.

Today, the farmstead includes the original barn and wooden granary, even as a new machine shed has replaced the original farmhouse. The shelterbelt planted by Wayne Stoskopf in the 1930s still stands, as do other windbreaks planted in the 1980s. The current farmhouse is framed by stones from the Henry Hull home (one-half mile west), including the decorative front and well-worn back steps. The farmstead serves as the hub for a Red Angus cow-calf herd. The main crops are wheat, grain sorghum, and alfalfa.

Stoskopf Family Barn, pre-1930

Stoskopf Farms, Inc. Barn – 2015

Daughter Julia (Stoskopf) Debes and her husband Josh joined the farm in 2014 with children Emaline and George. Dean and Mary Anne's son, Wayne Stoskopf, works in agriculture policy, currently in Washington, D.C. *Courtesy of Stoskopf Farms, Inc.*

STOUT BROTHERS FARM – Nestled comfortably in the rolling eastern Flint Hills is the Stout Century Farm. The Flint Hills is a massive picturesque grassland area, dominated by Blue Stem and Buffalo grass among lesser species. Covering almost 10,000 square miles, it is one of the largest grazing areas in the world. Once supporting large herds of American Bison, it now supports hundreds of cattle, many brought up from Texas to fatten.

Aerial view of the Stout farm

The Stout Farm has been owned by a Stout for 158 years. It is currently owned by two Stout brothers and their wives: Donald Stout, Sharon Stout, David Stout and Sue Stout. A third brother, Jerry and his wife, Jennifer, live on the place now. Keith Stout, my father, farmed the land after Jeptha Stout, my grandfather, quit farming.

There are two houses on the farm. The little house had one main room, which was built in the 1850s. Although it has been added to, the main room is still standing today. This is the house where five siblings and I were raised. The big house was built from a Sears and Roebuck blueprint in 1907. The barn was built a year later, using limestone quarried just south of the structure. The same stone was also used on the Sears foundation.

My grandfather, Jeptha Stout, bought the farm from his Uncle John Stout in 1898 after renting for several years. Jeptha was a Master Farmer and built a real showcase. The early farm raised Shorthorns for beef and milk, Poland China pigs and used Percheron horses for draft horses. The main livestock feed was corn fodder, which was shocked and fed to the animals out of the bundle. Today, the main crops are alfalfa, corn and milo.

Such a wonderful place to be raised, live and visit.

SUMP – Henry Sump, the original owner, homesteaded the Sump farm in 1868, 10 miles northwest of Randolph. Henry homesteaded 160 acres and then purchased an additional 520 acres. Henry died in 1886, leaving the farm to his son, Anthony. In 1938 Anthony retired,

leaving the farming operation to his son, Wesley. When Wesley retired in 1973 his oldest son, Tony, purchased the land. Tony and his wife, Sharon and three daughters: Staci, Gina, and Tanae lived on the original farm and continue to own it today although Tony passed in 2013. Sharon has recently moved to Manhattan but continues to own the land with Gina's family in charge of the cultivation of the cropland and pasture. *Submitted by Sharon Sump.*

SWANSON – In 1869 one Swedish settler named John Swanson sailed from the port of Malmo, Sweden and came to Kansas in search of good farmland. He was followed by his brother, Pete, in 1870 and his dad, Swan Eskelson and four daughters in 1871.

Harold Swanson and grandson, Matt Brock

The Santa Fe Railroad had every other quarter section, so the immigrants had the opportunity to purchase 80 acres and be given 80 acres. They had to stay five years, maintain the land and produce the goods the railroad could carry back and forth to the east. The early railroad track coming from the east was called the Interurban and came into Hutchinson in 1872. People could ride and get off at the Strandberg Station, the Kent Station and Swanson Station each a mile apart. Along that railroad started the 100-year Swanson legacy farm. Some immigrants worked for the railroad along with farming. The road by these sections was named Swede Road with all the Swedish families. They lived in a dug outs and talked of hearing the buffalos walking on top of their homes.

Swan Eskelson served as County Trustee and helped plot out roads throughout Clay Township. John Swanson and father, Swan Eskelson, helped start one room Obee School in 1873.

John Swanson and his wife, Kjersti, had four boys and one girl. John passed away when Edgar, the oldest child was 11 years old. Kjersti and grandfather Swan raised the children, farmed the land and continued forward with succeeding generations adding to the land. The four boys: Edgar, Clarence, Guy, Eskel and daughter, Ella, continued to grow the farm followed by another generation of Swanson boys: John, Harold and Donald who worked together and laughed together. The next generation - Kim, Hal and Guy - turned the farm over to grandson, Matt Brack and his family, who always wanted to follow in Grandpa Harold's footsteps.

Each generation has continued through droughts, wind, tornados and their faith in God. They've raised milo, wheat, corn, rye and cattle, participated in 4-H, and served in many civic and farm organizations.

Interurban railroad

John Swanson family with his father, Swan Eskelson

We are all working toward the common goal of keeping the farm intact. We've been truly blessed throughout all generations. *Submitted by Donna Swanson.*

SWEAT – On June 20, 1883 Solomon L. Sweat purchased 160 acres located in Smith County, Dor Township, Section 14, Range 15W. Wheat was the grain crop grown, along with livestock of cattle, hogs and chickens.

Original homestead

The farm passed from Solomon to Anna M. Sweat (daughter), to Lula Sweat (daughter), to Lennie and Dorthy Sweat (grandson and wife) and in January 2002 to John Sweat and Mary Sweat (great-grandson and great-granddaughter).

The farm retains the original 160 acres and now produces grain crops of wheat, milo, alfalfa and soybeans.

John, a widower, has four children: John, Joe, Jeff and Jeremy. Grandchildren are Konyr, Keoni, Kiara, Kenzie, Solomon, Cora and Kendra.

John retired in May 2010 but is still looking after things on the farm. *Submitted by John and Mary Sweat.*

SWINGLE/HOGG CENTURY FARM – The farm, located at SE 1/4 Section 21, Township 30, Range 9, West of 6 P.M., was purchased Nov. I, 1898 from the Charter Oak National Bank by Asa S. Swingle for a consideration of $800. He was a great-grandfather of Loretta M. (Swingle) Hogg. This was originally a 200-acre farm, which included 40 acres across the road.

Asa S. Swingle was a descendant of Hans Ulrich Zwingli, who emigrated from Switzerland to Pennsylvania in 1747.

Asa and his family traveled from Pennsylvania (1865) to Iowa then on to Torrent County, TX in 1869. In 1878 they moved to the Zenda, KS area where they homesteaded and farmed in several different locations. He served his community in a number of responsible positions and was a member of the Zenda Methodist Church.

Asa sold the 200-acre farm shortly after purchasing it to his son, Frank Swingle. Frank married Kate Osborn in 1904. They lived and worked the 200 acres all of their married life.

Donald Hogg and Loretta Swingle Hogg

About 1929 a new house was built on the property and the Homestead house was dismantled. The lumber was used to build a "Wash House" on an adjacent property around 1936 and it still stands. The "New" house fell in disrepair and was demolished in 2006. The property that the Wash House was built on was owned by Ira Swingle and his wife, Loretta "Etta" (Lewis) Swingle. Ira was the oldest son of Frank and Kate. An original batten board granary was moved to Ira and Etta's property and still stands.

Frank passed away 1951 and upon Kate's passing in 1964 Ira and Etta purchased the land from Kate's estate.

Ira and Etta's home was adjacent and north of the Century Farm for a total of 645 acres including the Century Farm.

Ira and Etta's daughter, Loretta M. Swingle, married Donald E. Hogg of Minneapolis, KS in 1959 at the Zenda Methodist Church and they reside in Wichita, KS.

Ira passed on in 1982 and Etta passed in 2005.

Upon Etta's passing Loretta M. (Swingle) Hogg and Donald E. Hogg purchased 160 acres of the original farm from the estate in 2006 and currently own the additional property to the north.

Currently the land is rented to others in the community for grain and livestock operations. *Submitted by Loretta M. Swingle.*

SYMNS – In 1881, Joseph Symns purchased land in Kansas and moved from West Virginia to farm. Today's Symns Family Century Farm in Doniphan County includes part of that original land. Joseph's brother, William Symns, married Maria Kent in 1872 and Perrin Kent (P.K.) Symns, one of their five children, was born in 1876. The family valued education and both P.K. and his wife-to-be, Helena Pincomb graduated from Kansas State University (KSU) in the class of 1901. P.K. and Helena were married in 1914.

In 1913, P.K. had purchased another tract of today's Symns Family Farm from William and Martha Armstrong. The price for that 252 acres was $12.70 per acre. The original home burned, but construction of the current brick home was completed in 1926. An original barn is still in use.

P.K. Symns farmed the land and raised prize-winning Shorthorn cattle until his death in 1939. Perrin Kent Symns Jr. and John W. Symns, the two surviving sons of P.K. and Helena continued to farm after their father's death. Perrin K. Symns, Jr. graduated from KSU in 1941. He married Barbara Stinebaugh in 1942, and they moved to the family home. Perrin taught classes in vocational agriculture for veterans of World War II, and he and his wife Barbara were supporters of 4-H and other youth and community activities. Perrin farmed the land and raised cattle both in partnership with his son, Perrin "Kent" Symns III from 1969-1980 and independently until Perrin's death in 1984. Barbara Symns, Perrin's widow, died in 2015. The farm currently is owned by Barbara Jane Stoecker, Kent Symns, and Richard Symns, as well as the six grandchildren of Barbara and Perrin. Matthew Symns, son of Kent Symns, currently manages the land on which corn and soybeans are rotated yearly on the tillable acres.

TALBERT – The Talbert Century Farm in Thomas County, KS is owned by The Morris C. Talbert Living Trust and The Dolores Talbert Living Trust.

My grandparents, James C. Emms and wife, Hattie Emms, moved from Kansas City, MO. They homesteaded the original quarter section of the farm's 800 acres in 1895. Mr. Emms outlived three wives and one daughter. He originally built a sod house and in 1905 he built a nice two-story farm house. He always said that he lived 100 feet from the best land in the world, which was well water. Life was unbelievably difficult as he had a plow team of six horses. He had things he enjoyed, one of which was playing a fiddle at barn dances. It would be late when it was over and he let the buggy reins go slack and the horse would find its way home. He also had a regular article in the *Oakley Graphic*, entitled Picked Up and Jotted Down. An occasional article was seen in the *Denver Post*.

James C. Emms and his work team of six horses.

He had 300 acres of cultivated land, 500 acres of grassland. He always had a cattle operation and raised wheat, barley, corn and feed crops. James also milked cows and sold cream in large cans. In later years he purchased a 15-30 International tractor on steel lugs. James died in 1939.

The land was passed on to the two surviving daughters, Myrtle Sim and Violet Talbert. Myrtle and John Sim sold their interest in the farm to Violet and Arthur Talbert. They jointly operated the farm until the death of Arthur Talbert in 1967, at which time Violet operated the farm until her death in 1991. The farm was passed to sons, Clinton and Morris Talbert. Clinton sold his interest to Morris and Dolores Talbert. *Submitted by Morris Talbert.*

TANGEMAN – The history of the Century Farm owned by Larry and Margaret Goering in Highland Township, Harvey County, KS begins with Larry's great-great-grandfather, Carl Heinrich Tangeman. Carl, a young farmer/soldier, left Prussia in 1847 to join his brothers in Ohio to find farmland. Cheryl Clay, a great-great-granddaughter, chronicled his journey in *Immigrant in Peril*. In Ohio he and his wife farmed and were raising their nine children. In 1878 their oldest son, Paul Tangeman and his new bride traveled to Kansas and purchased 80 acres in Highland Township, Harvey County. Paul persuaded his father, Carl Heinrich, to move the family, with Paul's five brothers and three sisters, to Kansas in 1880.

Pictured l-r: Henry William Tangeman, Henry Tangeman, Nora Tangeman, and a hired hand in front of the family home.

In 1900 Paul sold his land to Larry's great grandfather, Henry W Tangeman and moved his family to California. In 1919 Henry W. turned the farm over to his son and his wife, Henry and Lillian Tangeman, Larry's grandparents. In 1947 Henry and Lillian moved to town and rented the farm to their daughter and her husband, Bonnie and Eldon Goering, Larry's parents.

In 1973 Larry Goering purchased 80 acres of the original Carl Tangeman farm from Carl's grandson. In 1978 Larry and Margaret purchased the home place from Grandpa Henry Tangeman and moved their family to the farm. Their children - Sandy, Kevin and Suzie - grew up helping on the farm and taking care of their own 4-H and FFA projects. Kevin and Suzie both own land in Highland Township. Kevin and his family return regularly to help on the farm.

At various times since 1878 this farm has produced cattle, hogs, horses, Shetland ponies, chickens, Grade A milk, wheat, oats, corn, sorghum, alfalfa, sunflowers, brome and soybeans. Generations of children have been raised here with a strong faith, a strong work ethic and a desire to help make the world a better place.

THARP – This 160 acre Century Farm located in Marion County, KS has passed down through male Tharps for 146 years. On May 3, 1871 Joel H. Tharp (1831-1902), a Civil War veteran, applied for a homestead. He moved onto the land on May 20, 1871 and received the patent on Sept. 25, 1878. This farm is located in Catlin Township, 2-T215S-R-3E. There were 500 peach and apple trees, cattle, horses, chickens and hogs.

The old Tharp homestead, still standing in 2017

The farm passed to Atwell R. Tharp (1855-1931), to Oliver S. (1901-1967), to Max W. (1925-1993) to Blair R. (1954) owning 80 acres and David W. Tharp (1984) also owning 80 acres. Today the farm supports cattle and horses and produces hay.

The rock homestead house is still here. The original rock barn burned down in 1967 but there was an old time barn raising to replace it. The Tharp Cemetery is still located on the northeast corner of the farm.

In 1971 the Tharps were presented with a 100 year Century Farm plaque and held a family reunion of all living Tharps.

Max W. and Dalene Tharp had two children, Vincent J. and Blair R. Vincent and Patricia (Wood) have two children, Michael (Livia Osborn) Tharp and Paul (Annie McQuire) Tharp. Children of Mike and Livia are Shepherd, Silas, Noel River and Haven. Children of Paul and Annie are Elijah, Charlotte and Max.

Blair R. and Nancy Tharp have three children: Daniel (deceased), David W. and Rebecca Dalene (Tharp) and Herb Scherer.

Old time barn raising

David W. and Carla (Carlson) Tharp have three children: Jacob, Ryan and Claire. Abigail and Olivia are children of Rebecca and Herb Scherer. *Submitted by Dalene Tharp.*

THISSEN – Giles and Patricia Thissen are the third owners of a Century Farm located in Kingman County. In 1905, Anthony Gerber acquired 880 acres located in White Township, all of section 27 and part of section 28. Sheep, cattle, horses, chickens, wheat and oats were produced on this farm.

The Thissen house

The farm passed from Anthony to his daughter, Margaret and her husband, John Thissen. In 1986 the farm, now containing 330 acres, passed to their son, Giles Thissen. This farm now produces wheat and cattle.

Historic buildings remaining are the house, chicken house, two silos, granary and the surrey house. The house was one of the first homes in Kingman County to have electricity.

The Thissen house, silos, surrey house, and granary

For several years, the family has won the State Fair Champion Award for wheat.

Giles and Patricia have three children: Joyce (married to Brian Lies), John (wife, Kellie) and Joseph (Ji). They have seven grandchildren: Braley, Jennifer and Jerrod Lies; Cheli and Camron Thissen and Raleigh and Camellia Thissen. *Submitted by Giles and Patricia Thissen.*

THOES – The Peter Thoes farm was founded in 1854. Peter Thoes and his brother, Joseph, rode on horseback from Westport, MO in the spring of 1854 to Mill Creek area and selected the site of their homesteads. On March 1, 1855 Peter made his claim on his site and built his first log cabin on the banks of Mill Creek.

Big barn on Peter Thoes farm, 1975

The great importance of the Peter Thoes' place was that it was the location of the first Alma town site. Peter, Joseph and their brother-in-law, Ed Krapp, formed the Thoes Triangle with each of their claims bordering one another and it was in this location that all the settlers came, beginning in 1855. The Thoes' place was the center of a lively community with at least 40 German settlers making homesteads within a six-mile radius from the Peter Thoes farm. Peter had an "Indian" store and he was the first Postmaster. He gave the land and started a school by 1858. Ben Hankammer had a sawmill.

The Peter Thoes farm house, built 1869. The East wing was built in 1923.

"Alma" was the name of the Peter Thoes place. It was named for the river in Russia where Peter Thoes' father, Johanes, fought the Russians as a conscript in Napoleon's Army. Also because the hill (where later the Thoes house and barn were built) was called "The Alma," the name fitted the scene of the beautiful hill and valley of Mill Creek. This first Alma was also listed on the early maps of 1857 and 1859. (The present site of the town of Alma was not established until 1866)

The Thoes place was also in the contest for the site of the County Seat of Wabaunsee County. In 1860 it was incorporated as a town, according to the Incorporation Certification of Kansas Territory. However, the plotting and sectioning of a town was not carried out, probably due to the troubles of "Bleeding Kansas" and the approach of the Civil War.

Ernest Thoes, plowing

The Peter Thoes farm was operated by the family for 123 years. Peter, his son, Ernest and then Ernest's son, Elwin and his wife, Irene, from 1854 to 1977. Irene now lives in Wamego, KS as of this writing. The very famous "Big" barn is on the National Registry of Historic Places and the Kansas Registry of Historic Places. The barn is now owned by Martha Padgett Ross, a great-great-granddaughter of Peter Thoes. The barn is one of the very few "bank barns" surviving and for that reason was put on the registry and because of the historic importance of the Thoes Place. *Researched by Rev. Stanley D. Padgett, great-grandson of Peter Thoes. Stanley grew up on the Thoes Farm.*

THOMPSON – The Thompson Century Farm is located in Labette County, Kansas, and is currently owned by the Late Albert J Thompson SR and Dora L Thompson. George Thompson purchased the 160 acres from John Tibets on June 20, 1911; Mr. Tibets homesteaded the property. Albert's grandfather George purchased the farm at the age of 58. George's parents came from Ireland and were in New York City only one day before he was born in 1853. George always referred to himself as having been smuggled into the country.

First generation, George and Docia Thompson

Second generation, Arthur and Lucile Thompson

The Thompson Farm, Mound Valley, Kansas

In 1933 Grandfather George transferred title to Albert's dad Arthur James and his siblings George, Jack, Bill, Sam, Clyde, Susan and Button. Albert's father Arthur James acquired sole ownership when he bought out the interests of his siblings. In 1958 Albert and Dora purchased the land from his parents.

Albert and his twin sister Alberta were born on this farm in 1926 and in 1932 his younger brother Max was born. Albert lived there his entire life. Albert spent some time serving our country in Korea and when he was discharged he returned home to work with his family on the farm.

Third generation, Albert and Dora Thompson

The Thompson family has a deep faith in God. Church, friendships with family and with the community were and still are very important to the Thompson's. In the early years everyone always gathered at the Thompson's for birthdays and holidays to celebrate.

We raised chickens, swine, and cattle and always a big garden. Prairie grasses, wheat, milo, soybeans and corn were grown.

In 1989 Albert and Dora's youngest son Kenton moved one mile from the homestead and started farming full time. In 1995 Albert received an award from the Kansas Bankers Association and then in 2011 received Century Farm Recognition. Kenton and Albert farmed together until Albert's health made it impossible to help anymore. Albert passed away in March 2014.

If you drive past the farm today, you may see a barn that is the only remaining original building on the site. The home that Dora resides in was built in 1965. Today we still raise cattle, wheat, corn and soybeans.

The Albert and Dora Thompson family includes three children—Karen Thompson Phillips, Kenton Thompson and the late Albert J Thompson Jr.; grandchildren—Charity Phillips Frey, Caleb Phillips, Holly Phillips Miksch, Melody Phillips Kabrey, Treyton Thompson, Tyler Thompson, Tucker Thompson, Linsey Thompson Ackerman, Arthur Thompson, Jonathon Thompson; and 14 great-grandchildren.

TILLBERG – On July 31, 1886 Gustaf Tillberg purchased the east 80 acres of land located in Saline County, Ohio Township, 2-14-4 SE 1/4. On Dec. 6, 1892 he purchased the west 80 acres, making this a 160-acre farm. The farm produced grain crops of wheat, hay, barley and corn, and cattle.

Tillberg Farms, 2010

Andrew, Clarence, Carl, and Donald Tillberg

As with other farms, this farm has also been owned by a male Tillberg since it was originally bought. The farm passed from Gustaf to his sons, Andrew and Clarence. Next owners were sons of Andrew: Carl, Norman and Donald Tillberg. In 1996, Donald, son of Donald Tillberg, acquired acreage. In 2010, Alan, son of Carl Tillberg, also acquired acreage.

This farm now contains 120 acres and produces wheat.

Donald and Donnette Tillberg have three children: Mike, Kathy and Jimmy. Grandchildren are Adam, Jason, Lawa, Jessica, Sava and Annie.

Alan and Mindy Tillberg have two children, Jacob and Tyler. *Submitted by Alan Tillberg.*

VESTRING – The Vestring Century Farm in Butler County is currently owned by Joseph N. Hein (widower of Marilyn Vestring Hein) and Alan and Victoria Vestring Weldon, one of the great-grandchildren of the original owner, Johann Bernhard Vestring. Johann Bernhard "Barney" Vestring immigrated from Germany and was one of the original settlers of the community of Burns, KS in the late 1870s. The original deed to the farm states that on the 7th day of March 1900, Gottlieb Bauer and Christine Bauer of Essex County, NY sold to Barney Vestring the SE1/4 of Section 4, Township 23, Range 5 containing 160 acres in consideration of the sum of $1,200. This document was filed for record in Butler County, KS on April 14, 1900.

Vincent Vestring (far right), his sister Agatha (far left) and their brother Joseph (Bus) and his wife, Mabel Lovell, at the Vestring farmhouse in 1947.

Barney built a two-story home on the property where he and his wife, Elizabeth Uppenkamp, farmed and raised three sons. The oldest son, Joseph John and his wife, Mary Elizabeth Duffy, continued the family farming operation until 1949, raising cattle, hogs and chickens, as well as grain and hay. The farming operation was continued into the third generation by the oldest son of Joseph and Mary's six children, Vincent Bernard and his wife, Cleora Edwards, from 1949-1994. Vincent played a major role in promoting soil conversation practices and in 1993 received the Distinguished Service Award for over 30 years of service to the Whitewater River Watershed. For many years the Burns Lions Club Steak Feed, attended by hundreds of men from surrounding counties, was held on the farm. His son, J.C., later lived on the farm and continued to host this event well into the second millennium.

After his four children were grown, Vincent was assisted with the farm by his son, Vincent Bernard II "Bernie." The farm became the property of their daughters, Victoria Weldon and Marilyn Hein and their spouses in 2002. The land and buildings are currently rented to a local Burns family for their cattle, hay and grain business.

The descendants of Johann Bernhard Vestring currently includes three sons, nine grandchildren, 24 great-grandchildren, 41 great-great-grandchildren, 43 great-great-great-grandchildren and one great-great-great-great-grandson who live in Colorado, Florida, Illinois, Kansas, Louisiana, Missouri, Nevada, North Dakota, Oklahoma, Pennsylvania, Texas and Mexico.

VOGEL – On Sept. 21, 1889, George and Mary Vogel acquired 160 acres located in Marion County, Gayle Township, Section SW 1/4 11-19-3. Wheat and cattle were produced on this farm.

This farm, always owned by Vogels, passed from George and Mary to Frederick and Anna Vogel, to Ezra and Katherine Vogel and to Greg and Dina Vogel.

The farm now produces wheat, corn, beans, hay and cattle. Greg and Dina now live in the original stone house built by George and Mary.

George Vogel came from Russia and built a Russian style home. He traveled back to Russia and built an American style home. On this third trip back to Russia, George was kicked in the head by a horse and was buried in Russia. Mary passed away in Marion County, KS and is buried at Strassburg Cemetery.

Greg and Dina have three children: Jessica Vogel Fine (husband Zan), Jonathan Vogel and Jeremy Vogel. *Submitted by Greg and Dina Vogel.*

VOGTS – Johann "John" Vogts (March 13, 1866 – July 28, 1928) and Caroline "Carrie" Boesker (Sept. 23, 1867 – May 18, 1938) were married Sept. 14, 1892. On April 15, 1902 they acquired 160 acres located in the SW 1/4 of Section 16, Township 20, Range 1, McPherson County, KS. The farm produced wheat and corn as grain crops. Chickens, pigs and a milk cow were the livestock.

John and Carrie's son, Otto Henry Claus Vogts (Nov. 28, 1893 – July 31, 1960) and Susie Auernheimer (Dec. 21, 1898 – Dec. 19, 1958) were married Oct. 24, 1923. Their children were Roland (Dec. 19, 1924 – May 2, 2000), Mildred (Jan. 28, 1926 – Feb. 22, 2013), Ethel (Feb. 6, 1927) and twins, Richard and Robert (Sept. 23, 1928).

This farm retains 160 acres and now produces wheat, corn, milo and alfalfa. Also, feed out lightweight calves to feed lot stage.

John Vogts tied the cedar tees together, making an arch and also kept them trimmed. There were three sets of arches.

The Henry Volker family, front row: Henry & Mamie Volker with daughter, Iva, in between. Back row: Myrtle, Bill, Clyde Hamon (Mamie's brother), Nellie, and Opal.

The original house and barn are still being used today. Otto's five children were born in this house.

In 2010, this farm received Century Farm Award.

Richard Milton Vogts (Sept. 23, 1928) and LaVera Rose (Hensiek-Westerman) (Nov. 21, 1931) have four children: Jeffrey (Aug. 15, 1958), Kelly (March 20,1961), Verlin (Nov. 21, 1963) and Milton (July 20, 1966). They have eight grandchildren, 15 great-grandchildren, two step-grandchildren and five step-great-grandchildren. *Submitted by Richard M. Vogts.*

VOLKER/HARTNETT – The "Homeplace" is the farm where Martin and Myrtle (Volker) Hartnett lived their entire married life and raised their two children, Gerald Hartnett and Norma (Hartnett) Williams.

Henry Volker bought this farm from William R. Deniston on Dec. 14, 1906. Henry was an enterprising young man that had the goal of buying a quarter of ground with a house, for each of his six children. When Martin and Myrtle got married, the farm was given to them and they built a new house that still stands today. Henry died at the age of 53 but his wife, Mamie, continued his tradition of providing for their children as did Martin. "Mart" was able to provide a quarter and an 80 of ground with house for his children, Gerald and Norma.

The southwest corner of the farm has a grove of trees that had a meadow that was converted into a baseball field. Many community games were played there, complete with soda pop and a rope swing attached high in a GIGANTIC cottonwood tree. Mart had played minor league baseball as a young man. Mart and Myrt raised wheat, milo, alfalfa, sheep, cattle, work horses and chickens. When Grandpa decided to get rid of the chicken flock, Grandma told him she would never make chicken and noodles with a store bought chicken. To the grandchildren's dismay she was true to her word.

The farm was inherited by Norma Hartnett Williams, their daughter and then inherited by Norma's children: Jim Williams, Judy Algrim and Janet McCoy. Today John Hildebrand, Henry Volker's great - great grandson, farms wheat, milo, alfalfa and runs Angus Cattle on it. *Submitted by Gerry (Hartnett) Hildebrand.*

The Volker/Hartnett farm

VOPAT – The first owner of this land located in Ellsworth County, Sec. 22-14-10 was Jakob and Barbora Vopat. Jakob and Barbora came from Bohemia, Czechoslovakia in 1878 and acquired 80 acres where they raised cattle and wheat. Their family consisted of eight children.

The barn in winter

The second generation owners were Frank C. and Katherine (Lovci) Vopat. They had four children: Frank, Rose Diabal, Christine Bouska and Adolph.

Third generation owners were Adolph and Lila (Clark). They had two children, Thomas A. and Linde Lee.

The Vopat homestead limestone house

The fourth generation, Thomas A. and Christy (Paschang) Vopat established the Vopat Trust. They had two children, Thomas J. and Catherine A. Vopat, who are the fifth generation owners.

The Vopat Trust now owns a total of 900 acres, which produces cattle, wheat, milo and hay.

Buildings on the farm are the original homestead house made of limestone in 1882 and the 1900 hip roof barn.

The farm has received Century Farm status and the Governor Preservation Award. *Submitted by Thomas A. Vopat.*

WAHL – The Wahl Century Farm in Russell County, KS, consisting of 160 acres of pasture and tillable land, was purchased by Jacob Jr. and Anna (Hauschild) Wahl on Sept. 14, 1915 for $43.75 an acre. Beef cattle, hogs and chickens were raised. Wheat, grain sorghum, oats and Sudan grass were grown.

Jacob, 35 and Anna, 25, married on Dec. 18, 1912, southwest of Hays, reared a family of four boys and two girls: Philip, Earl, Alvin, Dean, Norma and Caroline. The sons all helped their father farm and graduated from Hays High School.

After high school, Philip remained home to till the soil and Alvin's interest was in quality beef cattle. Both of them are WW II veterans. Dean operated the farm with his father during the absence. Afterwards, Dean operated filling stations. Earl followed a career in sales. The girls both married men with farm backgrounds after college. Caroline Peterson, Clifton and Norma Strecker, Galatia are the present owners of the Wahl farm and operate their own farms.

The postrock barn, built in 1911.

The postrock barn built in 1911 by Ed and Jennie Corwin to stable work horses, store hay and grain for livestock and a room for the cream separator is being preserved. The large storm cellar with two dorms is still intact where canned meat, vegetables and fruits were stored as well as potatoes and newly gathered eggs.

The original three-room house with pantry has been torn down and replaced in 1949 with a five room, Bath Alladin catalogue home. It was shipped by Union Pacific Railroad in a boxcar and then constructed over a basement.

Their faith in God saw the family through the flu epidemic, the depression and the dust bowl days. The verse on an aged wall hanging, "The Lord will provide" continues to be the family's foundation.

WALLACE – Like many of the farms in Kansas, the Wallace farm was part of the Homestead Act and its first owner was a Union solider and he homesteaded 160 acres in 1867. There was one other owner before William Wallace and his wife, Mary Ann Claycomb come from Monmouth, IL and purchased the farm June 12, 1872. During their first years of ownership one of the relatives was shot by an Indian while working in the fields and is buried in the Lodd Cemetery nearby.

The Wallace home, built in 1911

Loil Franklin Wallace, their oldest child eventually took over the management of the farm. He married Louise Mott in 1887. They had six children: Leta, Elsie, Dee, Denzell, Frank and Melvin. All were raised on the farm.

Dee served in the US Army in France during WW I. Prior to going overseas he married Leah Marshall. She gave birth to their son, Dee Mott Wallace Jr., Nov. 30, 1918 in Erie, PA and she died 10 days later from influenza. After Dee Sr. returned from the war, he brought Dee Jr. back to Eureka where his mother, Louise, helped raise him. He also continued to farm and raise cattle. His siblings sold their interest in the farm to him in 1947 for $1.00 each.

The first Wallace home

Dee Sr. married Edith Painter in 1938 and Dee Jr. married Erma Porter the same year. Dee and Edith had one daughter, Grace Newby and Dee Jr. had two children, Roger and Deanna Born. Dee Sr. continued to farm and Dee Jr. moved five miles north and farmed on the Gray place.

Dee Sr. died in 1967. At that time, Dee Jr. went back to farming and raising cattle even though he was in the oil business. A couple years after the death of Edith in 1974, Dee Jr. bought Grace out and become sole owner of the farm along with his wife, Erma.

Erma passed away in 2009 and Dee Jr. in 2010. Their children, Roger and Deanna are now owners of the farm. The farmland and pastures are leased, as Dee Jr. had done some time prior to his death. Because of additional land purchases by the early owners, the Wallace farm now has 451 acres.

The Wallace's are proud to have five generations of ownership and hopes there are many more generations for years to come. *Submitted by Deanna Born.*

WARD – Great-grandfather, Andrew C. Dixon, was born in Glasco, Scotland Jan. 1, 1836. It is unknown when he came to Kansas, but he owned and operated a saloon in Montana, KS. He was a Civil War veteran.

Grandfather Andrew C. Dixon was born in Montana, KS March 22, 1877. In 1884, when he was 7 years old, they moved to the farm. Nov. 18, 1887 under the Osage Land Grant signed by President Grover Cleveland, great-grandfather homesteaded the southwest quarter of Section 23 in Township 32, south of Range 21, east of the South Principal Meridian in Kansas containing 160 acres. Wheat and corn were grown on the land.

The farm passed from Great-grandfather Andrew C. to his three children: Andrew C., James and Mary. The farm then passed to cousins Minnie Cook and Clarence Claxton. The next owners were Ernest and Opal (Dixon) Rigg. Barbara (Rigg) Ward, the present owner, inherited the farm at her mother's death in 2012.

Great-grandfather and his son, John, were skilled rock masons. They dug deep wells and lined them with rock. One is still on the home site. They helped build abutments for bridges crossing the Neosho River.

Barbara's grandparents retired in 1944. Her parents (Ernest and Opal) retired in 1971. They had purchased the land in 1945. *Submitted by Barbara (Rigg) Ward.*

WASINGER – Clemens Wasinger was the original owner of 160 acres located in 10-15-18. The farm produced wheat and livestock.

Through the years, the farm passed from Clemens Wasinger to Fidelis and Theckle Wasinger, then to owners William and Rose Wasinger and Leona Pfiefer.

The farm retains the original 160 acres and now produces wheat, milo and cattle.

William F. and Rosalie have five children: Linda, Robert, James, Francis and Brenda. They also have seven grandchildren and 10 great-grandchildren. *Submitted by William F. Wasinger.*

WEEDEN – The Weeden Century Farm in Cheyenne County is owned by a third generation son, Gaylen and his wife, Dorothy. George Weeden got the homestead patent on this 160 acres NE 22 5 42in 1892. He made some improvements before he brought the family out. We assume he built the house first. It was part wood and two more rooms of sod. He brought the family out later in a covered wagon. It was a long trip from Washington County, Ks. There was George Henry Weeden; his spouse, Lydia; children: Florie M., Lloyd V., Fred H, and Hattie M. They were listed in the 1895 Jaqua census.

Grandma planted an orchard and Grandpa built a barn, chicken house and milk house, all made from big rocks cut into blocks and stacked into walls with wooden roofs. It must have been a gigantic job to lift them in place. We do not know how they did it.

There was a school down the hill and our Mother (Vinnie Reish) came to teach there .She met and married our father Lloyd Jan. 25, 1908. For some reason Grandpa decided to move to Rock Island, TX. The whole family loaded their belongings on the train in a snowstorm and unloaded in Texas in the sunshine. They stayed one year. Our parents came back to Harrison, AA where their son, Vern, was born.

The Weeden farm in the early 20th century.

Grandpa and Grandma came back to Kansas and our parents came back also and bought the homestead from Grandpa March 5, 1913. Five more children were born to them: Virgil, Mildred, Audrey, Marie and Gaylen.

Our parents retired and I and wife, Dorothy, bought the farm in 1959. We have four sons: Terry, Dennis, Rod and Jeff. They married and our family includes three daughters-in-law, nine grandchildren and 10 great-grandchildren.

The farm raises corn and wheat. The big old evergreen tree Grandma planted is all that survives of the homestead. *Submitted by Gaylen Weeden.*

WEINERT/CAFFREY – Karl and Anna (Halbhuber) Weinert immigrated to the US; Karl from Germany and Anna from Austria in the 1860s. My grandfather, Charles J. Weinert, was born in Vliets, KS Jan. 11, 1880 the oldest child of Karl and Anna. At some point the Weinert family moved to near Frankfort, KS where Karl had a restaurant in Frankfort. In 1890 Karl, Anna and family moved to a farm near Wyoming, KS (later the town site was renamed Lillis, KS when the Union Pacific RR went thru the area in 1909). In 1901, at the age of 21, my grandfather, Charles J. Weinert, purchased 80 acres of land across the road from where his parents owned and lived. Later purchasing three more 80s, giving him a total of 320 acres. East 1/2 of 30-5-10, Cleveland Township, Marshall County, KS. In 1908 Charles J. married Catherine (Murphy). They had four children: Raplael J., Charles D. and two boys who died at a very young age. My grandparents had a fruit orchard, chickens, a couple milk cows, stocks cattle and hogs. My grandfather also had a thrashing machine and harvested wheat, oats and millet for neighbors. In 1922 they built a new home and in 1936 a new barn. We currently live on the first 80 acres my grandfather purchased and in the home he built. In 2002 we received the KFB Century Farm Award.

My grandfather Charles J. died at the age of 95 in July 1975. The land was passed down to my Dad, Raphael J. and my brother, James R. Weinert. Dad passed away in 1979 and my brother in 1984. Today I own the north half and my sister Mary (Weinert) Cygan owns the south half of the land once owned by our grandfather.

After my brother's death, my husband, Kenneth, and I quit our jobs in Topeka, KS where we had lived and worked for 25 years and went to farming. I grew up on the farm my great grandparents, Karl and Anna Weinert, purchased in 1890. My husband, Kenneth, also grew up in the Lillis area. We started out raising wheat, milo and soybeans, later corn and soybeans plus silage and prairie hay for our cow /calf herd. Our cattle herd is mostly Black Angus. For 50 years I was a Licensed Practical Nurse working part time as a geriatric nurse. We have one son, Tom. Currently we don't plant and harvest crops but still have the cow/calf herd. *Submitted by Martha Caffrey.*

WEIR – Daryl and Joyce Weir, present owners of the 100-year family farm purchased 80 acres from Grandmother Christina Weir and her five children in 1971. Later Daryl and Joyce purchased another 20 acres that had belonged to his grandparents. In 1980 Daryl received the Osage County Soil Conservation award.

They tore down the existing house and replaced it with a ranch style house. Original house was three stories, two stairways and eight sets of double pocket doors. Daryl used some these doors in the new home. All out buildings have been replaced with metal buildings. Daryl continues to raise row crops on this family farm.

Grandfather John Weir's family emigrated from Scotland and settled in Lonaconing, MD. After his father was killed in a mining accident John, his brother, Alex and his mother traveled to Kansas.

John and Christina Weir purchased the original 160-acre farm in February 1898. John raised livestock and row crops. Receipts have been found that he sold wool to Mid America Wool Company for $24.01. They also had a Bee Apiary in 1898. The honey was marketed as local honey. Inspection papers from June 20, 1950 stated two hives had been lost during the winter, in the spring had nine hives. Fifty pounds of honey produced in 1949. During this time period John served as Game Warden.

John obtained some of the farm from his brother, Alexander Weir. Other acreage was from John and H.E. Coffman.

Coal was strip mined on this acreage and still shows the scars. One train track bed was visible on the south end of acreage that went to Lawrence. It has been reclaimed and is farmed now. Another spur went to Carbondale to the Santa Fe railroad.

The acreage was granted by Patent to Private John Brown by President James Buchanan on June 1, 1860 and was sold many times until John Weir bought it. *Submitted by Daryl W. and Joyce Weir.*

WERNER – In 1869 a fine old Swiss gentleman, Jacob Werner Sr., came to Kansas and settled on 160 acres located in Section 16, Hanover Township, Washington County, KS. The stone house was built in 1879 at a cost of $1,000. And the stone barn was built in 1884, costing $600. In 1890, there was an orchard of 400 trees, including apple, cherry and peach plus a vineyard of over 500 vines. The farm had 16 head of cattle, 35 head of swine and a team of horses.

The Werner house that burned in May 1939

This farm passed to his son, Henry Werner, then to Conard Werner, on to Alfred Werner and to LaVonne Cook.

This 160-acre farm now produces corn, soybeans, milo and cattle. The original house burned on May 26, 1939. The second house burned in 1995. The stone barn is the only original structure remaining.

LaVonne and husband, Dale Cook, have three children: Debra Stout, Shelley Wichman and Alan Cook. Their grandchildren are Erin Stout (deceased), Carrie Oswald and Lesley White. *Submitted by Dale and LaVonne Cook.*

WESCH – The Wesch Century farm located in Decatur Co. KS was homesteaded by Reimer C. Wesch with the original 160 acre patent granted in 1899. Reimer and his parents along with several siblings emigrated from Holstein Germany in 1880. The family originally settled in Elk Township, near Malcolm, Lancaster County NE prior to mov-

Reimer Wesch home, 1915

Wesch Farm and Flag, 2016

ing to Decatur Co and Red Willow Co NE. The farm is located two miles north of Traer KS and has been in the Wesch Family since its origination. It consists of 480 acres, most of it purchased along with the original homestead.

Reimer passed away in 1925 leaving the farming burden to his wife Emma and her young family. The youngest child, Carl was 10 when Reimer died. They struggled during the dry depression days yet still were able to keep the farm. Emma, known as Granny always had sugar cookies for her grandkids. Reimer's Sons Bill and Mose (Francis) farmed the land until the 1980's and then farmed by Carl Wesch until it was purchased by Jim Wesch.

Wesch Farm during the Dirty Thirties

Today the farm is owned and operated by Jim and Pam Wesch. Jim is a grandson of Reimer Wesch and the son of Carl Wesch. The original milk house still stands and was built prior to 1900 by Reimer. The existing house was built in 1917 and is now 100 years old. The photo from 1915 depicts the Reimer Wesch family in front of the first house built on the property.

Since 1995, Jim along with the help of his wife and others have planted approximately 4000 trees and shrubs on the farm and Jim and His wife have received The Kansas Bankers Soil Conservation Award and the Windbreak award. It is quite a change from the treeless plains in the late 1800's.

WESTERMANN – In February 1917, Henry Westermann purchased 160 acres in Atchison County, 5-27-18, and raised corn, soybeans, wheat, milo, brome clover, cattle, chickens, and hogs. The farm passed from Henry, to his son Ernest Westermann, grandson Steve Westermann, and great-grandson Mark Westermann (and his wife, Marlene).

Mark and Marlene still manage the original 160 acres and raise corn, soybeans, brome clover, and alfalfa. An interesting fact: the farm house was originally the old Huron School prior to 1917 and was moved to the farm by Henry with a steam tractor.

Mark and Marlene have three children – Steve, Stacie, and Julie – and three grandchildren – Ella and Henry (Steve) and Ava (Stacie).

Mark and Marlene holding their Century Farm sign

WEST – George Samuel West and his wife, Margaret Everett West, originally came from Carroll and Columbiana Counties, OH. They lived briefly in Michigan and Illinois before settling in Linn County, KS. Joining them on their travel were their children — Menelvo, Delbert and Elmer — who were all born in Illinois. Born in Blue Mound were Grant, Chalmer and Wilber. Delbert, Grant and Wilber did not live beyond childhood.

The home of Juno and Clomer West, September 1962

From the Pike's, George H. and his wife, Sarah M., they purchased the north half of the southwest quarter of Blue Mound Township. The deed was dated July 22, 1863, filed May 14, 1866 and is recorded in book D, page 198.

In September 1898, George and Margaret deeded this property to the children of their deceased son, Chalmer, children Everett, Nelson Clomer, Archie and Hazel West, but reserved the use of it for the duration of their lifetime. The children were minors under the administration of their mother, Olive Murrow West.

The West family farm

When George died in 1910 and Margaret in 1911, the property passed to their grandchildren. Everett had died of a gun accident in 1903, so Clomer bought out the shares of his remaining siblings. In 1911 Clomer married Juno Steele and by 1913 they were living on the property. By 1925 they were able to settle the debt to Clomer's sister and brother. Clomer and Juno lived continuously on this land until her death in 1978 and his in 1983. The land passed to their daughter, Frances, until her death in 1992. Roland, the only surviving child, became the owner of the property. This land passed to me, Judy West Brownback, in 2006, thus keeping the property in the West Family, as it is today. George and Margaret West were my great-great grandparents.

WIENCK – In 1900 John and Hulda Wienck acquired 320 acres located in Lincoln Township, Washington County, KS, Section 21, Township 05, Range 005. Corn and wheat were crops raised to support cattle and chickens.

The 320-acre farm passed from John and Hulda to their son, Fred and Olga and to the third generation, Melvin and Frances. The farm now produces wheat, corn and soybeans. In years past, milo and alfalfa were grown. Dairy and beef cattle are raised on the farm.

All of the original buildings remain on the farm except the house, which was replaced after being destroyed by a tornado in 1960.

The Century Farm award was issued in 2009 for 100 years of Wienck ownership. In 2007 the Kansas Bankers Association Soil Conservation Award was presented.

Aerial view of the Wienck farm

Melvin and Frances Wienck have six children: Susan Arleen (Ron Graham), Linda Rae (Tim Poell), Dennis Allan (Lynn), Carol Jean (Doug Bickel), David Lee (Tonja) and Sharon Kay (Bradley Goering). Grandchildren are: Chris (Andrea) Graham, Travis (Korielynn) Graham, Nicholas (Stacy) Poell, Christina (Ryan Wisdom) Poell, Amanda (Travis Marfield) Poell, Joshua Bickel, Jewell (Pastor Justin Crisp) Bickel, Ariella Bickel, Evan Wienck, Garrett and Cassidy Goering. Great-grandchildren are: Jonathan, Olivia and Charlotte Graham; Chloe Jean Graham; Brandon and twins Lucas and Emily Poell; Isabel, Elijah and Evelyn Wisdom; Clayton and Madyson Marfield. *Submitted by Melvin Wienck.*

WIGGINS – The Wiggins Century Farm in Labette County is currently owned by Merrill and Norma Lee Wiggins Christy. The farm was originally purchased by her grandparents, John Marshall and Anna Bell Sargent Wiggins in 1894 and consisted of 80 acres.

Four generations, l-r: John Wiggins, Norma Lee Wiggins Christy, David Christy, Stephen Christy, and Edward Wiggins

Edward Wiggins I came to America from Dublin, Ireland in 1759, first settling in Virginia. Descendants moved to West Virginia and on to Ohio. John Marshall and Anna Bell initially moved from Ohio and settled in Iowa. From Iowa they moved to Kansas, eventually settling in Labette County.

The Wiggins Century Farm

The farm, located on the highway west of Oswego, eventually expanded to 440 acres and became known as "Wiggins Corner."

John and Anna Bell gave a two-acre tract of land for Maple Grove School. John and Anna Bell had nine children. Their son, Edward and several of his siblings each acquired 80 acre tracts of the original farmstead and continued their farming operations.

Edward was in WW I during 1917 and 1918, spending the majority of the time in France on the Western Front as a mechanic on the biplanes and tankers.

Edward and Mildred Tarkington Wiggins

Edward and Mildred Tarkington Wiggins were married in 1920. In 1924 their daughter, Norma Lee, was born and at the same time they built a new home in Oswego to be close to Mildred's elderly parents.

Later another 170-acre farm was purchased and Edward raised registered Hereford cattle on both farms. Ed put out a large garden each year and Mildred canned, which provided an abundance of food throughout the winter months.

Today there is the flood of rich memories that Ed and Mildred's three grandchildren have of spending summers with them and riding in Grandad's Ford truck out to the farm to help with chores. These wonderful remembrances are being passed on to future generations.

WIGGINS – William H. Wiggins and Rebecca White were married Nov. 4, 1868 in Michigan. Charles and Carrie were born there. They moved to Kansas Aug. 18, 1871, buying two city lots. Frank, George, Rachel, Anna, Fred were then born.

Nov. 16, 1877 the first 40 acres of farm were purchased, then March 3, 1884, 120 acres were purchased. Over the years more has been added, totaling 465 acres.

The W.H. Wiggins family, ca. 1905. Standing are Carrie, Frank, George, Charles, Rachel, and Fred. Sitting are Granma, Anna, and Grandpa.

William and Rebecca lived on the farm their entire lives. They had a fruit orchard and hog farm. He served on school board District # 65, was a man of many "hats:" doctor, veterinary, stonemason, blacksmith and inventor. He patented and manufactured the first one row horse drawn cultivator with cast iron seat bearing the name Wiggins and an alfalfa seeder shoe type drill which was restored by a grandson and is now in Eureka Historical Museum at Eureka, KS. The date on it is 1894.

During the Civil War, William served in Company 1, 2nd Ohio cavalry from 1861-1865, was taken prisoner at Jamesboro, and sent to Libby at Richmond and finally Salisbury in North Carolina. We have a journal he and a friend wrote later about it.

In the early 1920s oil was discovered on the farm with active production through mid-1960, then again in 2011 a small company drilled three wells with production.

Today my husband and I run a cow-calf operation with soybeans and hay. We continue to maintain the original stone house, which was completed in 1893. *Submitted by Leta A. Gray-Divine, great granddaughter.*

WILEY – The Wiley Century Farm in Rutland Township (Rutland), Montgomery County, KS is owned by Bill and Jay Wiley of Vaughn, WA.

Bill's great grandfather, William Franklin "Old Bill" Wiley, who was born in 1872 in Bourbon County, KS to the son of immigrants from Belfast, Ireland.

Left to right: Bill Willey, Homer Wiley and Jim Wiley, taken 1949.

Old Bill's family then moved to Rutland in 1875 where they homesteaded 160 acres just north of Havana, KS. The Wiley's were well known by the early settlers of Montgomery County.

Old Bill married Chloa Neel, who was born in 1871 in Radical Center, Montgomery County, KS in 1896. They had one son, James Homer Wiley "Homer," born in 1897. Old Bill farmed in Rutland for many years. One of the parcels of land that Old Bill farmed was the home of Dr. Tann of the Little House on the Prairie. During his lifetime Old Bill Wiley acquired 495 acres of land, all in Rutland. Chloa passed away in 1950 and Old Bill passed away in 1952. Their great grandson, Bill Wiley, still owns 416 acres of what Old Bill had accumulated during his lifetime.

After Old Bill's passing, Homer rented out the farm until his son, James H. Wiley Jr. "Jim" came back to Rutland in 1962 with his bride, Deloryse Anne Hough "De" and started farming the land again. Even though Jim had been at sea for many years as a merchant marine and also had been in the Nary for both WW II and the Korean War, his love of the farm in Rutland was very apparent. Jim and De moved back into the childhood home of Chloa Neel, Jim's grandmother.

Jim and De raised three kids: William Franklin (named after Old Bill), Anne Marie (Aspen, CO) and Jean Deloryse (Lawrence, KS.). De passed away in 2001 and Jim in 2007.

WINKLER – In 1867 Otto Winkler, alone at age 14, emigrated from Germany to America, settling in Lockport, IL with his uncle's family. In 1910 he read a Santa Fe Railroad advertisement for flat, fertile land with copious underground water and healthy air in Southwest Kansas. An offer of a free train ticket to view the land took him west to Gray County, KS. He decided to buy 1/2 of Section 6, Township 28, Range 29 and move his wife, Mary and their family west.

Otto and his three sons loaded two boxcars with two horses, a couple of cows, a bull, a moldboard plow, a wagon, farm equipment, personal items and lots of canned tomatoes. They dug a foundation for a house by hand and six months later when it was completed and a well dug, Mary followed to begin housekeeping.

Aerial photo of Winkler Century Farm, ca. 1940

The first years were spent plowing the virgin prairie and planting feed for the livestock and wheat to sell. Gradually the family added chickens, pigs, cattle and a dairy. Sons, Arthur and Irving, married sisters, Nellie and Lura Graves and they, too, made their homes on the original homestead. The family and farm survived the Depression and the Dirty '30s.

More acres were added over the years to increase efficiencies, production and provide a living for the multiple families. After Irving's death in 1991 his son, Riley, inherited the original homestead. Riley had been farming and ranching with his uncle and father since he was a young boy. He and his wife, Jackie, raised two children, Carol and David, on the property. Riley retired in 1998 but still oversees the property and production operations.

The family members were active in the community throughout the years, serving on school boards, church boards, extension services and the county commission.

WOHLETZ – The first family member to own the Brox Century Farm was Louis Wohletz. Louis was grandfather to Anton (Tony) Brox, current owner of the Brox Century Farm. Louis was born in Austria December 24, 1865. He came to America with his family when he was 16 years old and became a citizen of the United States in 1892.

During his ear y years in this country he worked as a farm hand. Shortly after his marriage in 1890 to Mary Keeler, Mr. & Mrs. Wohletz started farming for themselves one mile east of Huron, Kansas. About ten years later in 1900, they purchased a farm 1 mile south of Huron and moved there. In 1908, Louis purchased 160 acres of land 2 ½ miles east of Huron., which is now owned and occupied by Tony and Patricia Brox.

Louis Wohletz enjoyed hard work and was an honest, trustworthy citizen. In addition to his farm work he took considerable interest in the Huron Farmers Union and devoted much time and energy to the affairs of the organization.

Louis S. Wohletz Family

Tony, Pat & Family

Through industry and thrift Mr. Wohletz and his family accumulated 1,040 acres of land. This land was all in Atchison County, KS with most of it being in the Huron area except for two quarters north of Effingham, KS. Shortly after purchasing all of the land he died, leaving the land for the 4 sons to farm. In his will, a daughter, Anna Libel acquired the Brox Century Farm. Tony & Patricia purchased the land from Anna.

They have raised cattle and hogs along with farming the Century family owned farm since 1960. Their four children, Karen Shelly, Marcia Brox-Hughes, Nancy Koftan and Dennis Brox grew up here and helped with the farming operation. The children were all active members of 4-H.

WOLFF – Oscar Friedrich Wolff was born on Jan. 26, 1857 in Prussia (part of Germany). In 1885, at the age of 28, he travelled to Bremen, Germany and boarded the ship *Hohenzollern*, bound for the US and arrived at Battery Park in New York City on April 20, 1885. From there he migrated west to Quincy, IL and then travelled on to Meade, KS.

Oscar, Albert and Mary Wolff standing by the original farm house

In 1890 he came to the Caldwell, KS area where he met Mary Skvor Lebeda, who had travelled to Kansas in a covered wagon from Maribel, WI. Oscar and Mary were married on Nov. 14, 1891, 10 children were born to this union. They lived west of Caldwell until 1897 when they moved west of Renfrow, OK and in 1905 Oscar took over a Cherokee Strip plot and homesteaded it.

On Aug. 14, 1917, Oscar purchased 320 acres of Sumner County land from Labaron and Kittie Warlow and moved his family to that farm northwest of Caldwell, KS. Upon his death in 1939 the land went to his wife and nine living children. In 1946 240 acres of the original 320 acres was deeded to the youngest son, Albert and his wife, Lottie. That farm in now owned by Albert and Lottie's three children: Gary, Lary, and Patsy McCoy. Albert moved to the farm at age 2 and lived there his entire life.

The original farm house is still on the property but has been added on to and remodeled throughout the years and is still used as family housing. Throughout the years wheat has been the primary crop of the farm. Soybeans, milo and corn are rotated through on a regular basis. The pasture has been home to a shorthorn cow herd but is now used for prairie hay every year. The farm is located near the Chikaskia River so the family enjoyed many fun times at the river while growing up. Some of Oscar's great-grandchildren and great-great-grandchildren still spend many hours playing in the Chikaskia and helping out on the farm. *Submitted by Gary Wolff.*

WOMACKS – Samuel Claude Womacks, purchased the 320-acre farm March 12, 1907, located in Cowley County, SW ¼, S24, T30S, R6E and NW ¼, S25, T30S, R6E. On Nov 17, 1919 an additional 40 acres were purchased, NE ¼ of SW ¼, S25, T30S, R6E. The original crops were corn, wheat, and silage. Other crops over the years also included alfalfa, soybeans, and milo. Livestock included beef cattle, pigs, and sheep.

Lineage: Frances Marion Womacks immigrated to Kansas from Ohio in the late 1800's and settled near Hickory Creek, south of Leon, KS.

Samuel Claude (Georgia) Womacks – son of Frances drove some cattle to the Atlanta area one winter to feed them out. He had purchased some corn from someone in Atlanta. While feeding out the corn to the cattle, he discovered the 320-acre farm that he purchased in 1907.

Eugene Lloyd (Louise) Womacks – son of Samuel Womacks. Samuel died from appendicitis when Eugene was 13. Eugene took over the farm when he turned 18, 1926.

Larry Eugene (Ruby Ellen) Womacks – son of Eugene Womacks. After military service, Larry returned to Atlanta. He purchased a nearby farm and started farming and ranching also.

Samuel Claude Womacks family with their home in the background

Mark Eugene(Michelle) Womacks – son of Larry Womacks. When Eugene and Louise died, Mark purchased the homestead area and moved back to Atlanta. His family remodeled the original house, keeping much of its' original nostalgia, July/August 2001.

The original farm acreage is still owned by Larry (Ruby Ellen) Womacks and Mark (Michelle) Womacks. Cropland is leased to Hervey Farms. Recent crops have included corn, soybeans, milo, and wheat.

Mark Womacks still manages the grassland. We currently are contract grazing the grass on an intensive grazing program during the summer.

- The extended portion of the farm contains an old rock house (Originally part of the Ward Ranch).

- There has been evidence of Native American artifacts found in the crop fields along Timber Creek.

- A stone arch bridge crosses Timber Creek on the western edge of our property.

- We also operate a hunting and fishing guide service on the family farm. Whitetail Deer, Rio Grande Turkey, Bobwhite Quail, Waterfowl, Predators, and Fishing are the most popular species we offer.

WOOD – The Wood family began migrating from Illinois to Kansas in the 1870s. Several family members were instrumental in developing the new town of Cherryvale. John Henry Wood and his wife, Emily, bought the south half of a quarter section, 2.5 miles south of Cherryvale, in 1889. In 1898 they added the north half of the quarter section. That is the land that has been in the Wood family ever since.

Leonard Wood (1919 - 2005) and his work horse. Leonard's grandfather, I.H. Wood, bought the first section of the farm in 1889. Leonard's great-grandson, Jack Thomson, lived the first year of his life at the farm in 2011.

John and Emily's three sons – Albert, Will and Arch – attended Centennial School, located at the corner of the farm property. Centennial was the community focus for school, church and community meetings. Arch's son, Leonard, went to school there. Leonard's wife, Fern, taught school there and their three children - Mike, Sherry and Toni - attended 4-H meetings at Centennial. When Fern's and Leonard's granddaughter, Kerri Wood Thomson, moved to the farm with her husband, James, for a year in 2011, their son, Jack, represented the sixth generation to live on the land.

The farm economy has never been easy, especially with just 160 acres. Leonard and Fern, in the last half of the 20th Century, both needed full-time jobs to supplement the farm income. Their three children all attended college and moved to Kansas City, Ft. Worth and Wichita, with professional careers. With Leonard's and Fern's declining health, the ownership of the farm was in question. The three siblings formed an LLC, renting the pasture and crop lands to neighbors. In

2016 Mike and Margaret Wood, having retired from Wichita State University, moved back to the farm and built a new house.

Much has changed in 125 years, but not the natural rhythms of the land. Crops are still being planted and harvested and cattle are still grazing and birthing. In the undisturbed woods and creeks, one can still sense the spirit of the Osage Indians. *Submitted by Mike Wood.*

WOODWARD –This family of Woodwards have traced their ancestry back to Henry Woodward in England about 1730.

He came to Virginia in 1758, settling on land near George Washington. Rock from his land was used in building Mount Vernon.

The Woodward home and outbuildings

David T. Woodward, my great-grandfather, moved from West Virginia to Illinois in 1836 at 10 years of age, then to Kansas in 1856, taking a claim south of Clinton where he lived the rest of his life. Ezra Geary Woodward and Charles E. Woodward, twin brothers, were born in 1858 in the Apple Pie neighborhood on the farm homesteaded by their father.

On Sept. 7, 1891, Thomas B. Petefish, originally from Illinois, deeded the home place to my grandfather, Ezra Geary Woodward and Lotta Petefish Woodward. They raised nine children, he farmed and was active in all community affairs.

On Feb. 27, 1928, they deeded the place to their daughter, Linna Woodward Houk and Albert W. Houk, my parents. The farming was continued along with the livestock and additional rented farm ground.

After my Father was killed in a farm accident, we continued the farming until Oct, 22, 1979. My Mother deeded the property to us, Hayden K. Smith and Alberta Houk Smith. We continued the farming as before, with the crops and livestock.

Our family, all living close by, include the following three children: Gary, Randy and Jill, with seven grandchildren and eight great-grandchildren. *Submitted by Hayden and Alberta Smith.*

WOOLBRIGHT – This Century Farm owned by the William D. Woolbright Trust was obtained about 1996 from William's mother. Thomas W. Bell, William's grandfather, began purchasing land in Ford County during the 1880s. At the time of his death, he owned four sections – located in Township 28, Range 25. Cattle, wheat and milo were the original crops raised on the land.

The land located in Sections 3 and 11 (1/2 section) passed from Thomas Bell to G.M. Woolbright and then to William D. and Darlene. Wheat and milo are the crops raised today and 100 acres are rented for pasture. There are no original buildings remaining on the land. This farm received the Century Farm Award on Nov. 16, 2001.

William and Darlene have two sons, Terry L. and Timothy D, one grandchild, and three step-grandchildren.

William retired in 1996 and in retirement serves as Chaplain in Shawnee County Jail, assists Gideons, and is active in church.

WORREL – Robert Worrel purchased 80 acres in Manhattan, KS, in 1870, and through the years the farm has passed to his daughter Elizabeth Ann, to daughter Jesse White, to son Max White. Larry and Diane (Worrel) Hoobler bought the land at auction and is the fifth generation to own the farm, which they call West Fork.

When Jesse and her husband (Elmer) got this 80 acres in 1911, the first thing they did was build a barn. That original barn is still standing and in use today. The house was built at the same time and is still standing and in use today. There have been four families to live in that house. It still has original stained woodwork, pocket sliding doors, and leaded glass windows.

Current owners Diane and Larry have three daughters—Tammy, Tonya and Tina—one grandson and four granddaughters. Corn and soybeans are grown in rotation, and the land is irrigated. A pumpkin patch is used every year for the grandkids.

WORREL – Robert Worrel acquired this farm in 1870. Robert gave 80 acres to his son William Thomas in 1873. He would give 80 acres to each of his children. In 1892, he gave Levi 80 acres and William Thomas bought that 80 from his brother in 1902. This made up the 160 that then passed down the family of William Thomas. Larry and Diane Hoobler and Tonya Hoobler bought this land at auction.

William Thomas was known for growing a lot of different things and having a lot of different animals around. Like a lot of family farms of the time, he had a large orchard. He also had a Rumley tractor. Unfortunately, William perished in a fire that also destroyed his house.

Today the farm consists of 140 acres, and corn and soybeans are raised in rotation. The land is irrigated. Larry and Diane Worrel Hoobler have three daughters – Tammy, Tonya and Tina – one grandson, and four granddaughters. Larry retired from teaching Vocational Agriculture in 2002 to farm full time. Diane retired from Kansas Farm Bureau in 2013. When the Hoobler's bought this land at auction, Diane became the fifth generation and her daughter the sixth generation to own this land.

WRIGHT AND MONTGOMERY – The Wright and Montgomery Century Farm in Atchison County, near Cummings, KS, is currently in a family trust.

The original owner of this farm was Sara Wilkins (Jan. 4, 1832 – March 8, 1915). Second owner was Norah A. Donnelly (1874 – 1946). Third owner, Marguerita A. Wright (1903 – 1979). Fourth owner was Ralph G. Wright (April 26, 1924 – June 19, 2010). Fifth owner, Sherry A. (Wright) Montgomery (Dec 3, 1948 – April 4, 2003). Sherry's husband, Larry Montgomery, lives on his part of the farm. Sixth owner, Jodie N. (Montgomery) Thompson (July 31, 1972) and her four daughters live on the part of the farm they now own. Nikkie R. Thompson, Alicyn Macy Thompson, Claryssa Rylee Thompson and Brialyn Josie Thompson are the daughters of Jodie.

Original owners, Sara Wilkins and Norah Donnelly.

The rest of this farm is in a family trust. *Submitted by Larry Montgomery.*

YATES – Frank P. Yates purchased 120 acres located in Franklin County, Lincoln Township, Section 3, Range 19, in 1910. Dairy cattle, corn, wheat, oats and barley were produced on this farm.

In 1926 the farm passed to Frank's son, Wilbur F. Yates who retained own-

Aerial view of the Yates farm

ership until May 1974 when the farm passed to Frank D. Yates, present owner. Over the years, the farm grew to 640 acres that produces dairy cattle, corn, milo, beans, wheat, oats and alfalfa.

The original four-room house has been added to and in 1951 the size of the house was doubled. The barn built in 1926 has been used for milking since that time.

Frank D. and Mary Lou Yates have four children: Brian, Tammie, Robyn and Crystal. Their grandchildren are Loren, Brianna, Alyssa, Bradon, Morgan, Gavin, Ava and Ella. *Submitted by Frank D. Yates.*

YEAGER – On Dec. 22, 1874 Joseph Yeager purchased 140 acres located in Cowley County, 33 Township, Section 7, Range 3. The farm produced wheat and oats. On Feb. 28, 1898 the farm passed to Celia Patterson. On June 8, 1959 the farm passed to Glen Patterson and on Aug. 24,1988 from his heirs to Dale and Illene Hutchins.

The farm retains 140 acres and now produces wheat and soybeans.

Dale and Illene (McDaniel) Hutchins have three children: Gregory K. Hutchins, David D. Hutchins and Sheryl A. King. They have seven grandchildren: Luke Hutchins, Christine Webb, Tyler Hutchins, Jessie Lucero, Shelbie Fitch, Jennifer King and Spencer King. Great-grandchildren are Max and Lexi Hutchins. *Submitted by Dale and Illene Hutchins.*

YOUNG – In 1876 John Bird Young was one of the first settlers in Evans Township in Kingman County. He came from Illinois and homesteaded 160 acres. His document (or certificate) shows he was owner of his land by Feb. 20, 1880. It is signed by President R.B. Hayes. The farm was named the Shady Grove Farm. He had a large orchard of around five acres, mostly apple trees. He piped water from his windmill to his orchard by way of underground pipes. He also sold nursery stock of fruit trees and shrubs.

Mr. Young had two tanks about 15 feet high outside his home, which could be filled with water and warmed by the sun. Many enjoyable baths were taken in the then-modern bathtub in the house.

Mr. Young was also a postmaster for a few years. The Post office was called Lorette. The pasture on this farm is still virgin land. There are several buffalo wallows in the pasture where the buffaloes rolled.

On March 1, 1879, Mr. Young married a neighbor girl, Catherine Laura "Fanny" Moxley. There were five children born to this family. A girl died in childhood.

In the '20s after the death of J.B. Young and his wife, Frank L. Young, their son bought the land from his sisters and brother. The old house was torn down to make room for a new home. The basement was dug and the foundation made when the Depression came. A neighbor wanted to sell his land so he could go to California. The Young's bought the land instead of building a new home. In the early 1930s farmers were having a very hard time financially. Along with six men in the area, Frank L. young helped form the Kingman Production Association. Each man put in money for a share of stock in the Association. It was organized to help the farmers.

J.B. and Fanny Young in back, with their daughter Pearl Young driving

The Young's milked registered Jersey cows. During the '40s their cattle were rated the best west of the Mississippi. Also Mr. Young was a Master Farmer. Their first tractor was an Avery. The Young's did some threshing. They usually had a crew of around 11 men. There was a cook shack on wheels, which included a stove and tables. There were usually two women hired to do the cooking. The women would sleep in the shack but the hired men usually slept in the barn. Their first car was a model T bought around 1916.

In 1959, F.L. Young, a grandson of J.B. Young and his wife, Nyla (Arents) Young, moved onto the original homestead. They have four living children: Linda, Rhonda, Jason and Bryan. Brenda passed away in 1973 at the age of 11. They had a dairy herd for eight years then went to raising beef cattle. A new home was built in 1966 where the Young's live. It is on the same location where J.B. Young built his home. *Submitted by F.L. and Nyla Young.*

ZERFAS – In 1900 John and Mary Zerfas (my great-grandparents) with their five children immigrated to Ellis County KS from Bukovina Austria. They were seeking a better life for their family away from wars, etc. It has been stated that my Great-Grandmother Mary cried when she saw the bare prairie, not the trees and lush farmland that was so prevalent in Bukovina.

They applied for a homestead on 160 acres, namely the SE 1/4 of 30, 11, 19 in northwest Ellis County, KS. They proved this acreage in 1908 for the cost of $8.00 and also acquired 40 acres of SW 4 of 30, 11, 19 from a neighbor and 160 acres of NW 4 of 31, 11,19 from the Union Pacific Railroad. They built a stone house and other out buildings, also drilled a water well on the quarter of the homestead. Crops raised were wheat, sorghum, cattle and a large garden.

They moved to the town of Ellis in 1918. Several of their children and grandchildren lived in the stone house at various times. John passed away in 1931 and Mary took ownership of the acreage. She had various tenants until her death in 1960. The land was then sold, Ernest Zerfas, an uncle of my father, purchasing the homestead quarter.

My dad bought the Railroad Quarter and the 40 acres so that he could have access to the water. Of this 200 acres only 20 acres were farm ground, the rest is pasture land. The water from the windmill was accessible to the 40 acres. My father also farmed the homestead for his uncle.

My husband, Tom McClelland and I began farming the land in 1980. Upon my father's death in 2002, my husband, Tom McClelland and I inherited the land. The farm ground is enrolled in CRP. The rest of the acreage is in pasture land. The cattle are watered by Rural Water District No 2. We are the fourth owners of this property and plan to keep it in our family for many years to come. *Submitted by Martha L McClelland.*

ZIMMERMAN – In June 1898 Benjamin and Mary Ann Zimmerman purchased 160 acres of land located in Ellis County, Lookout East Township, 36-15-19. They raised wheat, feed and cattle on the farm.

The 160-acre farm then passed to Rudy and Remigia Zimmerman. In 1972, the farm passed to Jacquelyn (Zimmerman) and her husband, Francis.

The land is now farmed by renters and produces wheat and milo.

Jacquelyn and Francis have two children, Simone Werth and Shawn Braun. Their grandchildren are Cameron Werth, Leighton Werth, Nathan Braun and Breanna Braun. *Submitted by Jacquelyn Braun.*

A Roster of Century Farms by Kansas Farm Bureau in Celebration of 100 Years of Continuous Family Farming, 2000-2017

The roster is listed alphabetically under the county where the farm is located, listing the recognized name under which the award was given, the year awarded, and the original year the farm was purchased in italics. If different than farm location, member's county Farm Bureau is listed in parentheses.

Allen County

Frederick & Denise Apt, Jr., 2002, *1875*
Melvin L. Patricia I. Bruenger, 2006, *1889*
Ron & Frances Daniels, 2005, *1883*
Marie Martin Goodspeed, Matthew, Paul & Heather, 2010, *1857*
Leon Harris & Sharon Harris Harvey, 2004, *1867*
HBS Farms, Tracy & Teresa Smith, 2017, *1865*
Sue Hurst, 2009, *1902* (Neosho County)
Lloyd Krone, 2009, *1856*
Patricia J. Marshall, 2000, *1891*
James G. & Marilyn B. Miller, 2002, *1901*
Glenn E. Roberts, 2000, *1878*
Paul & Dorothy Setter and Mike & Pam Setter, 2014, *1912*
Strong Family, 2003, *1874*
Deloris & Dale Wiles, 2006, *1902*
David Wilson, Lucy Thompson, June Toland, Clyde Toland, 2017, *1892*

Anderson County

John Blaufuss, 2002, *1880* (Coffey County)
Rick & Kathy Feuerborn, 2000, *1894*
Gwin Farms-Charles Gwin, H. Floyd Gwin & Shirley Campbell Roeckers, 2006, *1906*
Wayne & LaVerne Howarter, 2014, *1898*
Gail & Debbie Kueser, 2016, *1891*
Mark & Angie Larson, 2013, *1912* (Allen County)
Danny & Diane Martin, 2016, *1876*
Andy McAdam, 2001, *1907*
Gary and Sharon Minckley, 2017, *1892*
Jim Mulcahey, 2007, *1902*
Earl & Roberta Poss, 2004, *1880*
Carl & Iona Sobba, 2001, *1890*
Robert O. & Joan L. (Kulp) Warren, 2002, *1877*
Charles & Rita Wilper, 2003, *1886*

Atchison County

John Junior Armstrong, 2000, *1883*
Jeannette Bishop, 2009, *1886*
Anton and Patricia Brox, 2012, *1908*
Alan Cummings, 2010, *1888*
P.J. and Ramona Cummings, Alan Cummings, 2015, *1901*
Raymond & Helen Erpelding, 2003, *1868*
Jerry & Janis Falk, 2014, *1882*
Harvey & Nancy Jo Fasse, 2002, *1895*
Mildred Flattre, 2003, *1872*
Sam & Karen Flory, 2002, *1878*
Mr. and Mrs. Robert F. Hall, 2011, *1901*
William W. Handke, 2015, *Homestead*
Mr. and Mrs. Jason Harden, 2011, *1893*
William Joseph & Joanna Hines, 2006, *1894*
Milton F. Hines, 2006, *1903*
Donald Hoffman, 2001, *1810*
Sarah K. Jensen, 2007, *1884*
Miriam Kaighin, Emogene Kautz, JoAnn Kautz and Yvonne Ricklefs, 2009, *1881*
William & Dorothy Kloepper Family - Sally Ellerman, Patricia Snyder, Larry Kloepper, Linda Chartier, Emma Warren and William Kloepper, 2013, *1883*
Kreider Farm, 2003, *1901*
Leonard W. & Nancy J. Lange, 2007, *1880*
Dan and Christine McCoy, 2017, *1913*
Melvin & Lois Myers and Kevin & Lisa Myers, 2008, *1904*
Daryl J. and Emma Lou Nieman, 2002, *1878*
Niemann Farms, 2003, *1903*
Albert Nolting, 2001, *1900*
Melvin & Emily Nolting, 2003, *1892*
The Perdue Farm - Joan Olga Perdue Nelson, Helga P. Nelson Huckaby, 2014, *1883*
Kenny & Charlotte Rathert, Jack & Kathy (Rathert) Ryser, Geoff & Vicky (Rathert) Hope, 2015, *1913*
Jeff & Beverly Reichart, 2001, *1864* (Jefferson County)
Jeff & Beverly Reichart, 2001, *1875* (Jefferson County)
Willard & Ruby Royer, 2000, *1876*
Willard & Ruby Royer, 2010, *1909*
Daniel & Lucy Schletzbaum, 2008, *1877*
Ron and Miriam Scholz, 2011, *1882*
Dwight R. Scholz, 2014, *1907*
Troy & Angie Scholz, 2015, *1915*
Dennis B. Schrick & Loretta L. Schrick, 2010, *1907*
Mike & Terry Slattery, 2014, *1860*
Richard & Gerri Taliaferro, 2013, *1890*
Thomas Wagner, 2001, *1895*
Walt & Eileen Wohletz, 2012, *1912*
Mark and Marlene Westermann, 2017, *1917*
Ralph Wright & Sherry Wright Montgomery, 2000, *1884*

Barber County

Dillman Ash, 2010, *1885*
Edna (House) Boyd & Calvin Boyd, 2007, *1907*
Wes Cline, 2005, *1884*
The Davenport Family Trust, 2015, *1900*
L. Wayne Davis, 2003, *1880*
The DeGeer Ranch LLC, 2008, *1886*
Harvey DeMint, 2001, *1888*
Albert & Fern Heublein Trust, 2009, *1908*
Kelly D. Hrencher and Carolyn F. Hrencher, 2009, *1908*
Mary Ann Larkin Jones, Fred Stump, Eddie Stump and Kenneth Larkin, 2013, *1913* (Wyandotte County)
Donald W. Lonker, 2014, *1880*
Steve Lukens, 2014, *1910*
Paul & Sue Marshall, 2009, *1909*
Michael L. Miller, Martha Spicer, 2006, *1884*
Michael L. Miller, Martha Spicer, 2006, *1884* (Comanche County)
Margaret Mills Reiman, 2003, *1890*
Charles & Donna Rowe, 2000, *1892*
Gary & Fayetta Rugg, 2010, *1901*
John & Sandy Smith, 2012, *1904*
Ralph Stout, 2001, *1884*
Thompson Angus Farm, 2003, *1903*

Barton County

Raymond & Catherine Bahr, 2001, *1892*
Richard & Debbie Batchman, 2000, *1887*
William J. & Marcia R. Berchek Jr., 2007, *1905*
Eugene H. & Virginia Bitter, 2005, *1898*
Brian & Melissa Bitter, 2016, *1900*
Robert L. Debes, 2007, *1884*
James J. Debes, 2007, *1884*
Warren Deckert, 2003, *1897*
Elnita Ehler Jezek, 2002, *1897* (Ellsworth County)
Doris Evers West, 2001, *1894*
Reuben & Jean Flagor, 2004, *1900*
Marion & Rose Marie Frieb, 2001, *1884*
Daniel & Teresa Frieb, 2001, *1895*
Louis Hickel, Jon Hickel, Brenda Black, Barbara Gilmore, Janie Manning, Chris Kelly, Sue Luketich and Bernie Hartig, 2008, *1887*
F. Lawrence & Elinor Hiss, 2005, *1899*
Richard & Sharon Hiss, 2005, *1905*
Anthony A. & Caroline A. Hoffman, 2002, *1883*
Eveyln Johanning Hoffman, 2002, *1892*
Gerald L. Lonnon and Marcus & Caraleta (Lonnon) Huslig, 2009, *1884* (Sedgwick County)
Barbara Jordan, 2000, *1889*
Michael F. & Rosalie A. Kinzel, 2003, *1883*
Dale Knop, 2002, *1886*
Helen Koester, Richard Koester, Ken Koester and Duane Koester, 2008, *1907*
Eugene & Mary Jo Maneth, 2000, *1884*
Anetha (Soderstrom) Mauderly, 2006, *1887*
Mary McKinney, 2002, 1893
John H. Miller, 2012, *1912*
Loretta Morgenstern, 2007, *1882*
Morgenstern Family Corp., 2010, *1910*
Karen P. Neuforth & Royce Neuforth, 2000, *1877*
Neuell Nicolet, Kenneth, Celia & Darren Nicolet, 2007, *1885*
Randy Oeser, 2007, *1897*
Gerald Oetken, 2002, *1899* (Pawnee County)
Peter Family - Rollie W. & Sue Ann Peter, Normalee (Peter) Isern, 2012, *1876*
Donald J. & Eudora Reif, Marilyn Hitz, Terry L. & Anita Reif, 2007, *1905*

Mrs. Orville (Katherine) Rusco, 2000, *1870*
Rita Schartz, 2001, *1898* (Washington County)
Mary Lou Schenk, 2009, *1899*
Rosalia Ann Schick, 2010, *1886*
Marvin & Marguerite Schneider, 2001, *1881*
Keith Schneider, 2005, *1902*
Steven Schneider, 2006, *1882*
Leonard D. & Virginia L. Sharp, 2009, *1903*
James & Judi Sharp, 2012, *1903* (Sedgwick County)
Marvin G. & Marilyn Skolaut, 2010, *1910*
Dean & Mary Anne Stoskopf-Stoskopf Farms, Inc., 2006, *1900*
L.R. Stoskopf, Brad & Brian Stoskopf and Matt, Greg, Bryce & Kevin Stoskopf, 2008, *1901*
Ronald Stoss, 2002, *1891*
Alan Ray Stueder, 2011, *1909*
Edward J. Susank & A.L. Chenoweth, 2006, *1891*
Ellen Skolaut Haberman Family, 2012, *1896*
Kay K & Joyce J. Vanek, 2003, 1887
Kay K & Joyce J. Vanek, 2003, *1891*
Margie & Howard Wallace, 2009, *1888*
Larry Weathers, 2001, *1896*
Glenn & Kathryn Yager, 2001, *1889*
Virgil & Diane Zink, 2014, *1878*

Bourbon County

Charles Bruner, 2000, *1867*
Dewaine & Beverly Geiger, 2009, *1873*
Floyd G. & Helen L. Harper, 2004, *1900*
Meaurice Hennessy, John F. Hennessy, John J. Hennessy Sr., John J. Hennessy Jr., 2007, *1874*
Clarence Earl Holeman & Daisy Bonita Holeman, 2014, *1883*
James Joseph King, 2000, *1880*
Joe & Ann Ludlum, 2009, *1909*
Bill & Connie Perry, 2001, *1893* (Allen County)
Warren & Irene Schooley, 2013, *1907*
Rudy Simpson, 2001, *1870*
Doris Williams, Diana Williams Almasi and Brandon Pitts, 2014, *1914*
Steven C. and Marci Williams, 2017, *1911*

Brown County

Jerry C. & Jeanette R. Aller, 2006, *1906*
Harlan & Virginia Bailey, 2001, *1900*
Virginia M. and Harlan L. Bailey, 2011, *1906*
Charlott Loyd Blanton, Virginia Loyd Higginbotham, Dorothy Loyd, 2005, *1895*
Jere D. & Patty S. Bruning, 2002, *1902*
Don A. Cashman, Maurice R. Cashman, Jr., Rachel E. Rademacher, Alison A. Cashman, 2009, *1876*
Mary Chadwell, Lyle Chadwell, Gary Chadwell, Steven Chadwell, John Chadwell and Jeanne Dowell, 2013, *1913*
James Ray & Patti Kay Deaver, 2006, *1880* (Nemaha County)
Delzeit/Geiger Farm, 2003, *1871*
Wayne A. Finger & Vickie K. Finger, Edward M. Finger & Agnes L. Finger, Brenda K. Weih & Justin Weih, and Janice M. Schroeder & Tim Schroeder, 2013, *1901*
James & Gladys Gall, 2000, *1869*
Ethel Geiger, 2003, 1878 (Nemaha County)
Edward Hageman, Sr., 2001, *1884*
Austin N. & Rhoda M. Heise Trust, 2003, *1872*
Glenn & Linda Hennigan, 2001, *1881*
Robert, Rise, Christopher & Rick Hoffman, 2012, *1865* (Doniphan County)
Thomas & Beth Hooper, 2004, *1904*
Jacobsen Farms, 2013, *1913* (Johnson County)
Marcia E. (Plamann) Kilgore, 2009, *1900*
Frank Jr. & Arlene King, 2000, *1860*
Frank Kopp, 2000, *1859* (Jackson County)
Herbert & Rosalie Pape Koppa, Gene & Eilene Koppa, 2003, *1903*
Randall & Karen Korthanke, 2001, *1885*
James J. & Lois E. Larson and Van J. & Dawn R. Lawson, 2008, *1907*
Tom & Joyce Lockwood, 2009, *1892*
John & Gerol Moore, 2009, *1871*
Plamann/Meyer, 2014, *1874*
Roderick Phillips, 2001, *1900*
Roger & Liz Ploeger, 2016, *1889* (Nemaha County)
Diann A. Schuetz, 2008, *1900*
Eugene & Nancy (McCoy) Swearingen, 2002, *1873*
Eugene & Nancy Swearingen, 2002, *1898*
Tom & Judy Torkelson, 2003, *1901*
Raymond E. & Marilyn J. Wehrman, 2003, *1903*
Dale K. & Karla L. Williams, 2003, *1893*
George M. & Marjorie M. Wilson, 2002, *1895*
Mildred L. Winter, Mike Winter, Ed Winter, Alan Winter, 2017, *1916*
Walt & Eileen Wohletz and Bob & Dedra Bunck, 2013, *1859* (Atchison County)

Butler County

Beverly A. (Goodson) Baker, Kristin K. (Goodson) Thiessen, Kara S. (Goodson) Nice and daughters, 2009, *1899* (Sedgwick County)
James & Loriene Brewer, 2004, *1881*
Frank & Elsie Brown, Byron Brown, Raymond E. & Catherine Brown, Mark and David Brown, 2014, *1902*
Richard J. Chase and Faralane Chase, 2008, *1907*
Alvin & Katherine Claassen, 2004, *1891*
Harold & Sandy Cooper, 2001, *1900*
David R. & Betty Corbin, 2000, *1869*
Olen F. & Juanita P. Deichman, 2005, *1894*
Olen F. Deichman and Ronnie K. Deichman, 2008, *1894*
Robert N. Diller, 2000, *1872*
Thomas & Sarah C. Dixon, Charles T. & Mary E. Dixon, Charles R. Dixon, 2015, *1907*
Cecil Faulconer, Vernon Faulconer & Don Faulconer, 2007, *1906*
Lyle & Shirley Gray, 2008, *1898*
Larry & Deborah Gray, 2008, *1898*
Richard & Glenda Harder, 2007, *1891*
Gary & Linda Harms, 2004, *1898* (Harvey County)
Dr. Luman A. Harper, Mercy N. Harper, Flora E. Harper, Susie H. Snodgrass, Elizabeth A. Snodgrass, 2010, *1873*
Robert L. & L. Donnese Harsh, 2008, *1872*
Carl D. Humbolt & Frances M. Humbolt, 2002, *1880* (Sedgwick County)
J. Larry & Judith L. Jackson, 2006, *1906*
Tom & Linda Klaassen, 2005, *1903* (Harvey County)
The Marshall Family, 2015, *1911*
Eldon & Debbie Marshall, 2010, *1899*
Velma Fay Meyersick, 2009, *Homestead*
Louvina M. (Regier) & Herman C. Miller, 2004, *1876* (Harvey County)
James & Opal Nuttle, 2001, *1883*
David Oursler & Frank Ousler, 2002, *1902* (Harvey County)
Elizabeth Parsons, 2016, *1909* (Sedgwick County)
The Chalmer J. Paulson Family, 2004, *1895* (Harvey County)
Patrick K. Regier Family, 2003, *1876*
Virginia M. Sluss, 2000, *1892*
Edwin F & Susie H. Snodgrass, Philip H. & Mary M. Snodgrass, Elizabeth A. Snodgrass, Almira S. Slusser and Glen E. Snodgrass, 2010, *1908*
Mary Millbourn Stackley and Arlan & David Stackley, 2000, *1878*
Paul Stephens, 2002, *1879*
Varner Farms, Inc., 2002, *1900*
Fred H. Weber Jr. & Mrs. Laura J. Weber, 2014, *1914* (Harvey County)
Marilyn Vestring Hein and Victoria (Vicki) Vestring Weldon, 2013, *1900*
Shawn & Kristin Wiebe, 2004, *1882* (Harvey County)
Richard S. & Margaret E. Wiebe, 2005, *1884*
The Benson Farm - Gregory Scott Wiley, 2009, *1878*
Zack G. & Jean H. Wilkerson, 2014, *1912*
Robert & Mary Wilson, 2002, *1898*
Elwood H. & Norma J. Zuercher, 2002, *1892*

Chase County

Glendon E. & Eula M. Barrett, 2003, *1890*
Brenda Bruch, 2000, *1891*
Campbell/McNee Family Farm, LLC; Maurine McNee Campbell, Alex N. Campbell, Jr., John B.T. Campbell III, 2007, *1907*
Elder Family Farms, 2006, *1906*
John Sharp, John Sharp Jr., Lulu Kitty Sharp Galentine, Howard F. Galentine and Brian C. Galentine, 2009, *1860*
Robert Glanville, 2001, *1901*
Donald ImMasche, 2002, *1901*
Charles McCabe, 2001, *1888* (Lyon County)
Patricia Potts, Shawn Potts, Seth Potts, 2011, *1882*
David J. Rettiger, 2009, *1891*
John Rogler, 2001, *1870*
Pat Sauble, 2001, *1856* (Marion County)
Jaqueline L. & Michael L. Schmidt, 2013, *1906* (Lyon County)
Marvin Dale Schwilling, 2000, *1872*
David K. Stout & Donald E. Stout, 2004, *1861*
The Yeager Family, 2016, *1859* (Lyon County)

Chautauqua County

Kenneth & Beverly Aiken, 2001, *1881*
Earl & Norma Arbogast, 2002, *1873*
James W. and Wilma M. Bell, 2017, *1890*
James W. and Wilma M. Bell, 2017, *1915*
Orson D. Phillips (Original Homestead Owner), Doyle & Ruth Call (Trust), 2010, *1889*
Kenneth & Marjorie Cooper, 2014, *1883*
Maxine Dale, Donald Dale Jr., Margaret Dale Hudgins, 2006, *1884* (Cowley County)
Paul Jeanneret, Martha Jacot, Floyd A. Jacot, Paul Jacot, John Jacot, 2006, *1903*
Gary A. Kelley, 2007, *1873*
Mabel P. Kinnaman, Sharilyn K. McAlister, Connie L. Hey, Pat L. McAlister, Gary J. McAlister, Kathy S. Spurlock, 2003, *1902*
Reta Kirchner Pipher, 2015, *1872*
JD Rector and Norma J. Pomeroy, 2000, *1874*
Dale Ramsay, 2001, *1875*
Rhonda J. Reed and Paula M. Sneed, 2010, *1906* (Riley County)
Ronald L. Rogers Trust, 2012, *1891*

Cherokee County

Marion W. Atkinson, 2009, *1884*
Earl & Thay Ross Burns, 2000, *1870*

Evan D. Capron, Virgil J. Capron, Kenneth M. Capron, 2004, *1890*
Edward & Carol Cassell, 2002, *1898*
Doris M. Christiansen & Edna E. Lassman, 2000, *1873*
Adelia Clark, 2006, *1870*
Judith A. (Chase) Clugston, 2002, *1902*
Randy N. and Sibyl Sue Commons, 2005, *1870*
Bonnie (Sartin) Courtney, 2000, *1897*
Brenda A. & Jim C. Dooley, 2009, *1909* (Jewell County)
Dorothy L. Dugger Daniels, 2002, *1902*
Pyne Farms, Inc. - Arlie & Rosalie Pyne, 2014, *1902*
Fred D. Fox, 2000, *1891*
Charles E. Frank, Lafana G. Frank, 2016, *1888*
Merle & Alta Mae Frobish, 2010, *1908*
Malcolm and Linda Goodrich, 2017, *1900*
Robert N. Goodrich, 2005, *1882*
Bob D. Guy and Betty L. Guy, 2009, *1886*
Ray Hale, Tim Hale, 2017, *1916*
Donals & Wilma Hefley, 2000, *1884*
Lloyd Ray & Vera E. Helwig, 2012, *1909*
Mike & Rhonda Hibbard, 2011, *1873*
Gerald Jones, 2001, *1860*
Hazel Carlson Kresyman, 2006, *1906*
Warren C. Lyerla, 2000, *1857*
Warren C. Lyerla, 2004, *1865*
Francis R. Martin, 2006, *1895*
Anna Marie (Leo) Martin, Gregg Martin, Jeff Martin, Gina Peak and Becky Medved, 2006, *1899*
Larry & Carolyn McKee, 2004, *1904*
Clarence C. McNemar, 2000, *1872*
Jim O'Malley, 2004, *1870*
Mary O'Malley, 2014, *1914*
Rosalie Rennie, Kevin Rennie, Kyle & Leslie Rennie, Kregg & Janice Rennie, Ed & Kelli Vanderbeck, and Richard & Kara Parker, 2013, *1881*
Alice M. Roberds, 2000, *1886*
Dorothy Roth, 2000, *1887*
Henry & Saundra Stricklin, 2003, *1902*
Joan Sullivan, Dennis E. Sullivan, Rose M Sullivan, 2011, *1889*
Linda Swaim, 2005, *1880*
John R. & Judy Thompson, 2004, *1904*
Donald C. Watson, 2002, *1902*

Cheyenne County

Royce Cook Farm, 2003, *1898*
Bruce & Sharon Feikert, 2015, *1911*
Hurb & Lola Graves, 2004, *1886*
Alice Isernhagen, Glenn and Louanne Isernhagen, 2007, *1907*
Elton & Ella Mae Keller, 2001, *1887*
The Family of Keith & Jean Loyd, 2009, *1906*
Dean E. Smith, 2000, *1892*
Gaylen Weeden, 2001, *1892*
Robert & Gertrude Zweygardt, 2014, *1899*

Clark County

E. Vance Baldwin, 2000, *1892*
Dean L. Beebe & Carma Lee Beebe, 2009, *1909* (Ford County)
Roberta Cusick, 2002, *1885*
Jim Harden, 2000, *1898*
Willis M. Harden, 2000, *1898*
Willis Harden & Paul Harden II, 2000, *1898*
Ellen Harrington, 2005, *1905*
John & Carla Huck, 2001, *1860* (Wabaunsee County)
John L. Bair, Alice A. (Bair) O'Brien, Linda L. (Bair) Mahieu, Mark & Michelle Mahieu, 2006, *1905*
Bruce C. McKissick, 2005, *1905*
Lynn & Wanda Tedford, 2003, *1903*

Clay County

Clifford & Evelyn Bergmeier, 2004, *1904*
George E. Munson, Robert D. Munson, F. Eugene Munson, and Patricia M. Bowen, 2008, *1880*
Harry & Marjorie Brumbaugh, 2002, *1881* (Shawnee County)
Harry & Marjorie Brumbaugh, 2002, *1890* (Shawnee County)
Harry & Marjorie Brumbaugh, 2002, *1890* (Shawnee County)
Buenta Burger, 2016, *1907*
Ronald & Beverly Carlson, 2001, *1877*
Anne & Bill Clark, Elizabeth Clark, James A. Schaffner, David J. Schaffner, Linda R. Shaw, 2004, *1871*
Harold & Mildred Fehlman, 2004, *1882*
Wanda E. & Wayne Hafner, 2004, *1800s*
Hanson Family Farm, 2015, *1867*
Dolores Heigele, 2001, *1862*
Vernie Heigele, 2001, *1875*
Rodney Hofman, 2001, *1884*
Areata J. Hofmann & Lee A. Dalrymple, 2004, *1884*
Dallas & Alice Iseli, 2002, *1876*
Verl & Myrna Johnson, 2004, *1877*
Verl & Myrna Johnson, 2004, *1882*
Orcena Annabelle Jurey, 2003, *1880* (Washington County)
Warren & Bernice Leidig, 2004, *1900*
Dorothy J. Lilley, 2002, *1878* (Shawnee County)
Dorothy J. Lilley, 2002, *1882* (Shawnee County)
Walter & Evelyn Mugler, 2012, *1912*
William Dale & Anna Mae Newell Trust, 2004, *1869*
Ruth A. & Clifford C. Owens, 2002, *1870*
Lawrence & Eunice Pacey, 2002, *1873* (Ottawa County)
Roy Rau, John W. & Roger Forsythe, 2007, *1900*
Ferdinand Schwab, 2002, *1880*
Christian Steinbach, 2009, *1908*
Mark & Debra Taddiken, 2004, *1869*
Henry Sr. & Lois Tillinghast, 2004, *1874*
Douglas & Patricia Tippin, 2001, *1877*
Marvin & Darlene Yarrow, 2014, *1913*
Vernon & Kathleen Yenni, 2004, *1891*
Alice (Adams) Iseli and Dallas W. Iseli, 2017, *1887*

Cloud County

Norman G. Andersen, 2011, *1897*
Dallas & Anita Breen Family, 2007, *1880* (Ottawa County)
Keith Buss, Steven Buss, 2010, *1909*
Sharon Carlgran, 2000, *1890*
David Carver, 2002, *1872* (Ottawa County)
JoAnn Cool Albán, 2014, *1879*
Lee W. Doyen, Ray A. Doyen, Ross O. Doyen, Mark M. Doyen, 2007, *1877*
Lee W. Doyen, Ray A. Doyen, Ross O. Doyen, Mark M. Doyen, 2007, *1894*
Steven J. & April F. Elstrom, John A. & Bethany Elstrom, Kristi D. & Mike Hubert, and Angela & Mark Savoy, 2013, *1876*
Feight Farms, 2005, *1880*
Dr. & Mrs. Warren S. Freeborn, Jr., 2002, *1893*
Rick & Alice Faye Hanson and Noel & Gloria Hanson, 2012, *1870*
Dean & Dorothy Hanson Family, 2016, *1876*
Gerald Hart, 2001, *1883*
Jarold & Linda Hayden, 2009, *1901*
George Kocher Trust; Helen Kocher & Kurt Kocher, Trustees, 2012, *1910*
Colleen A. Larsen, Kane R. Gray, Kirk R. Gray, Robert E. Gray II, Frances Burdick, 2006, *1882*
Vernon & Sylvia Mai, 2000, *1900* (Washington County)
Catherine L. Newland and Mary F. Gibbens, 2000, *1879*
Stanley & Corda McMillan, 2005, *1888*
John W. & Linda L. Morgan, 2003, *1893*
The Swanson & Gram Families, Richard Swanson for Swanson Family (owners) and Mark & Theresa Gram for Gram Family (renters), 2014, *1879*
Richard L. and Denise A. Swenson, 2017, *1879*
Donald & Donna Tasker, 2007, *1881* (Ottawa County)
John A. Tobald, 2015, *1889*
Jason & Amanda Townsdin, Marquerite Townsdin, 2010. *1889*
Carl & Verna Trost, 2005, *1878*
Max & Mary Ann Widen, 2000, *1874*
Walter Zimmerman, 2004, *1870* (Mitchell County)

Coffey County

The Bazil Family, Butch, Brenda, Jarrod and Megan, 2017, *1894* (Franklin County)
Melvin Bunge, 2000, *1885*
Gale D. Combes and Christine A. Combes, 2013, *1877*
Charlene L. Danford, Lois A. Nelson, and Larry W. Roth, 2008, *1884*
John Decker, 2002, *1902*
Roger L. Douglass, 2000, *1865*
Catherin Yates, William H. Yates, George W. Hess, Laurence N. Hess, Marilyn A. Eccles, 2007, *1884*
Francis A. & Jimmie D. Force, 2000, *1883*
Roger & Sheila Gilbert, 2004, *1902*
Arlie Grimm, Sarah Grimm, Ilene Rogers, 2002, *1877*
Herbert F. Hammond, 2011, *1904*
Dennis Homer Hatch, 2000, *1891*
Larry E. & Linda L. Hermon, 2004, *1884*
David A. & Connie K. Kunkel, 2007, *1907*
Gene Morrow, 2000, *1860*
Charles Nickel, 2002, *1897*
Donna Nickel, 2002, *1902*
Dixie Lee George Potter, 2009, *1874*
James Raaf, 2002, *1889*
Virgil Ryan, 2003, *1883*
William G. Shoup, 2000, *1868*
Levan Skillman Family, 2002, *1902*
Johnnie R. & Retha I. Sleezer, 2002, *1889*
LaVerne Smith, 2001, *1886*
Mary M. Smith & Wilma L. Hess, 2000, *1879*
Philip & Judith Stukey, 2015, *1915*
Newell E. Thomas, 2013, *1903*
Kerry O. Trostle, 2009, *1854*
Richard G. & Carole J. Wilson, 2008, *1895*

Comanche County

Roy & Sarah Beeley, 2008, *1885*
Brown-Todd, LLC - Judith Bowman & Marilyn Dudley, 2010, *1910*
Donald and Elin Bryant, 2011, *1911*
Charles Burt, Jr., 2003, *1903*
The Crossed J Ranch, Inc., 2002, *1885*
Kurtis J. & Adrianne Dale, 2006, *1887*
Mike and Jane Nash Deewall, 2007, *1892*
Larry Ellis, Judith Kerstetter and Carol Thies, 2013, *1909*
Connie Mae (Uhl) & Ralph John Keller & Velma J. (Uhl) Smith Etal, 2005, *1902*
John and Phyllis McNeely, 2017, *1915*
Cleon Parkin, 2004, *1884*
Patricia Sanders, Joyce Veatch, 2010, *1885*
Kenneth & Jeanne Webb, 2009, *1898*

Cowley County

Stan Ahlerich, 2001, *1870*
John Baird, 2001, *1873*

Gene Beard, 2004, *1899*
Sheryl A. Dowler Cary, 2004, *1900*
Conrod Farm, 2008, *1900*
Kent A. Crain & Pamela K. Crain, J. Elmo Crain & Barbara A. Crain, 2016, *1916*
John C. Dicken, Diana K. Dicken, Sharon Hosier, Kaylene Dicken, Pam Schmidt, Shelly Prichard and Susan Hamilton, 2013, *1876*
Donald & Wanda Donley, 2005, *1898*
Frank L. Baxter & Carrie Baxter; John & Vesta Webb; Jimmie L. & Patricia Webb; Mark & Jancie Glover, 2004, *1892*
Connie S. & Robert W. Gottlob, 2006, *1906*
Justin Groom, 2001, *1870*
Duana Hale Guinty & Steven K. Guinty; Children- Chelsea R. Guinty, Velva R. Guinty, Ethan Hale Guinty, Charles Steven Guinty and Chance H. Guinty, 2012, *1872*
Barbara Lowe Hartpence, 2005, *1900*
Arthur G. Hoyt, 2004, *1880*
Connie Hoyt, Rex & Shirley Hoyt, 2014, *1896*
Dale & Illene Hutchins, 2004, *1874*
Lewis & Charlotte Johnson, 2014, *1871*
Pauline Jones, 2001, *1876*
Lee Kelly, 2005, *1884*
Royal & Naomi Kennedy, 2001, *1886* (Chautauqua County)
Royal P. Kennedy, 2014, *1873* (Chautauqua County)
Royal P. & Naomi Kennedy, 2014, *1886* (Chautauqua County)
George Liermann, 2002, *1902*
Shirley Lindly, 2002, *1901*
Bertha Williams Maddox, 2004, *1872*
Bill K & Patricia Marker, 2005, *1887*
Jack R. and Catherine Meyer, 2008, *1907*
Maureta J. Meyer, 2010, *1907*
Philip & Alene Miller, 2007, *1906*
Morton Meadows LLC (Mike H. Morton & Sharon L. Gordon), 2010, *1909*
William K. Mueller, 2002, *1898*
Argenia P. O'Neil, 2004, *1904*
Helen Phillips, *2005*
William A. (Bill) and Neewannah Ramsey, 2004, *1875*
William A. (Bill) and Neewannah Ramsey, 2004, *1897*
David & Martin Ray, 2002, *1897*
Patrick & Larisa Sarntee, 2001, *1901*
Patrick N. Sarntee, Larisa V. Sarntee, 2005, *1901*
James C. Seal, 2004, *1904*
Eugene G. & Marjorie M. Sellers, 2004, *1877*
Neil Shaffer & Nancy Taylor, 2005, *1891*
John Warren & Gloria L. Sickles, 2005, *1898*
Robert Tatum, 2001, *1896*
William C. and Lou E. Tharp, 2017, *1903*
Dorothea E. (Nixon) Thompson, 2003, *1897*
Dorothea E. (Nixon) Thompson, 2003, *1901*
Allene Vaughters, 2001, *1875*
James A. Wheeler, 2014, *1905*
Larry E. Womacks and Mark E. Womacks, 2007, *1907*
David W. Wood, 2005, *1870*

Crawford County

Richard W. Bennett, 2008, *1897*
Raymond L. Bezinque, 2012, *1908*
Janet & Steve Burt, 2001, *1881*
Randy and Marilyn Chiartano, 2017, *1917*
Maybelle Circle, 2010, *1881* (Marshall County)
Dennis & Barbara Clutter and their children Pam, Brad and Janet, 2013, *1910*
Floey Collier, 2001, *1900*
Dean & Mila Dixon, 2000, *1883*
Galen A. Gaston, 2000, *1881* (Cherokee County)
LaVerna Nadine Gemmell, 2000, *1883*
Keith E. & Lisa A. Harris, 2009, *1892*
Joe & Frieda Hofer, 2004, *1880*
The Hofer Family Jim, Loretta, & Children LaDonna, Mary, Charlotte, David, Christopher, Bernadette, Marta, Michael, Rita, & Father Andrew, O.P., 2004, *1880*
Ray Hofer, 2004, *1880*
Wayne & Kay Huning, 2000, *1881*
Esther Jagels, 2002, *1886*
Leonard E. Lucke, 2000, *1898*
Barbara McClaskey, George McClaskey, Kay Martinous, 2015, *1893*
James L. Mechling, Nancy A. Mechling, 2009, *1883* (Labette County)
Gordon & John Michael, 2000, *1892*
Richard A. & Ramona R. Murphy, 2015, *1910*
Steve & Martha Murphy, 2000, *1889*
Warren Murphy & Marjorie Smith Murphy, 2000, *1872*
O'Brien Cattle Co., Inc., 2002, *1868*
Margaret R. Phillips, 2000, *1881*
Daniel C. & Alice C. Renn, 2015, *1883*
Daniel C. & Alice C. Renn, 2015, *1878*
Gary & Colleen Renn, 2015, *1910*
L.D. Herlocker and Joseph Marsh, John Herlocker, Glen Herlocker, Madrienne Herlocker and Kevan Schupbach, 2010, *1880*
Orval Smith, Alice Mussa & Lela Grant, 2000, *1875* (Cherokee County)
John & Donna Tersinar, 2001, *1875*
Clarence Tiedemann, 2000, *1897*
Rick & Jane Weimer, 2013, *1911*
Steve & Ruth Zimmerman, 2009, *1988*

Decatur County

Lora & Lee Arnold, 2002, *1887*
Elmoin & Marguerite Bailey, 2002, *1899*
Darrel and Alice Barnett, 2005, *1892* (Norton County)
John E. Bremer, 2003, *1891*
Doyle & Kay Brown; Kent & Robin Brown, Jordan & McKEnzie; Gordon & Kara Kuntz, Katherine & Zachery; Kurt Brown, Millicent & Dashiell; Phillip & Keli Kleidosty & Syndey, 2012, *1885*
Doyle B & Kay (Tacha) Brown, Kent & Robin Brown, Gordon & Kara Kuntz, Kurt Brown, Phillip & Keli Kleidosty, 2015, *1882*
Elder & Lois Carlson, 2002, *1895*
Dennis Euhus, 2001, *1889*
June Fortin, Jason Fortin, 2004, *1904*
Edith E. & Andrew L. Foster, Foster Family Trust, 2009, *1880*
John & Ann Gallentine, 2001, *1893*
Gene and Kristi Gallentine, 2011, *1897*
Demmer Farms - Connie, Dan and Greg Grafel, 2014, *1887*
Dean & Peggy Groneweg, 2006, *1902*
Ivis Hanson, 2006, *1888*
Floyd & Ruth Laidig, Glenva Nichols, Janice Mehl, 2005, *1893*
Kathleen Marcuson, Ronald Marcuson, David Waldo, Daniel Waldo, Barbara Eskew Bradley and Elaine Marcuson, 2017, *1916*
Walter & Betty May, 2000, *1886*
Arthur and Alice May, 2007, *1907*
Robert and Anita Montgomery, 2007, *1894*
Loyd and Fern Diehl Moore, 2007, *1894*
Gary Orr, 2001, *1892*
Louise Pollnow, 2000, *1889*
Pansy Price, 2001, *1879*
Virgil & Rowena Cathcart, Lynn & Joyce Rippe, 2007, *1906*
Erlene Schroer, 2003, *1900*
Ralph & Voilet Shaw, 2002, *1900*
Roger L. & Dorothy M. Shimek, 2002, *1899*
Glen & Carolyn Strevey, Bob & Kay Strevey, 2005, *1904*
Gary Torluemke, 2006, *1887*
Evan J. & Sheryl S. Unger, 2002, *1900*
Donald Wasson, 2003, *1878*
Jim & Pamela Wesch, 2014, *1900*
Eugene W. Wurm, 2005, *1904*
Don & Beverly Stacey Zumbrunn, 2005, *1904*

Dickinson County

H. Robert Altwegg, 2010, *1893* (Ottawa County)
Francis & Clema Anderson, 2001, *1866*
Royal Barten, 2006, *1897*
Raymond & Janet Bielefeld, 2014, 1881
Bistline Estate: LaVerna Ernst, Gloria Bistline, Robert Bistline, Carolyn Beetch and Lois Hines, 2004, *1890*
Phyllis E. Britt, 2007, *1894* (Geary County)
Wayne E. Cook and Dorothy Cook, 2002, *1871*
Lowell Davidson, 2002, *1868*
Monte & Donneta Felbush, 2003, *1901*
Karl D & Yasu Gfeller, 2002, *1896* (Atchison County)
Kevin & Rosemary Harris, 2001, *1883*
Hilton & Patricia Helbach, Mindy Helbach Foote, Michael Adam Helbach and Margo Helbach Tinney, 2010, *1910*
Lester Hill, 2003, *1876*
Gene & Cynthia Hoffman, 2007, *1901*
V. Dean and Janice L. Holdsworth, 2017, *1917*
Kandt Family, 2013, *1866*
George Kandt, 2009, *1863* (Riley County)
Brian & Joyce Lady, 2006, *1906*
John & Doris Larson, 2002, *1881*
Paul & Dorothy Lehman, 2003, *1902*
Robert & Leola Linder and Tom & Debra Linder, 2008, *1906*
Maurice Lorson, 2004, *1904*
Vincent & Laurel McBoyle, 2003, *1883*
John Jr. & Peggy Mueli, 2002, *1899*
Leon J. Mills, 2010, *1874*
Pray Family, 2001, *1872* (Shawnee County)
Dean & Marilyn Roles, 2002, *1881*
Melvin & Stella Rufener, 2003, *1872*
Deb and Tim Sanders, 2017, *1911*
Armin & Evelyn Scripter, 2002, *1882*
Ron, Beth & Jon Scripter, 2012, *1902*
Shelly A. Sexton, Gregg E. Sexton, 2017, *1886*
William J. Shippy, Charlotte A. Barten, Kevin D. Shippy, Darrell A. Shippy, Brian W. Shippy, 2009, *1879*
Jack D. & Roberta M. Sleichter, 2000, *1883*
Mary Evelyn Rakestraw Stillwagon, 2008, *1908*
Dean Underhill, 2004, *1871*
Eldon & Velma Wuthnow, 2004, *1902*
Evelyn Herr York, 2003, *1880*
Don & Beverly Stacey Zumbrunn, 2005, *1870*

Doniphan County

Ronald & Linda Benitz, 2010, *1887*
Beryl E. (Smith) Blevens, Kenneth Nolan Blevins, Brandon Geiger, 2013, *1908*
William W. & Julia C. Franken; William A. & Ruth J. Fuhrman; Al, Jeff & Chuck Fuhrman and Eric Fuhrman, 2013, *1913*
Bettie J. Herring Fritch, 2009, *1887*
William R. Gordon, Susan (Gordon) Holguin and Mary Jo (Gordon) Abbot, 2009, *1886*
David G. Halling, 2012, *1860*

Milton Hartman Jr. & Patricia Hartman, 2014, *1911*
Michael J. Holt, 2000, *1864*
Raymond Hoverson, Nellie Goforth, Deanna Wood and Wilma Nurski, 2013, *1869*
Adam & Mildred Huss, 2001, *1897*
Joy Elizabeth Jeffers, Trustee Gene E. Jeffers Trust, 2004, *1872* (Brown County)
Steven S. Jones, Betty S. Gleason, Mark D. Jones, Sharon A. Shouse, Andrew Clements, Cindy L. Hoverson, and Matthew B. Clements, 2013, *1854*
Franklin C. Kinsey, 2000, *1860*
Kenneth & Mary McCauley, 2014, *1900* (Brown County)
Charles & Jean Nimtz, 2000, *1887*
Anna Nimtz, 2002, *1902*
Luetta Petersen and Brenda Petersen, 2009, *1857*
Ricklefs Family, 2009, *1877*
James & Mary Ruddy, 2001, *1875*
Becky Swinford, Charles Swinford, Brian Swinford and Curtis Swinford, 2013, *1857*
Barbara E. Symns, 2013, *1881*
Dave Taylor, 2000, *1878*

Douglas County
Mrs. William (Marcella) Anderson, 2000, *1871*
Charlene & Robert Bierly, 2005, *1879* (Shawnee County)
Churchbaugh Family Farm, 2009, *1885*
Fred & Margaret Clark, 2000, *1887*
Nora Cleland, 2002, *1877*
Geraldine E. Coffman, 2000, *1879*
Margaret Counts, 2009, *1899*
Richard Eckman, 2001, *1890*
Charles C. Fawl, 2000, *1860*
Adrian & Debbie Fishburn, 2002, *1896*
David & Laurie Shuck and Mike Forth, 2010, *1905*
Chip & Jan Hornberger, 2004, *1872*
Marvin & Norma Jardon, 2002, *1887*
Harold C. Jehle, 2000, *1898*
Evelyn Griffith Johnston & Paula Johnston Aiken, 2000, *1855*
Jim & Frank Johnson, 2014, *Homestead*
Norman & Ralph Leary, 2000, *1869*
William R. Lothholz, 2000, *1868*
Dr. George Badsky Maichel, D.V.M. and Jacqualine Lea Boeka, 2009, *1900* (Osage County)
William Meairs Jr., 2000, *1854*
Family of Phillip Metsker, 2004, *1860*
Gail and Ginny Neis, 2011, *1909* (Johnson County)
Roy & Martha Parker, 2001, *1868*
Charlene Pohl, 2000, *1884*
Heffner Farm owned by Pratt Family, 2013, *1898*
Don & Jane Schwartz and Greg & Kelly Schwartz, 2008, *1856*
William H. and Catharine Dude Schaake, Edward C. and Magdalena Altenbernd Schaake, John H. and Irene Schaake Vogel, Nancy S. Vogel, Jerry M. Vogel, 2007, *1800* (Leavenworth County)
Wells Family Trust, Alice Wells Lassman, Opal Wells Hesselbarth, Wilma Wells Johnson, Eleanor Wells Curtis, and Annette Wells Gaston, 2016, *1866*
Allen K. & Charlene F. Winter, 2000, *1879*

Edwards County
The Elmer Boehme Family, 2009, *1905*
Robert Burcher, Jr., 2001, *1876*
Butler Family Trust, 2010, *1909*
Richard and Judy Dockendorf, Andy and Shawna Dockendorf, 2017, *1903* (Kiowa County)
Eugene Fisher, 2001, *1901*
Marcella Wortmann Heit, 2010, *1888*
William Herrmann, Albert Herrmann, Ronald J. Herrmann and David S. Herrmann, 2013, *1908*
Bret & Debbie Lancaster, 2006, *1903*
Duane Lancaster Trust, 2013, *1884*
Earl & Fern McBride, 2007, *1885*
Jefferis & Zelma Mead, 2003, *1903*
John & Gina Mundhenke, 2008, *1908*
Ploger Farms - Mark Ploger, 2011, *1910* (Pratt County)
Doris Schultz, 2001, *1893*
Rean & Jerri Stegman, 2010, *1906*
Paul and Mary Janice Strong, Paul Robert and Beth, 2007, *1907*
Jane Michaelis Wenstrom and Richard James Wenstrom, 2013, *1906*

Elk County
Wilfred L. Bauder, 2000, *1896*
Dorine Benest, 2001, *1880*
Janet S. Black, 2005, *1905* (Pratt County)
Raymond & Phyllis Divine, 2000, *1900*
James F. Hogan and Kenneth A. Hogan, 2013, *1882* (Montgomery County)
Dennis & Harriet Horner, 2014, *1884* (Montgomery County)
Kenneth & June Kramer, Kristine Kramer, Joni Kramer Steffen and Thomas Kramer, 2014, *1912*
Paul & Izola Moore, 2001, *1870*
Pat Moore, Paula Pearson and Doug Moore, 2013, *Homestead*
Lynn L. Perkins, 2000, *1898*
Maurice & Jackie Perkins, 2013, *1871*
Don & Shelby Perkins, 2013, *1908*
Pat Preston, 2000, *1885*
Orlin & Aaron Russell, 2005, *1896* (Wilson County)
Owen Neal & Wilma Smith, Randy & Sandy Smith and Jacob & Kim Smith, 2010, *1910* (Wilson County)
James Wiseman II, 2001, *1873*

Ellis County
Paul J. & Lucy Baier, 2000, *1897*
Francis & Jacquelyn Braun, 2002, *1898*
D. Ralph & Margaret Chrisler, 2002, *1894*
Roy & Helen Davis, 2005, *1883*
Robert L. Dickinson, R. Kirk Dickinson, 2002, *1890*
Elmer Dortland, 2001, *1878* (Russell County)
Jim & Opal Flinn, 2002, *1893*
Leo N. Groff Family Trust, 2002, *1887*
Harold E. Hall, Steven W. Hall, Donna M. Frenzl, 2017, *1901*
Betty L. Holt, 2000, *1897*
Betty L. Holt, 2000, *1898*
Arnold & Hanora Homburg, 2001, *1894*
Edna May Homburg, 2001, *1901*
Daniel H. Johnson 2000, *1897*
Imelda M. Karlin/Norman J. Karlin, 2002, *1888*
Donald Kraus, 2001, *1895*
John & Esther Kraus, 2001, *1880*
Warren C. & Rosemary Kraus, 2002, *1890*
Paul M. Leiker, 2017, *1883*
Thomas M. & Martha L. McClelland, 2003, *1903*
Gilbert Meis, 2002, *1901*
Alexander & Anna Pfannenstiel, C.A. & Philomena Pfannenstiel, Cliff J. Pfannenstiel, 2015, *1915*
Leona Wasinger Pfeifer, 2001, *1894*
Emery Schmidt, 2002, *1885*
Eugene J. Schmidt, 2000, *1897*
Virginia Schmidt, Norma J Braun, Kathleen Schmeidler, and Mary Jane Schmeidler, 2004, *1901* (Rooks County)
Gary & Roxanne Solomon, Gordon & Kelly Solomon, Roger Solomon, 2007, *1891*
Leo & Betty Walter, 2002, *1897*
William & Mary Walz and The Walz Family Trust, 2013. *1910*

Ellsworth County
Dale & Diane Barta, 2002, *1897*
Calvin Boyd & Carla (Schepmann) Boyd, 2005, *1884* (Barber County)
Steve Dlabal Family, 2002, *1902*
Haase Farm, 2002, *1886*
Raymond Lee Hanken, 2010, *1896* (Barton County)
Dianne M. Hemry, 2007, *1906* (Rice County)
Stuart Paul Janssen, 2006, *1879*
Thelbert W. Janzen, 2007, *1907*
Emma L. Doherty, Evelyn E. Parker, Erlene K. Fox, Ernest D. Jezek Jr., 2007, *1887*
LeRoy & Janice Huseman, 2010, *1904* (Sumner County)
Richard E. & Grace A. Johnson, 2000, *1892*
Arthur & Kathryn Kohls, 2002, *1891*
John Kueser Family - John & Katie Kueser, Nancy Thornburgh, Terry & Nancy Jane Kueser, 2016, *1899*
Ernest J. & Darlene A. Mattas, 2002, *1890*
Marlene & Ron Montgomery, 2001, *1897*
Joel & Janet Nelson, 2014, *1910*
Ronald Piester & Brenda (Schepmann) Piester, 2005, *1887* (Pratt County)
Jack & Delores Rathbun, 2002, *1887*
Jack Rathbun, 2008, *1887*
Louis Rolfs, 2002, *1892*
Randle C. Rolfs Trust & Vada R. Rolfs Trust, 2002, *1890*
Ronald E. Schepmann, 2002, *1884*
Donald F. Schepmann, 2005, *1889*
Schultz Farm, 2002, *1893*
Joseph & Libbie Sebesta, 2001, *1885*
Mark Sebesta and Susan Kurz Sebesta, 2017, *1913*
Siemsen Family Farm, 2005, *1895*
Marilyn Soukup, 2002, *1878*
Melvin Vern Splitter, 2002, *1877*
Joseph Truhlar, 2002, *1882*
Adolph J. Vopat, 2000, *1876*
Laura Wamhoff, 2001, *1880* (Russell County)
Triple J Partnership - Ona L. Wright, Joe L. Wright, Jon L. Wright, 2009, *1900*

Finney County
Pearl L. Greathouse, 2005, *1886*
S. Fielding Hands, 2015, *1913*
Aloysius Ohmes, 2017, *1917*
Walter & Esther Rundell, 2015, *1910*
Elsie Stone, 2001, *1895*

Ford County
Carma Lee Beebe & Dean Beebe, Charles Lawrence Deniston, 2009, *1907*
Timothy & Lynda Burghart, 2013, *1904* (Edwards County)
Scott & Marilyn Butler, 2012, *1910* (Edwards County)
Larry & Kathy Feikert, 2000, *1884*
Walter Goff, 2002, *1890*
Jerry Hager, 2006, *1902*
Jerry A. Hager, 2009, *1905*
Richard Hamilton, 2010, *1891* (Gray County)
Floris Jean Snyder Hampton, Marvin Hampton, 2003, *1891*
William Herrmann, Paul Herrmann and Austin A. Herrmann, 2013, *1904* (Edwards County)
Charles V. Imel, 2000, *1887*
Charles V. Imel, 2000, *1900*
Charles V. & Yuvonne Imel, 2006, *1902*
John Indiek, 2009, *1905*
Robert Issinghoff, 2001, *1885*
Mabel L. Keller, 2007, *1905*
George and Jean Leaming, 2017, *1917* (Clark County)
Philip & Barbara Lutz, 2008, *1907*
Kirk Molitor, 2010, *Homestead*
Morris Family Trust - Ford Farm, 2016, *1905* (Ness County)
Ira Paulin, 2001, *1899*
Earl E. Plattner, Sharene Oldham and Connie Evans, 2008, *1905* (Kiowa County)

Grant & Bonnie Powers, 2001, *1892*
David E. & Maryalys Robb, 2007, *1905*
Dorothy Roesener, 2007, *1905*
Todd Edward Scheib, 2003, *1888*
Todd Edward Scheib, 2003, *1897*
John & Bernadine Stegman, 2001, *1900*
T.H. & Anna Stein, Thomas & Marcella Stein, and Locke M. Stein, 2005, *1905*
Richard & Eva Torline, 2001, *1901*
Zelma I. VanHorn, 2000, *1885* (Franklin County)
Louise VanNahmen, 2014, *1903*
Willis B. Warner, 2000, *1884*
Jack W. Warner, 2000, *1898*
John B. Weber, 2015, *1914*
Bill & Darlene Woolbright, 2001, *1899* (Shawnee County)

Franklin County

Katherine E. Alexander, 2000, *1860*
Edward & Terry Altic, 2003, *1898*
Gerald & Mary Anderson, 2006, *1894*
Harley and Janet Scott Broers, 2007, *1855*
Dean and Alexis Chambers, 2002, *1893*
Charles Conley, 2002, *1865* (Linn County)
Wendell & Claire Davis, 2006, *1905*
Charles Albert Dunbar Family, 2008, *1908*
James Conard & Nancy Roofe Dunn, Roy Conard & Deborah Hall Dunn, 2004, *1858*
Robin Dunn, 2007, *1907*
Robin Dunn, 2011, *1904*
Roy S. Edwards, 2000, *1876* (Anderson County)
Robert & Roberta Eichenberger, 2000, *1882*
Jim Eichenberger & Carl Eichenberger Trust, 2004, *1904*
Merle & Lois Fleming, 2001, *1883*
John Adam and Cressie B. Fisher, 2015, *1912*
Robert D. Herring, 2000, *1893*
Edwin and Mary Horstick, 2008, *1908*
Tom & Linda Horstick, 2010, *1910* (Linn County)
Charles Gary & Terri Lou Hower, 2000, *1864*
Clarence Keith, 2001, *1865*
Donna Kempin, 2001, *1875* (Johnson County)
Douglas & Patsy McCrea Kershner, 2004, *1857* (Johnson County)
George Lickteig, Hubert Lickteig, Leroy Lickteig, 2016, *1914*
Patricia L. Martin Gretencord, 2005, *1901* (Johnson County)
Raymond W. Mason & Jeanne Dagenette Mason, 2005, *1890*
Gene & Pat Maxwell, 2003, *1866*
Rex Montague, 2002, *1901*
Valeta N. Parmely, 2009, *1866* (Coffey County)
Valeta N. Parmely, 2009, *1878* (Coffey County)
Harry & Juanita Peckham, 2014, *1914*
Lawrence W. Reh & Kathleen E. Corbin, 2000, *1882*
Mary Shelton, 2001, *1898*
Rex & Rosalee Simmons, 2000, *1895*
Frank & Mary Lois Yates; Brian & Susan Yates, 2012, *1910*

Geary County

Ray & Sharon Allaman, 2007, *1902*
Stanley & Dorothy Ascher, 2001, *1883*
Harold J. Boller, Peggy A. Boller, 2006, *1866*
Daniel P. and Dorothy Boller, 2008, *1907*
Dorothy Simmons Boyd, 2005, 1886 (Morris County)
Richard & Alice Cross, 2006, *1860*
Harvey B. & Patricia H. Deatherage, 2014, *1866*
Harris & Vada Ramsour, 2003, *1884*
Judith Russ, Alona Hedstrom, 2003, *1897* (Rooks County)

Gove County

Abell Family, 2014, *1907*
Philip & Paula Beesley, 2007, *1902*
Mildred E. Bentley, 2006, *1884*
Bill & Kerry Briggs, Bart & Amy Briggs and Joss & Megan Briggs, 2011, *1909*
John Feldt, Anton & Anna Feldt, Anthony & Mary Ann Feldt, 2007, *1898*
Lyle & Janice Glick, 2001, *1889*
Keith & Naomi Hockersmith, 2011, *1907*
Shirley and Bill Jones, 2007, *1885*
Thomas & Barbara Kinderknecht, 2013, *1906*
Terry L. Mendenhall and Patricia R. Mendenhall, 2013, *1906*
Irwin & Betty Ann Porter, 2006, *1886*
Roy Thornburg, 2001, *1897* (Ness County)

Graham County

Everett A. Anderson, 2005, *1880*
Tim & Brenda Benoit, 2011, *1908* (Rooks County)
Betty Jean Elliott, 2007, *1891*
Laurel T. Goddard, 2005, *1884*
Kendal & Barbara Grecian, 2006, *1892*
Connie Griffith, John Griffith & Nancy Bollig, 2000, *1879*
George & Lois Gustafson, 2001, *1888*
Dwayne & Rose Hardman, 2008, *1887*
Joe, Paul & Wayne Heskett, 2005, *1904* (Logan County)
Hobbs Ranch, 2002, *1887* (Trego County)
Delmer & Evelyn Hoeting, 2002, *1890* (Mitchell County)
Everette & Matilda Keith, 2003, *1820*
Ben Keith Farm, 2007, *Homestead*
Paul, Christine, Kenneth Kohart, 2004, *1903*
Robert Morin, 2004, *1888*
Kathryn A. Davis Mollhagen, 2006, *1901* (Ellsworth County)
Ron & Betty Nickelson, Don & Kim Nickelson, Steve & Saundra Nickelson, 2006, *1890* (Trego County)
Ron & Betty Nickelson, Don & Kim Nickelson, Steve & Saundra Nickelson, 2006, *1885* (Trego County)
Mark & Janet Niehaus, 2011, *1879*
Don & Rea Paxson, 2002, *1899*
Janet Shaw, 2012, *1903* (Sheridan County)
Jacob Spies, August Spies, August Spies Jr. and William Spies, 2013, *1903*
Carol Trexler, 2004, *1903*
Floyd & Elizabeth VanLoenen, 2005, *1905*
Roy VanLoenen, Helen Miller, 2009, *Homestead*
Charles E. Webb & Etta M. Webb, Charles E. Webb Jr. & Elizabeth C. Webb, 2006, *1906*
Jean (Moore) White, 2003, *1898*
Allen Worcester, 2005, *1880*
Joe Zohner, 2000, *Homestead*

Grant County

Lindley Cox, Madeline Hegi, Julie Cox, Joanne Black, John Cox, Cheryl McLaughlin, Beverly Long, Terry Cox, Larry Cox, 2004, *1887* (Anderson County)
Lindley Cox, Madeline Hegi, Julie Cox, Joanne Black, John Cox, Cheryl McLaughlin, Beverly Long, Terry Cox, Larry Cox, 2004, *1887* (Anderson County)
Keith & Kathy Goff, 2013, *1895*
Larry & Virginia Kepley, 2013, *1910*
Helen Leigh, 2001, *1901*
Robert D. Moore, LaVon R. Wright, Lynette S. Metcalf, Oralea L. Harn and Linda C. Leggette, 2014, *1913*
Donna M. E. Walker, 2007, *1907* (Stanton County)

Gray County

Love & Love Farms, 2013, *1910*
Wiley McFarland, Sara McFarland and Greta McFarland Huebert, 2014, *1911*
Steve and Margo Sayre, 2011, *1890* (Ford County)
Riley Winkler, 2010, *1910*

Greeley County

Lyle J. Dixon, 2011, *1910* (Riley County)
Ella & Lyle Griffin, Julia Griffin Fisher, Rita Griffin Shumate, Ben Fisher, 2015, *1915*
Max D. & Thelma Liggett, 2008, 1904, Kiowa, Greensburg (Check – think this is Kiowa County)
Walter Ernest Rauch, 2006, *1896*
Steve Schmidt, 2001, *1900*

Greenwood County

Kenneth & Evelyn Baker, 2004, *1871*
Gary D. Barg, 2003, *1892*
Halverson - Breech Farm, 2003, *1862*
William Roberts Browning, Patricia Browning Berner, Susan Browning Pogany, 2007, *1906*
Jack & Virginia Connell, 2006, *1905*
James & Dorothy Cox, 2005, *1885*
Leta A. Divine & James W. Gray (Gray-Divine Farms), 2004, *1877*
Larry J. Fankhauser & Mary M. Fankhauser, 2016, *1891*
Jimmie D. and Frances A. Force, 2002, 1895 (Coffey County)
Heather and Cory Fuesz, 2011, *1899*
Gayla Garner, 2014, *1913*
Charles & Cathy Gowen, Richard & Cheryl Gowen, 2005, *1902* (Lyon County)
Billy D. Greenwood, 2004, *1884*
John Grundy & Norman Grundy, 2004, *1902*
Charles Willis Hall, 2003, *1871*
Charles Willis Hall, 2003, *1871*
Wallace & Anita Hall, 2004, *1881*
Vernon & Sally Hawthorne, 2005, *1905*
Vernon & Sally Hawthorne, 2005, *1989*
Larry & Cindy Hibbard, 2009, *1870*
Larry P. Hibbard, Mary Alice McNitt and Gordon V. Hibbard, 2009, *1885*
Donald J. & Dixie R. Huntington and Donelda R. Perkins, 2009, *1901*
Irlene and Clinton Huntington, 2011, *1884*
Pamela Huntington Cowley, 2001, *1869*
F. Duane Knapp, 2005, *1884*
Murray K. Knox, 2011, *1909*
Daryl & Glenn Lewis, 2004, *1877*
Maxine Pritchard & Marilyn Miller, 2004, *1904* (Allen County)
Thedda Olson, Olson Family Farm, 2007, *1907*
Bonnie Ott & Gerald Waddell, 2004, *1903*
The Tom Perrier Family, 2003, *1868*
Ron & Nichelle Rader, 2001, *1896*
Lloyd & Billie Rice, 2001, *1895*
William "Bill" Frank & Hazel L. Russell & sons; Steven Frank Russell; Allen Kent Russell and Bryan Lee Russell, 2011, *1911*
Dorothy Sanderson, 2001, *1884*
Harold & Martha Sauder, Sharon Sauder Barnhard, Derek & Kelly Barnhard and Staci Barnhard Elliott, 2010, *1909* (Montgomery County)
Bruce A. & Julie M. Smith, 2000, *1890*
Celia L. Smith, 2004, *1869*
Keith & Rita Stapleton, 2016, *1914*
Marshall & Trudy Stauffer, 2010, *1908*
Eldon & Lynnie Teter, 2011, *1881*

Sand Road Farms, LL- Alleen S. Van-Bebber, Richard L. Diklich, Kitty G. Shinkle, and Neva A. Diklich, 2014, *1871*
Wallace-Born Farm LLC, 2014, *1872*
William, Don, Jim, Bert, Dwight Wiggins, 2004, *1885*
Harlan C. Wingrave, 2004, *1879*

Hamilton County

Elsie Buck, 2002, *1892*
Grady Cook & Rick Cook, 2006, *1906* (Stanton County)
Oliver C. Hines, Stanley E. Hines, 2004, *1882*
Clarence & June Nickerson, 2010, *1907*
Alma Lou Plunkett, Harmon J Plunkett, Harlon R Plunkett, Wesley M Plunkett, Whiney L Plunkett, Lucas T Plunkett, Jeron Plunkett, Devan Plunkett, Richard C Plunkett, Aaron Plunkett, Austin Plunkett, 2009, *1906*
Mary Frances Wright, 2002, *1891*
Mary Frances Wright, 2002, *1900*

Harper County

Bruce Bell & Sandra Lee Ball, 2007, *1900*
Titus - Beisel, 2004, *1884*
Wilma Burgess, Margaret Wright, Max Yocam, Gary Yocam & Madeline Banta, 2007, *1884*
Lee Burgess, Jessie Smith, Brenda Foil, Carolyn Lynch, Mike Gammill, Tim Gammill, 2007, *1882*
Harold W. Carr, 2005, *1901*
Milford F. Coady, 2000, *1892*
Karen & Neil Coffman, 2001, *1882*
Frederic & James Coleman, 2000, *1882*
Innis Croft, 2006, *1902*
Irene Duncan, 2000, *1892*
IrmaLee Duncan, 2000, *1892*
Elsie Jane Gardner, 2004, *1881*
R. Douglas Gill, 2000, *1885*
Donald L. Gill, 2006, *1870*
Eugene F. Grabs, 2004, *1904*
Marion Hamilton, 2005, *1885*
Leonard Lee (Larry) and Gloria Bradshaw, Ronald and Rebecca Bradshaw Holt, 2016, *1916*
Brian & Tammy Hoopes, 2004, *1898*
Ralph & Doris Hunt, 2007, *1902*
Karen Sue Ioerger-Francis, 2003, *1884*
Karen Sue Ioerger-Francis, 2003, *1899*
Frances Laird Johnson, 2006, *1906*
Arthur & Georgia Joyner, 2002, *1887*
Howard L. & Jeanne Kastens, 2004, *1883*
Anna Marie Kiser, 2010, *1888*
Carol V. Drouhard-Long and W. C. Long, Jr., 2005, *1890* (Sumner County)
Glenn E. Mandevill, 2000, *1886*
Lloyd E. Mathes Trust, 2004, *1890*
Andrew O. McIntire, 2000, *1892*
Richard & Lois McKee, 2000, *1883*
Rogene McKnight, 2001, *1888*
Mrs. George Miller, 2002, *1902*
Cleo & Dorothy Pearl, 2001, *1882*
James N. & Patricia A. Robb, 2008, *1899*
Harold D. Ryan, Linda Ryan Lesher & Jennifer Schoenhofer, 2007, *1907*
Pauline Salser, 2000, *1899*
John Schnackenberg, Jr., 2002, *1882*
Puterbaugh/Senff Farms, 2016, *1901*
Kathleen Shaffer, 2001, *1883*
Robert Neil & Joyce M. Snodderly, 2006, *1905*
Don Vornauf, 2001, *1899*
Merle & Mary Ellen Walker, 2002, *1884*
Bob Warner, 2009, *1907*
Herbert L. & Carole Wilcox, 2006, *1906*
Charles S. Wood, 2001, *1882*

Harvey County

Milton C. Boyle, 2000, *1886*
James and Geneva Chapple, 2005, *1885*
Wesley and Amy Chapple, 2007, *1885*
R. Dwight & Connie J. Claassen, 2000, *1877*
Irvin Dey, 2001, *1896*
Anabel & James Downey, 2001, *1880*
Willard & Edna Entz-Larry & Carol Entz, 2004, *1901*
James & Freida Everley, 2000, *1891*
Elda Gatz, 2001, *1879*
Larry & Margaret Goering, 2001, *1897*
LuJean Graybill, Jerry M. Short, Allen L. Short, Philip D. Short, 2009, *1902*
LuJean Graybill, Jerry M. Short, Allen L. Short, Philip D. Short, 2009, *1872*
Kent D. & Myrla Cole Haury, 2004, *1874*
Donald & Harriet Hiebert, 2001, *1893*
Terry & Dana Knott, 2004, *1900*
Charles A. Koerner, 2000, *1898*
James R. Lohrenz, 2000, *1884*
Don Joe McGinn, 2000, *1898*
Stella Miller, 2001, *1880*
Melvin O'Keefe, 2000, *1881*
Shaun, Kaitlen, Kalyn & Megan Ortman, 2015, *1915*
Carol Penner, Roberta Paulsen and Naomi Andreos, 2017, *1916*
Wilbur Prouty, 2001, *1882*
Kenneth Regier, 2015, *1889*
Rita Schlagowsky, 2000, *1899*
Darleen Schrag Bartell, 2001, *1891*
Howard & Eldred Schroeder, 2008, *1908*
Judith & Dana Shifflett, 2015, *1911*
Clarence Shirk, 2000, *1891*
Robert Dean Suderman, Jr., 2011, *1877*
George R. TenEyck, 2000, *1885*
Thomas Veazey, 2001, *1882*
Phillip Eugene Wendling, 2000, *1883*
Pat & Bonnie Wendling and Norma Wendling, 2013, *1911*
Neal & Sheila Williams and Robert & LaVeta Williams, 2000, *1900*
Shirley Wilson Sheffler, Arlene Wilson Smith & Marjorie Wilson, 2000, *1882*

Haskell County

Garetson Family Farm, 2002, *1902*
Russell Kells, Darcy Kells, Nathan Kells and Vance Kells, 2008, *1893*
Paul Weidner, 2000, *1885*

Hodgeman County

Paul Bradshaw, 2000, *1869*
Ethel Dresie, 2002, *1900*
Douglas & Barbara J. Ewy, 2006, *1884*
Arylene Jackson, Roland Jackson and Marcia Dian Hall, 2009, *1878* (Ness County)
Kenneth F. Regnier, 2000, *1873*
Ron & Losis Sebes, 2005, *1899*
Virginia Shortridge, 2000, *1891*
James R., Sr. & Mary Joann Sorem, 2000, *1886*
Andy Stairrett, 2002, *1896*
Virginia Stramel-Strong, 2015, *1910*

Jackson County

Kathryn Arnold-Weckman, 2005, *1864*
Richard A. Bone, Robert Bone, Ronald Bone, Nancy Bone Johnson, 2004, *1903*
Lester & Dorothy Randel Bottom, 2007, *1892*
John Braum, 2001, *1891*
Eugene Brockelman, 2000, *1893*
Fred S. & Nancy S. Coombs, 2000, *1858*
Joe L. & Mary E. Doyle, 2000, *1869*
Vera & Everett Frazier, 2001, *1866*
Darin & Elizabeth Holaday, 2008, *1868*
Ideker Farm, 2014, *1904*
Morris Jenkins Family Farm, 2003, *1895*
Joseph L. Kennedy, 2007, *1907*
Michael C. Kennedy and Patrick C. Kennedy, 2009, *1907*
Al & Sarah LeDoux, 2013, *1913*
James Lewelling Family Farm, 2003, *1898*
James McAdam, 2001, *1901*
Tom and Delores McAsey, 2009, *1887*
Dean, Larry, Ross & Roy Montgomery, 2000, *1896* (Atchison County)
William V. & Eula F. Moulden & Linda Totten, 2006, *1881*
Ross M. Nelson, 2016, *1905*
Glenn & Doris Oxandale, 2005, *1905*
George W. & Sarah J. Phillips, 2008, *1881*
Brad Porter, 2004, *1904*
David C. and Lou Ann Porter, 2011, *1911*
Diane K. & RW Ruppert, 2007, *1903*
Rodney & Beth Stadel, 2002, *1878*
Rex E. Stapel, 2004, *1903* (Shawnee County)
Kevin & Laura Stauffer, 2016, *1880*
Marvin D. & Mary L. Stous, 2000, *1860*
Stous Family, 2007, *1882*
William & Phyllis Zibell & Gary & Becky Zibell, 2006, *1895*
Mark Zlatnik, Linda Zlatnik and J. Laverne Zlatnik, 2010, *1910*

Jefferson County

Roger E Barnes, 2015, *1857* (Shawnee County)
Daryle and Donna J. Bates, 2010, *1908*
Jermiah Croley, Wm F. Croley, Anna Cornelison, Jerry Cornelison, 2015, *1867*
Danny L. & Anne C. Dailey, 2000, *1883*
Thelma Farris, Thomas E. Farris Sr., Ronald Grant Farris and Lonnie Dean Farris, 2009, *1868*
John & Connie Gering Family, 2015, *1901* (Atchison County)
Charles Hamon, 2001, *1900*
Charles E. & Mary Jo Hamon, 2006, *1906*
David J. Hemme, 2000, *1868*
James D. Lobb, 2003, *1854*
Don & Aldiene Myers, 2001, *1884*
Don & Aldiene Myers, 2001, *1893*
Scott & Lori Navinskey, 2004, *1856* (Atchison County)
Daniel J. & Kay F. Noll, 2004, *1891*
Everett Nottingham, 2000, *1872*
Margaret Rathert, 2001, *1864*
Jeff & Beverly Reichart, 2001, *1898*
John & Zelma Sanders, 2005, *1881*
Harold J. Schrick, 2013, *1905*
Dr. Wilfred & Gail Schuler, 2015, *1905* (Atchison County)
Clarence & Lucy Seaver, 2001, *1883*
Mary Jane Tucker, 2003, *1854*
Paul & Doris Schneider Weishaar, 2013, *1903*
Beverly (Zinn) White, 2011, *1904* (Jackson County)

Jewell County

Newell E. Alexander & Sons, 2000, *1897*
Robert & Betty Collins, 2010, *1868* (Mitchell County)
L. Joan Dean, 2001, *1876*
Walter Dillon, 2002, *1900*
Jerry Durham, 2002, *1871* (Mitchell County)
Marilyn Endsley & family, 2000, *1885* (Mitchell County)
Sandra Jean Gillett Franco, 2002, *1897*
Samuel D. Gillett, 2000, *1883*
Larry Gillett, 2002, *1902*
Norman Greene, 2002, *1886*
Fern Griffitts, 2001, *1880*
Ferne Harris, 2002, *1869*
Lynda Harris Schreck, Richard Harris, Thomas Harris, 2017, *1878*

Gerald Rea Hart, 2000, *1880*
Margaret Hayman, 2002, *1883*
Marcia Gillett Litzenberger, 2003, *1887*
Donald W. McElroy, Beulah M. McElroy, 2017, *1915*
Daisibel McIntyre, 2002, *1900*
Don A. and Mary E. Modlin, 2002, *1883*
Ila M. Nelson, 2000, *1898*
Ila M. Nelson, 2000, *1898*
Jim & Sharon Nelson, 2000, *1878* (Mitchell County)
Roger & Barbara Oplinger, 2005, *1890*
Peckham & Brooks - Charlotte Peckham Walters, 2002, *1872*
Philip C. & Stephen J. Pennington, 2005, *1871* (Leavenworth County)
Donna Porter, 2004, *1904 (*Mitchell County)
George Edward Shute, 2000, *1886*
Elwyn Topliff, 2001, *1880* (Mitchell County)
Eugene & Irene Wood, 2003, *1882*
James A. and Sherryl L. Zoltenko, 2017, *1917*

Johnson County

E.N. DeVault, 2002, *1893*
Byron & Maribeth Finley, 2000, *1864*
Donald Wayne Gay, Gerald Edward Gay, Bernard Lee Gay and Jacelyn Dee Gay, 2013, *1882*
James L. & Bernice E. Harmon, 2003, *1882*
David B. Hayden, 2008, *1856*
Richard & Linda Knabe, 2000, *1900*
Roberta Knabe, 2008, *1894*
P. Franklin & Barbara J. Lane, 2015, *1907*
Edward N. & Margaret J. McCarthy, 2004, *1864*
E. Neal & Joan McCarthy, 2014, *1864*
Lorena Meyers, 2001, *1871*
Gus & Kate Moll, Richard & Opal Jensen, Kenneth Garnett Moll, 2015, *1901*
Charles Pretz, 2009, *1887*
Joe & Margaret Schwartz, 2001, *1899* (Miami County)
Mick & Linda Seuferling, 2012, *1893*

Kearny County

Donald & Rhonda Litton, 2016, *1916*
Maddux Farms, 2006, *1906*
Vernon M. & Pauline Neff, 2004, *1886* (Grant County)
David M. & Anna Summers, Henry D. & Alice R. Summers and Donald E. Summers, 2009, *1907*

Kingman County

Ralph & Lucille Albers, 2004, *1903*
Stanley E. & Betty I. Benson, 2009, *1907*
Peter & Loretta Clouse, 2007, *1907*
Robert L. & Geraldine L. Cole, 2000, *1882*
Theda Davis, 2003, *1884*
Ellis/Converse Family, 2015, *1914*
Roland & Mary Jane Elpers, 2015, *1908*
David & Karen Fieser, 2003, *1878*
Bernard & Mary Gillen, Mike & Janene Gillen, 2005, *1905*
Paul & Dorothy Handkins, 2005, *1898*
Marcelline Henning, 2008, *1902*
Loretta M. (Swingle) Hogg & Donald E. Hogg, 2006, *1898* (Sedgwick County)
Duane Hufford, 2003, *1885* (Harper County)
Robert G. & Donna L. Keller, 2002, *1900* (Reno County)
John & Cynthia Kuszmaul, 2004, *1900*
Paul M. & Carole M. Meng, 2007, *1887*
Martin Dale & Elizabeth Ann Sheldon, 2008, *1900*
Scott & Karlene Sparks, 2003, *1899*
Arland Stephens, 2000, *1882*
Martha E. Stuart, 2006, *1882* (Sedgwick County)
Giles & Pat Thissen, 2006, *1902*
Charles Werner, Dennis Werner, Vera Werner, 2017, *1905*
F.L. & Nyla Young, 2001, *1880*
Mary Jo, Steve, Susan, Chris, and David Ziegler, 2014, *1914*

Kiowa County

Clifton D. & Betty Jo Clark, 2010, *1901*
Arnold Fast, John Fast, Jean E. Clark, 2005, *1893*
Ki & Kimberly Gamble, 2009, *1909*
John & Janice Haney, 2009, *1901*
Keith E. Hinshaw, 2008, *1907*
The Hubert Farm, Jim & Joyce Keith, 2009, *1895*
Allen & Juanita Schmidt; Rex & Janie Schmidt, 2011, *1911*
Schwarm Ranch, 2009, *1899*
Mark A. Schwarm, DVM and Velda J. Schwarm, 2008, *1897* (Reno County)
Lawrence & Pauline Schwarm, 2013, *1913*
Paul W. & Shirley Unruh, 2015, *1908*
Sloan Family, 2013, *1903* (Riley County)

Labette County

Howard & Lois Baker, 2000, *1882*
Charles & Diana Buckley, 2000, *1898*
Merrill Christy, 2000, *1895*
Merrill W. and Norma Lee Christy, 2017, *1904*
Norma Lee Wiggins Christy, 2000, *1900*
Raymond & Mary Cook, 2001, *1869*
Robert & Barbara Evans, 2000, *1867* (Cherokee County)
Betty & Elmer Gittins, 2000, *1899*
Arthur Graves, 2002, *1869*
Arthur Graves, 2002, *1871*
Charles & Shirley Greenwood, 2001, *1873*
Vera Maybell Hanigan, 2012, *1910*
Terry & Velma Hayden, 2006, *1904*
Gary & Luella Hucke, 2000, *1899*
William & Connie Journot, 2001, *1895*
C.L. & Mary E. Landrith, Clifford I. & Arlene Landrith Family, Mike, Melanie, Brenda, Nancy, & Lynette, Mike & Brooka Landrith, 2015, *1914*
Stan & Marty Lips and Fred Lips, 2010, *1910*
Kenneth & Mary Jean Manners, 2001, *1880*
W. Leon Maxson, Jean Maxson Paul, Carole Maxson Brown, 2004, *1898*
Jeanne Allen McKinley, 2001, *1883*
Ronald Wayne Neidigh, 2009, *1909*
Andy & Regina Nelson, 2000, *1869*
Darrell & Deborah Nelson, 2001, *1883*
Mike Nixon, 2001, *1869*
Thomas O'Connor, 2002, *1884*
Karen (Quathamer) & Thaddeus W. Osgood, 2014, *1914*
Leon W. & Geneva F. Pearce, 2005, *1894*
Mark & Charlene Piper, 2000, *1868*
Opal Rigg, 2001, *1887*
Earl Stegmeir, 2001, *1870* (Montgomery County)
Jim, Grant, Clyde, Earl N. Stigenwalt and Johanna B. (Miller) Stigenwalt, 2013, *1867*
Joseph & Charlotte Stottmann, 2003, *1872*
Cora Edna Thompson, 2000, *1900*
Albert J. & Dora L. Thompson, 2011, *1911*
Jonathan M. Tower, 2004, *1869*
Ed & Maureen Wassom, 2003, *1874*

Lane County

Paul Prose, Amy Prose Rhodes & Audrey Prose Barron, 2000, *1886*
Judith Collins, 2001, *1901*
George Cramer, 2000, *1892*
Don & Tracey Hagans, 2000, *1895*
Walter Owen & Patrycia Ann Herndon, 2010, *1905*
Don & Betsy Hineman, 2000, *1898*
Edwin McLeish, 2008, *1885*
Elgin McWhirter, 2000, *1885*
Ethel Mudd, 2000, *1879*
Jim and Dora (Ross) Nelson, 2002, *1885* (Trego County)
Aaron & Diana Paris and Robert & Delores Paris, 2013, *1912*
Janis & John Reed, 2001, *1900*
Bessie West, 2001, *1883*
Lee R. Zink, 2000, *1888*

Leavenworth County

Robert Clark Aaron and Cecilia Joanne Clark, 2010, *1882*
Robert & Karla Beying and Virginia Mowery, 2010, *1906*
Charles E. Christ Sr. & Carol A. Christ, 2015, *1890*
Walter Lee & Nila Denholm & Families, 2005, *1885*
Edward E. & Theresa L. Eberth, 2000, *1882*
Floyd A. Eberth, 2011, *1909* (Wyandotte County)
Jimmy R. & Edna I. Elder, 2000, *1874*
Donald Ernzen, 2014, *1913*
Ronald H. & Alvina K. Ewert, 2010, *1888*
James Heintzelman, Tom Heintzelman, Angela Hund-Nelms, 2015, *1880s*
Bonnie J. Hobbs, 2013, *1885*
Robb Family Farms: Nancy Hudson, Connie Gummelt, Eva Prince and Kathy Kurpeikis, 2009, *1902*
Irene M. Kiester, 2006, *1881* (Johnson County)
Robert E. Klinkenberg and Charlotte L. Klinkenberg, 2009, *1870*
Anthony & Sherry Kramer and Eugene & Eileen Kramer, 2000, *1898*
Edwin & Elizabeth Lamborn and family, 2003, *1877*
Helen E. Lohman, 2015, *1872*
Leonard M. Meyer, 2000, *1884*
Daniel & Blanche & Richard Murphy, 2004, *1900*
Timothy Dale Ney, 2003, *1874*
Michael & Angela Oelschlaeger, 2006, *1869*
Philip C. & Stephen J. Pennington, 2005, *1902*
Brian and Renee Potter, 2007, *1907*
Herman J. & Kimberly J. Sachse, 2010, *1900*
Rainbow Hill Farm -Sloan Family, 2002, *1856*
Douglas Tayrien, Tyler Tayrien and Paige Comer, 2009, *1862*
William E. Torneden, 2003, *1878* (Douglas County)
Elaine & Randy Watson, 2001, *1869*

Lincoln County

Marylyn Kroenlein Bell, 2000, *1882*
Roland & Bernice Bolte, 2007, *1907*
John A. Buttenhoff, 2013, *1886*
Frances Clark, 2001, *1901*
Crawford Farms, Inc. - Vera M Crawford & Royce K. & Marilyn J. Crawford, 2005, *1894*
Mary Jane (Meili) Day, 2000, *1899*
Anna Mae Feldkamp, 2014, *1887*
Carlene Friesen, Carol Lingenfelter and C. Kay Gabelmann, 2000, *1898*
James F. Gabelmann & Sylvia S. Goss, 2002, *1900*
James F. Gabelmann & Sylvia S. Goss, 2004, *1904*
Pauline Hartzell, 2002, *1873*
Louise D. Heiser, 2005, *1903*
Loren E. and Kay A. Heller, 2002, *1884*
Larry & Janet Hiitter, 2016, *1892*
Dorothy Hogan Hollenbeck and Janet Hogan Marshall, 2014, *1904* (Johnson County)
Roy Huhl, 2013, *1884*
Louis & Tracy Hynes, 2010, *1910*

Beatrice Kroenlein, 2000, *1889*
Daniel Metz, 2002, *1898*
Forrest Meyer, 2001, *1868*
Daniel Merle Millburn, 2014, *1906*
Ronald & Jodi Nelson, 2001, *1879*
Martha Alice Patterson Nelson, 2002, *1882* (Mitchell County)
Richard Plinsky, 2005, *1905*
Melvin & Luella Reinert and Bill & Joan Detmer, 2009, *1898*
Glenn P. Ringler Jr & Sharon L. Ringler, 2010, *1871*
Glenn P. Ringler Jr & Sharon L. Ringler, 2010, *1875*
John & Shelley Thaemert, 2010, *1910*
Hulda Ulrich, 2004, *1901* (Mitchell County)
Bob Vanek, 2001, *1891*
Cecil L. and Ardith I. VonFange, 2002, *1887*
Cecil L. and Ardith I. VonFange, 2002, *1898*
Gene and Adis Wacker, 2002, *1894*
Max D. and Tammy Wallace, 2002, *1892*
LaVern "Doc" Weber, 2012, *1912* (Mitchell County)
Lyle G. Weber, 2009, *1898*
John and Jody Wiebke, 2007, *1889*
John R. and Margaret Wilbeck, 2011, *1907*
Evergreen Farm - Mary E. Greene Wilson & Thomas W. Wilson, 2002, *1879*
William (Bill) & Barbara Wolford, 2004, *1886*

Linn County

Earl & Dale Avery, 2003, *1902*
Clayton E. & Barbara A. Baird, 2004, *1904*
Tom & Judy (West) Brownback, 2009, *1863*
Larry Cox & Terry Cox, 2000, *1858*
Donnie L. & Connie Fultz, 2009, *1894*
Reta N. & Kenneth M. Kennedy, 2002, *1868*
Carson Leasure, 2002, *Homestead*
Dorothy Markley & Robert E. Markley, 2002, *1900*
Glenn McGee, 2002, *1902*
Kenneth D. McGee & Kathryn N. McGee, 2005, *1892*
Charles Edwin McGrew & Ruth E. McGrew, 2006, *1866*
Richard C. Morrell, 2015, *1883*
Donald Stainbrook, 2001, *1893*
Donald Stainbrook, 2001, *1899*
Stites Farms, Inc., 2003, *1891*
Jeff Stroup, 2002, *1890*
Stroup Farms, 2002, *1891*
Robert & Iva Teagarden, 2006, *1894*
Joe & Vicki Teagarden, 2007, *1881*
Tim and Caryn Tyson, 2017, *1871*
Hilma Ungeheuer, 2000, *1890*
Mary Delores Warta, 2011, *1878*
Robert, Wanda, James & Charles Weickert, 2002, *1893*
Robert West, 2001, *1869*

Logan County

Velda Bollinger, 2001, *1885* (Scott County)
Glennis & Lyle Boyd, 2002, *1892*
Helen Jordan, 2003, *1892*
Loren & Reta McDaniel, 2001, *1885* (Scott County)
Morris C. and Dolores Talbert, 2004, *1895* (Rooks County)
John Fred & Eva Margurette Wassemiller and Fred J. & Martha L. Wassemiller, 2014, *1910*

Lyon County

Ron Bathurst, 2015, *1887*
Ailene Johnson Bodquin, Barbara Johnson McAntee, 2003, *1883* (Douglas County)
Winifred Boline, 2003, *1899*
Boyce Farms - Michael & Connie Boyce, 2015, *1915*
Sylvester & Sarah Brown, 2004, *1900*
Don & MaryLou DeLong, 2014, *1913*
Thomas & Ruth Fowler, 2007, *1900*
Iola Hall, 2003, *1883*
Gene & Jackie Harder, 2001, *1885*
Page Nicklin, David Hill, 2007, *1871*
Willis & Helen Huston, 2008, *1878*
Marvin & Sylvia Klumpe, 2014, *1914*
Gerald K. Laws, 2000, *1900* (Coffey County)
Roberta Lewis & Jim Lewis, 2006, *1882*
Ross & Betty Ludy, 2006, *1864*
Earl W. Lyddane, 2009, *1883*
Duane N. Ogleby, 2012, *1872*
Osborn Family Ltd. Partnership, 2003, *1883*
Lavern Schallehn, 2005, *1896*
Vera Kathleen Schesser, 2008, *1878* (Sedgwick County)
Ivan & Dorene Smith, 2012, *1874* (Coffey County)
William E. Smith & Linda Kay Smith, 2011, *1855*
Jan Spade Family, 2017, *1891*
John & Mary Beth Watts, 2000, *1893*
Donna Williams, 2002, *1870*
Carl & Naomi Wingert, 2006, *1906*

Marion County

Royden Albrecht, 2016, *1916*
Fred & Carol Bichet, 2001, *1858* (Chase County)
Scott and Sherry David, 2017, *1904*
Norman F. & Rosella M. Epp, 2003, *1894*
Warren D. and Paula Fike, 2007, *1907*
Edla Freeland, 2003, *1902*
Frank J. Heath, Jr., 2000, *1879*
Janzen Family Farms - John Janzen, David Janzen, Sara Janzen Regier, 2009, *1898 (Harvey & Butler Counties)*
Dennis P. Krause, Velma Krause, 2002, *1901*
Carrol Logan, 2000, *1878*
Murphy Farms, 2002, *1902* (Dickinson County)
Naomi E. Nelson Estate - David Nelson & Thomas Nelson, 2015, *1914*
Herbert D. & Wilma Regier, Arnold W. & Edith Regier, 2002, *1885*
Wm. P. Robinson, 2005, *1904* (Chase County)
Robert and Leah Sellers, 2008, *1878*
Robert and Leah Sellers, 2008, *1886*
Rex L. Siebert, 2006, *1896*
Gary & Carole Spohn, 2002, *1887*
Ronald Suderman, 2001, *1880*
Greg and Dina Vogel, 2017, *1903*
Randal Vogel, 2003, *1893*
Lawrence Winkley, William L. Winkley, Ruby McElfresh and Jerry Winkley, 2013, *1887*
Lawrence Winkley, William L. Winkley, Ruby McElfresh and Jerry Winkley, 2013, *1909*
Helmut and Carol Wituk, 2010, *1883*
Evan & Rebecca Yoder, 2004, *1904*

Marshall County

Billie L. & Betty Jo Abitz, 2007, *1901* (Pottawatomie County)
Howard & Louise Anderson, 2002, *1901*
Martha & Kenneth Caffrey, 2002, *1901*
Gerald & Irene Donahy Trust, 2009, *1884*
Chester Jr. & Mary Ann Fincham, 2002, *1901*
Gwynn & Marilyn Hazen, 2009, *1864* (Nemaha County)
Lyle & Ann Howell, 2009, *1908*
Roger & Konnie Jones, 2002, *1872*
Gary & Sharon Kabriel, 2010, *1902*
Eunice Larson, 2014, *1880*
Brice E. & Bruce M. Leonard, 2000, *1900*
Esther, Theodore & Charles Musil, 2000, *1900*
Warner Pape Jr., 2010, *1876* (Brown County)
Arlyn & Charlotte Peterson, 2005, *1900*
Dan Petr, 2010, *1907*
Jean Roepke and Dwayne Roepke, 2010, *1881*
Dan and Kim Schmidt, 2017, *1907*
Brandon M. Vering, 2010, *1910*
Norman & Loren Volle, 2000, *1876*
Kenneth & Dolores Wanklyn, 2001, *1860*
James A. & Marjorie C. Wilson, 2006, *1900*
Samuel S. & Patti J. Wilson, 2009, *1901*
Lee & Shirley Wohler, 2001, *1874*

McPherson County

Ann & Dean Becker; Michael & Sarah Becker, 2011, *1893*
Stanley & Jenise Ediger, Delbert & Lillian Ediger, 2005, *1898*
Randy and Janie Ediger, 2013, *1874* (Harvey County)
K. Duane Embers, 2000, *1898*
Dale & Genevieve Frownfelter, Gary & Amber Frownfelter, 2007, *1900*
Gary, Amber & Karson Frownfelter, 2007, *1906* (Rice County)
The Dale S. & Bertha Galle Family: Nelson & Marilyn Galle, Dorothy & Ralph Claassen, Edith & Larry Dahlsten and June & Perry Krehbiel, 2008, *1894*
Ernie & Florene Goering, Ernest & Vera Lynn Goering, 2003, *1900*
Euncie Graalman, 2000, *1886*
C.F. Norstrom and Gustaf Helberg, 2012, *1868*
Howell Family Trust, 2011, *1902*
Duane & Rosemary Kasey, 2002, *1883*
Kaufman Dairy Farm, 2009, *1903*
Brad Kaufman, Cindy Kaufman Bartell & Allen Kaufman, 2006, *1902* (Harvey County)
Gene Knackstedt, 2002, *1900*
Craig & Jackie Larson, 2010, *1880*
Orvin Lundquist, 2003, *1869*
Don & Virginia Mitchell, 2003, *1878*
Don & Virginia Mitchell, 2003, *1883*
Don & Virginia Mitchell, 2003, *1883*
Kermit M. Nikkel, Christopher Nikkel, Suzanne Meier, 2009, *1882*
John E. & Evi B. Otte, 2002, *1876*
John E. & Evi B. Otte, 2002, *1878*
Linn & Mary Peterson, 2001, *1900*
John H. & Judith L.T. Reimer, 2014, *1902*
Vincent & Christine (Johnson) Rocco, 2002, *1902*
David A. & Charlene S. Rupp, 2000, *1878*
Pearl Sandahl, 2001, *1891*
Jonell & Jerald Schenk, 2009, *1901* (Mitchell County)
Mr. & Mrs. Fred Schlender, 2000, *1898* (Harvey County)
Norman Schroeder, 2009, *1890*
Emma Creek Farms - Jim & Bruce Schmidt, 2000, *1876*
Russell & Phyllis Swanson, 2001, *1882*
Richard M. & LaVera R. Vogts, 2010, *1901*
Ella Vogts Trust, 2010, *1908*
Verda & Daton Wedel, 2003, *1892*
Abraham Woelk, Heinrich A. Woelk, Eldon & Melba Woelk, Janice L. (Woelk) Yartz, Eileen (Woelk) Stoddard, 2009, *1893*

Meade County

Victor L. Bohling, 2010, *1907* (Cloud County)
William Bolan, Jr., 2001, *1892*
John, Richard & Steven Fleming, 2011, *1911*
Leila M. & Donald H. George; Delbert & Weymouth Pope; James K. & Lula Pope, 2007, *1907*
Ed & Pat Hissom, 2005, *1902*
Carl Dean Holmes and Marcellene Whinery, 2013, *1912* (Seward County)
Joe Mertens, 2014, *1913*
William J. Painter, 2000, *1899*
Russell Vail, Joyce Knott, Vail Fruechting, Ann Metzger and Julee Schmidt, 2013, *1913*

Gordon & Eldeana Winfrey, 2007, *1907*
Darrell Lee Winfrey, Vera Mae Winfrey, 2007, *1907*

Miami County

Geo W. & Doris J. Allen, 2003, *1903*
Charles W., Jr. & Rose Mary Atwood, 2000, *1878*
Vernon L. & Carol E. Bartlett, 2000, *1877*
Steve & Ann Boehm, Mildred Beets, 2007, *1854*
Helen M. Kennedy Carter, 2000, *1879*
Roy Dakin & Cecilia Kueser Dakin, 2000, *1860*
Violet Davidson, 2001, *1891*
Raynard & Helen Endecott, 2001, *1885*
Ardith A. & Howard E. Frerichs, 2004, *1883*
Mark & Dorothy Gardner, 2001, *1885*
Loree P. Alpert Gardner, 2002, *1895*
Gerald & Susan Hall and Alvin & Rosalie Hall, 2000, *1857*
Robert W. Harper Sr., 2015, *1907*
Robert J. & Wilma M. Hays, 2000, *1897*
Aisla Higgins, 2001, *1871*
Billy Hurt, 2001, *1881*
Kenneth E. Kaiser and Merle R. Kaiser, 2014, *1908*
Erma McCaskey, 2002, *1901*
Vernon Edward and Donna Lou Cole Prothe, 2009, *1886*
Daryl & Marie Rohrer, 2004, *1858*
Cecil & Marlene Shannon, 2001, *1901*
Larry & Janis Tenny, 2003, *1860*
Sara J. Maddox Walter, 2003, *1896*
Warren Family, 2002, *1880*
Karen S. (Butel) & Derrick C. Weigel Family- Jake, Alix, Nick & Joe and Annetta E. (Wolken) & Anthony J. Butel, 2014, *1910* (Shawnee County)
David W. & Barbara S. Wilson, 2008, *1879*

Mitchell County

Donald A. & Patricia A. Abercrombie, 2010, *1908*
Eugene E. & Kris Baird, 2000, *1886*
DeWayne, Gloria & Jed Black, 2004, *1900*
Esther Walter Boehner, 2006, *1906*
Rex & Eva Borgen, 2002, *1870*
Joan Pinkell, Carol Moffatt and Leo Brown, 2000, *1891*
Lila M. Bryan, 2006, *1880*
Clifford Campbell, 2000, *1884*
Robert D. Cather, 2004, *1885*
Jack & Linda Clover, 2000, *1877*
Edward A. & Alberta A. Cordel, 2004, *1885*
Ruth Tice Courter, 2002, *1897*
Donald & Barbara Cox, 2002, *1883*
Rosemary Deneke, 2006, *1906*
Robert E. Dubbert, 2000, *1881*
Robert & Mary Gradig Dubbert, 2013, *1904*
Sam Eilert, 2001, *1893*
Bill & Evelyn File, 2004, *1898*
JoAnn Hansen and Bill & Kris Hansen, 2010, *1904*
Charles & Josephine Hartman, 2001, *1872*
Thelma Hauptli-Hendricks, 2001, *1893*
Harold Heller, 2001, *1880*
Andrew John Heller, 2007, *1892*
R.D. (Rex) Hodler, 2002, *1891*
Keith & Debra Houghton, 2002, *1872*
George Jordan, 2002, *1887*
Michael D. & Joyce E. Jordan, 2012, *1880*
Michael & Beverly Kadel, 2009, *1905*
Martha Koster, 2001, *1877* (Osborne County)
Marjorie J. Ludwig, 2006, *1885*
Helen Mann, 2000, *1879*
Daryl & Charles W. McConnelee, 2002, *1874*
Charles D. & Vickie Mears, Jennifer Mears Standley & Alex Mears, 2006, *1884*
Mehl Homestead, 2002, *1891*
Elva Tice Michal, 2002, *1897*
Raymond Mick, 2000, *1884*
M. Louise Nelson, 2002, *1879*
Gregory L. Nicholson, 2006, *1896*
U B Farms LLC - W.K. Oliver, Donald R. Oliver, 2002, *1883*
Joyce Pertl & Nadine Lee, 2005, *1873* (Russell County)
John A. Prewett, 2013, *1908* (Lincoln County)
R. H. Quakenbush Trust, 2002, *1862*
Darrell Remus, 2001, *1899*
Schmidt Family Farm, 2002, *1878*
Harold Seely, 2001, *1897*
Joe C. Simmons, 2004, *1885*
Alan & Julie Snyder, 2000, *1872*
Mike & Shirley Stewart, 2002, *1885*
Smyth Farms LLC- Karen Siskosky and Vaughn Studer, 2009, *1877*
Vernon & Ann Thiessen, 2001, *1894*
Marilyn Thomas, 2009, *1908*
Richard Williams, 2001, *1880*

Montgomery County

Anita Etter Atherton, 2000, *1894*
Florence Babcock, 2010, *1885*
Florence Blaes, Mike Blaes, Wayne Blaes & Carol Gerstenkorn, 2011, *1910*
John & Joyce Coons, 2001, *1877*
John L. and Joyce Coons, 2017, *1877*
Harold Ewing Trust- Jerald & Debbie Ewing, Aaron & Kristi Ewing, Adam & Kristi Ewing, and Angela & Kevin Edwards, 2012, *1907*
Robert D. Faler & Helen (Faler) Maitland, 2003, *1883*
Sally & Ralph Holliday, 2000, *1899*
Merle and Frances Mavers, 2007, *1875*
Leoda Meyer, Charlene & Harold Swearingen, 2005, *1904*
George & Reba Ellison Miller & family (Fawn Creek Farm), 2006, *1879*
Jack and Marcialee Mitchell, 2017, *1916*
Virginia Noble, George Copeland, 2009, *1891* (Labette County)
Billie D. & Bonnie Oliver, 2000, *1893*
Harry & Jeroldine Pittman, 2009, *1873*
Jesse Wallick, 2002, *1873*
Evelyn Ward, 2015, *1913*
Bill & Jay Wiley, 2010, *1901*
Five Woods Farm, LLC - Mike & Margaret Wood, Toni Wood, Sherry & Dick Ruddell, 2016, *1889*

Morris County

J. Max Amos, 2001, *1899*
Gene and Juanita Barr, Ben and Rachael Barr, 2017, *1879*
Clover Hill Farm, 2011, *1886*
Jon D. Comp and Juanita K. Comp, 2007, *1904*
William, Michele & Patricia Cunningham, 2011, *1880*
Gladys Diekmann, 2008, *1906*
Kevin & Karla Effland; Korby, Mary & Ian Effland; Kelly Effland and Kaleb Effland, 2009, *1909*
J.B. & Helen Kendall Haage, 2005, *1864*
Milan & Darrell Harkness, 2000, *1873*
Pete Unruh, Charles Keiffer, & Ed Keiffer, 2007, *1906*
Gail Lee, 2001, *1882*
Evan & Theresa McCreath, 2010, *1904*
Dean & Marilyn Miller, 2010, *1906*
Dean & Marilyn Miller, 2010, *1910*
Kenneth Muller and Richard Muller, 2008, *1874*
Eldon Nelson, 2002, *1884*
Galen Peterson, 2001, *1883* (Dickinson County)
Charles L. & Beverly Y. Rodman, 2010, *1884*
Marilyn Mitchell Sharp & Jim Sharp, 2000, *1872*
Marilyn Mitchell Sharp & Jim Sharp, 2005, *1872*
Dwayne & Delores Stroda, 2014, *1912*
Leland W. and Linda K. Viar, 2011, *1909*
Janice Wilkison, 2001, *1895*
Kirby & Cindy Zimmerman, 2004, *1901*

Morton County

Floyd O. and Maxine Turner Coen, Robert and Patti Coen Grieshaber and Brynn Claes, Linda Coen Passmark and Niklas, and Celia Coen and Shannon, 2008, *1908*
Charles Milburn and John Milburn, 2012, *1905*
Stacy Trahern & Lance Trahern, 2000, *1893*
Patricia Trahern, 2001, *1893* (Stevens County)
Phil & Betty Yoder, 2013, *1911*

Nemaha County

Richard & Diane Allison, 2002, *1860*
Harry C. M. Burger, 2000, *1883*
Roy Claycamp, 2004, *1900*
Dexter F. Davis, 2013, *1913* (Jackson County)
Leslie & Lola Droge, 2001, *1874*
Clyde Eisenbarth, 2004, *1878* (Jackson County)
George Grossenbacher, 2000, *1878*
Dolores Grossenbacher Aul, 2000, *1875*
John and Janell Haug, 2017, *1916*
Lester A. & Edith E. Haug, 2002, *1890*
Al & Janet Haverkamp, 2002, *1902*
Francis & Silvia Hecht, 2000, *1880*
Roy and Dorothy Heiman, 2017, *1895*
Daniel J. & Karen Henry, 2002, *1900*
Hickman Family Farm, 2010, *1862* (Marshall County)
Roger and Jolene Holthaus, 2017, *1879*
Galen J. & Susan A. Huninghake, 2004, *1903*
Kenneth & Diane Keegan, 2000, *1879*
Larry D. King & Joella K. King, 2015, *1912* (Jackson County)
John, Lucille & Jerry Kissel, 2009, *1909*
Joseph Koch, 2001, *1880*
Leonard & Clara Krogman, 2000, *1887*
Donald J. & Alice A. Macke, 2000, *1892*
Larry & Judy Macke, 2003, *1880*
Kenneth & Alice Meyer, 2001, *1892*
Robert & Janet Meyer, 2011, *1881*
Aurel & Jeanette Montgomery, 2001, *1901*
David & Donna Nordhus, 2000, *1899*
William & Wilma (Heinz) Perkins, 2003, *1902*
Claude, Betty (Capsey) Rieschick, 2004, *1872* (Jackson County)
Michael L and Rosalyn Stallbaumer, 2017, *1917*
Joseph J. Strahm, Arthur J. Strahm and Craig A. Strahm, 2014, *1913*
Duane & Jane Tanking, 2000, *1870*
William & Julia Thompson, 2008, *1903* (Brown County)
Darryl & Wilma Thompson, 2008, *1908*
Jim & Kelli Uphaus, 2015, *1873*
Gordon & Faye Weyer, 2013, *1892*
Lee & Esther White, 2001, *1895*

Neosho County

Joe & Helen Arthur, 2002, *1885*
Mark R. Blackburn & Tammy J. Blackburn, 2007, *1873*
Philip Joseph Blaine, 2002, *1882*
Marion Tom Burress, 2013, *1876*
J&R Cheyney Farms, Ronald Cheyney, Rhonda Cheyney-LaRue, 2015, *1903*
William A. and Jean M. Cook, 2017, *1915*
Wayne and Joyce Erikson, 2017, *1909*

John & Mavis Gard, 2001, *1882*
Neill & Brian Gildart, 2000, *1886* (Wilson County)
Max & Bonnie Grossnickle, 2003, *1902*
Greg & Rebecca Hunn, 2013, *1874*
Ralph E. Janssen, 2010, *1901*
Todd Lindberg, 2017, *1917*
Richard L. & Patricia A. Lock, 2011, *1885*
Randall & Judith Moyer, 2001, *1901*
William Neely, 2001, *1871*
Cathy J. Rydquist and Nels E. Olson, 2012, *1912*
Richard & Linda Reinhardt, 2004, *1904*
Clinton D. Swiler, 2000, *1881*
Jerry & Dorothy VanLeeuwen, 2003, *1883*
Jerry & Dorothy VanLeeuwen, 2009, *1883*
Jessie Waltermire, 2000, *1886*

Ness County

Courtney & Paula Atwell, 2015, *1894*
Merritt Atwell, 2002, *Homestead*
Daryl Barricklow, 2015, *1901*
Dean & Olinda Barricklow, 2003, *1901*
Melvin Boese, 2005, *1901*
Mary Brady, LoRee Logan, Verneta Matt and Linda May, 2009, *1904*
Walter Duane Brown, 2005, *1887* (Leavenworth County)
James Burdett Family Trust, 2011, *1890*
Laverne & Martha Carlson, Neil & Sharon Carlson, 2010, *1887*
Myra G. Coker, 2008, *1892*
Vernon & Luzella Evel, 2003, *1890*
Fred Goodman, 2003, *1897*
Goodman Family Trust, 2009, *1909*
Clyde L. & Marla K. Goodman Family Trust, 2010, *1900*
Orville & Willa Hair, 2005, *1884*
Valdah Harkness Benson, 2000, *1885*
Kaye and Gail Harris and Prairie Farms Partnership, 2014, *1913*
Janis Haynes, 2002, *1888*
Richard & Kay Horchem, Todd & Brenda Horchem, Brad & Cindy Horchem, 2015, *1904*
Rexford (Rex) D. Huxman, 2000, *1886*
Arylene Jackson, Roland Jackson and Marcia Dian Hall, 2009, *1878*
Darryl & LaVina Johnson, 2015, *1895*
Alan G. & Peggy J. Knotts, 2005, *1879*
Dale L. Leighty, 2014, *1914*
Harlan S. McNinch, 2000, *1896*
Dean & Harriet McVicker, 2000, *1894*
Linda Shomo, Wade & Gayla Michaelis, Gwenda & Doug Philbrook, Gary & Deborah Lamoree, 2015, *1902*
Dennis & Janell Michel, Larry & Donna Michel, 2004, *1899*
Robert & Patricia Mishler Family, 2011, *1911*
Jon & Maxine Nuttle, 2003, *1888*
Douglas R. & Harriet J. Petersilie, 2015, *1888*
Travis & Laurie Petersilie, 2015, *1899*
Jared & Erin Petersilie, 2016, *1915* (Rush County)
Berniece F. Pfannenstiel and the Berniece Pfannenstiel Trust, 2017, *1917*
Glenn Richardson, 2001, *1893*
Carl B. Ruthrauff, 2015, *1890*
Harry E. & Bettie Shank, 2010, *1905*
W. Lance & Pamela Stum, 2015, *1907*
Roy Thornburg, 2001, *1898*
Gerald Whipple, 2002, *1882* (Hodgeman County)
Rex Whipple & Donna Whipple, 2015, *1915* (Lane County)
Keith Wierman, Francis Wierman & James Wierman, 2007, *1898*
Keith and Wanda Wierman, 2007, *1898*
Verlyn F. & Melba Witthuhn, 2005, *1990*

Norton County

Merle Applegate, 2002, *1900*
Rosella & Henry Austerman, 2001, *1884*
James and Shanna Brooks, 2012, *1902*
Maxine Brooks, 2000, *1897*
Robert & Sally Clydesdale, 2016, *1910*
George Samuel Colip Homestead, 2006, *1889*
Robert & Betty Cotton, 2008, *1902*
Brock & Kim Ellis, 2004, *1896*
Eugene & Florence Fisher, 2001, *1881*
Doyle & Arlene Foss, 2000, *1882*
R. Harold & Glenda D. Hahn Trust, 2011, *1879*
Kenneth & Faith Hays, 2002, *1881*
Dennis Leichliter, 2001, *1885*
John & Becky Linden, 2007, *1901*
Jon & Elaine Lofgreen, 2001, *1899*
James A. & Suzann Maddy, 2002, *1894*
Bob & Donna McClelland, 2002, *1877*
Rusty A. Miller, 2005, *1902*
Patricia C. and Joseph C. Miller, 2005, *1905*
James & Peggy Ninemire, 2006, *1898*
Todd & Amber Nuttycomb, 2010, *1906*
Arnold M. Schoen Trust, Ruth C. Schoen Trust, Carmen Schoen Rowley, Trustee and Barton, Dorinda and Gavin Rowley, 2010, *1907*
Arthur Shelton, 2001, *1865* (Decatur County)
Merle & Zelma Sleffel, 2002, *1899*
Mr. & Mrs. Robert Smith, 2002, *1875*
A.F. and Ruth Vahling, 2012, *1886*
Bruce & Cindy Wegener, 2009, *1891*
Whitney Farm, 2012, *1893*
Duane & June Wray, 2006, *1905*
Lloyd & Linda Zierlein, 2006, *1906*

Osage County

Roger & Garlena Bartz Anstaett, 2002, *1900*
James S. Badger, Jeffrey D. Badger, 2015, *1880*
Ruben M. and Hedy M. Bauck, 2009, *1903*
Loren & Clara Bryan, 2011, *1906*
Charles F. & Donna L. Bull, 2000, *1879*
W. Dean Burkdoll, 2000, *1878*
Douglas & Roxie Bush, 2007, *1905*
Casten Farms, 2003, *1891* (Franklin County)
Larry & Jeannine Colstrom, 2005, *1902*
Sue Lyon DuBois, 2000, *1866*
Ronald Fredrickson, 2001, *1881*
Jack Gilmore, 2000, *1858*
Robert L. Hendrickson, 2004, *1888*
Esther Hollar, 2003, *1896*
Edward & Robert D. Hug, 2000, *1868*
Charles A. & Hannah J. Hutchison, 2004, *1900*
Lee Kaff, 2005, *1881*
Roezetta & Larry Keim, 2013, *1911*
Wanda Kinney, 2001, *1860*
Dean, Martha & Ken Kuykendall, 2000, *1897*
William & Joan Lieber, 2016, *1903*
Carl F. Meyer & Earl H. Meyer, 2003, *1873*
Carl F. Meyer & Earl H. Meyer, 2003, *1903*
Fred & Lucile Michaels, 2004, *1890*
B. Neil & Kathy A. Mickelson, 2002, *1896*
Lewis L.W. and Laura Louise Miles, 2002, *1902*
Carrol Niles Henderson, 2000, *1872*
Sahlberg Farms, 2004, *1877*
Donald D. & Anita L. Remington, 2000, *1882*
Jerry L. & Ann D. Sand, 2000, *1886*
David Shoup, 2013, *1891*
William D. and Joan Skonberg Lieber, 2017, *1908*
Harold & Pauline Supple, 2002, *1869*
Jim Ullery, 2011, *1911*
Dale A. Vawter, Donna M. (Vawter) Peoples, 2007, *1868*
Daryl & Joyce Weir, 2005, *1900*

Osborne County

Kenneth & Wilma Bartholomew, 2001, *1878*
Kenneth Bates, 2002, *1882*
Edmund Brant, 2002, *1894*
Bryan & Arleta Byrd, 2006, *1890*
Deryl & Wilda Carswell, 2002, *1890*
Everette A. Dibble, 2000, *1871*
Lyle J. & Carol R. Dixon, 2005, *1904* (Riley County)
Ernest J. Heitschmidt & G. Dorine (Heitschmidt) Elsea, 2004, *1901*
Earl & Ula Emerson, 2000, *1885*
Alan & Debby Guttery, 2000, *1879*
Lowell Guyer, 2000, *1874* (Sherman County)
Clifford & Iris Hahn, 2003, *1872*
Clifford and Iris Hahn, 2007, *1907*
Gary & Jean Krier, Janice Motes, Wayne Krier, Steve & Tammy Krier, Roger & Pam Krier and Ron Krier, 2010, *1901* (Mitchell County)
Gerald & Veda LaRosh, 2002, *1887*
Gerald & Veda LaRosh, 2002, *1891*
LaRosh Farms - Jhan & Marsha LaRosh, 2011, *1900*
Ethel Limbocker, 2001, *1900*
Carol McNeal Dixon, 2005, *1882* (Riley County)
Marvin McReynolds, 2003, *1889* (Rooks County)
Wanda C. Meyer, 2004, *1878*
Wanda C. Meyer, 2007, *1906*
Mitchell Farms, 2005, *1895*
Muths Family Partnership, 2002, *1887*
Wilma Nichols, 2001, *1887*
Douglas C Norris, 2003, *1901*
Linda S. Rice and Janice A. O'Toole, 2002, *1902*
Pruter Farm, 2006, *1905*
Laurence Reinert Family, 2002, *1883*
Arnold Reinert, 2002, *1887*
Scott & Nadine Sigle, 2013, *1892*
Louise Smith, 2002, *1887*
Winder Farms, 2013, *1880*
Oid L. Wineland, 2002, *1890*
Janice O. Zamecnik, 2014, *1914*

Ottawa County

Florence Ablard, 2002, *1869*
Thomas E Austin Sr., Thomas E Austin Jr., 2016, *1873*
Steve Baccus, 2001, *1880*
Frank Blaha, 2008, *1908*
Lucy Buck Carlson, 2002, *1899*
Clyde F. & Alma L. Christenson, 2009, *1901*
Kathy & Bill Connelly and Bill Fischer, 2013, *1912*
Joe & Maxine Culley, 2001, *1862* (Saline County)
John S. Darrow, 2010, *1890*
Bob & Sharon Flegler, 2010, *1882*
Margaret & John Fleming, 2009, *1909*
James & Crystelle Ganson, 2003, *1900* (Republic County)
Kenneth & Muriel Greene, 2003, *1897*
LaDonna M. & Maurice D. Hill, 2015, *1898*
Gary L. Kay, Francis G. Kay, James P. Kay, Robert S. Kay, 2016, *1916*
Howard Oscar Larson, 2002, *1885*
Joyce C. Lyne Marrs, 2009, *1893*
Herman A. Mende, 2006, *1900*
Hubert & Ruth Mull, 2002, *1896*
Ruth Mull, 2003, *1901*
Sean M. Myers, 2012, *1906*
Ron & Karen (Ziegler) Seyfert, 2003, *1903*
Everett & Doris Shroyer, 2004, *1862*
Robert & Sara Coatney Thompson, 2004, *1884*
Jane Hanes Thurston and Nancy Hanes Milum, 2013, *1881*
Riley D. Tibbits, 2010, *1908*
William Jay Wagner, 2012, *1868*

Pawnee County

Charles J. Aldrich, Charles E. Aldrich, Stanley D. Aldrich, 2015, *1904*
Ralph H. Baird, 2004, *1904*
Richard and Kim Bennett, 2017, *1956*
Fern Bruntzel, 2002, *1890*
John Lorraine & Wilma Couchman, 2003, *1898*
John Lorraine & Wilma Couchman, 2003, *1901*
Phyllis Deege, 2002, *1888*
J. Timothy Dryden, 2002, *1895*
Roger G. Fox & Family, 2002, *1898*
Eugene & Marjorie Louise Grizzell, 2005, *1904*
Albert & John J. Hays, 2003, *1903*
William L. Horner, 2005, *1884* (Barton County)
John & Jane Kelty, 2000, *1886*
David Lupfer, 2002, *1878*
Margaret Lupfer Casady, 2002, *1882*
Harold H. Miller, 2000, *1879*
Nairn Farm, 2002, *1870*
Barbara & Robert K. Petro, 2006, *1885* (Shawnee County)
Charles L. Pfenninger, 2005, *1889* (Rush County)
James & Joan Rainbolt, 2003, *1888*
James & Joan Rainbolt, 2003, *1893*
Marvin Saxton, 2002, *1890*
Edgar L. Schadel and Karen M. Schadel, 2009, *1902* (Rush County)
Gary & Nina Seibert, 2001, *1900* (Stafford County)
Hattie Belle Zahn, 2002, *1902*

Phillips County

Martin Jr. Bayens, 2011, *1889*
Lloyd A. & Deloris Becker, 2005, *1880*
Gerald Bredemeier, 2002, *1888*
James Burton, 2002, *1872*
Cecil Chester, 2002, *1872*
Winston Churchill, 2002, *1885*
Wanda Dusin, Dan Cole, Connie Sylvester, 2002, *1889*
Wanda Dusin, Dan Cole, Connie Sylvester, 2002, *1890*
Dolores & Richard Fischli, 2002, *1898*
Keith Eugene & Partricia Ann Forssberg, 2000, *1885*
Fred & Marie Flanigan & Gertrude Erickson, 2000, *1872*
William & Diana Greving, Colby & Cher Greving, 2001, *1891*
Luroy Gruwell, 2002, *1896*
Elda Mae, Alan & Connie Hanke, 2005, *1886*
Tom & Kendra Highley, 2006, *1890*
Cornelius Hopson, 2002, *1887*
L.P. Johnson, 2002, *1873*
Neil Johnson, Wayne Johnson, Juanita Gould, 2003, *1896*
Krafft Family Farm - Blaine, Candy, Jason & Tina, 2002, *1882*
Denton, Sharon, Jodie, Eric & Jenna Krafft, 2002, *1882*
Audrey Tompkins, Sharon Krafft, Dennis Merklein, 2006, *1898*
Carl & Esther Merklein, 2005, *1896*
Paul W. Merklein, 2007, *Homestead*
Kenneth Molzahn, 2001, *1890*
William Molzahn, 2001, *1900*
Beverly J. Morgan, 2004, *1903*
Forrest R. & Marlene R. (Smith) Peters, 2008, *1898*
Norma F. Preuss, Larry Preuss, Mary Harders, 2002, *1884*
Stephen & Karen Reese, 2001, *1881*
Keith & Charlene Ross, 2014, *1883*
Lewis & Janene Schneider, 2000, *1895*
Pauline Shoemaker Mankin & Andra J. Shoemaker, 2002, *1883*
Russell W. Smith, 2008, *1908*
Billie & Hilda Stapel, 2004, *1903*
Robert & Betty Still Family, 2005, *1905*
Kenneth & Diane Stockman, 2002, *1872*
Mike & De Suchsland, 2002, *1885*
Marion and Joyce Tien, 2011, *1889*
Bernard Townley, 2003, *1901*
Conrad & Barbara Van Kooten, 2005, *1887*
Donald J. VanRoekel, Marcia A. Shott, John L. VanRoekel, Ronald R. VanRoekel, Dale VanRoekel, 2016, *1916*
Eugene Vogel, 2009, *1905*
Edna L. (Selbe) Wood and Hazel I. Tracy, 2002, *1887*
Robert & Harriet Young, 2006, *1902*
Andrew C. & Betty M. Zillinger, 2002, *1893*

Pottawatomie County

Wayne & Margaret Bairow, 2002, *1897*
Galen F. Bosse, 2000, *1891*
James & Marian Bradley, 2001, *1870*
Darold Brunkow, 2007, *1892*
Cleo E. Brunkow, 2005, *1905*
Gary J. Burgess, 2000, *1864*
C.A. & Anna Ebert, Kenneth & Dorothy L. Ebert, 2007, *1902*
Kenneth L. & Janice Goehring, 2000, *1880*
John W. & Patricia A. Huey, 2013, *1913*
Lowell & Evelyn Jones, 2007, *1894*
Bruce, Ann, Branson & Keyanna Kocher, 2004, *1901*
Kenneth L. & Denise L. Melvin, 2005, *1859* (Jackson County)
Basil & Betty Moran, 2000, *1880* (Shawnee County)
Dennis Nelson, 2003, *1885*
Rezac Farms - Jay and Stacy Rezac, 2017, *1899*
Joan L. Rezac Family, 2011, *1887*
Ron & Mary Sue Roller, 2016, *1896* (Dickinson County)
Lee & Beverly Stowell, 2010, *1898*
Ken Teske, 2002, *1876*
Larry Wahl, 2009, *1899*

Pratt County

Dwight Adams, 2004, *Homestead*
Stephen and Carol Aschenbrenner, 2009, *1889* (Wichita County)
Benny Bailey, 2001, *1887*
Benny Bailey, 2001, *1901*
Wilma Beck, 2001, *1887*
James Bergner, 2002, *1894*
Portia Blair Terry, 2003, *1884* (Harvey County)
Berry K. Bortz, Brandon S. Bortz, Darnell Bortz, 2017, *1864*
Berry K. Bortz, Brandon S. Bortz, Darnell Bortz, 2017, *1900*
Jim & Dorris Brant, 2001, *1900*
Pam Dietz, 2001, *1888*
Marjorie Fincham, 2002, *1902*
Lester Goyen, 2001, *1901*
Corlis & Mildred Goyen, Jr., 2002, *1889*
Jack & Dorothy Grier, 2001, *1879*
Mary Jane Hawver McEntire and Stephanie McEntire, 2013, *1908*
Lewis & Sharon Hillard, 2001, *1892*
Luke & Tammie Hoeme, 2007, *1887*
Hoener Family, 2004, *1903*
Glen & Sue Honeman, 2005, *1902*
Bill & Jean Howell, 2003, *1899*
Sandra K. Hetrick and Vicki A. Hetrick Howell, 2009, *1894*
Dorothy Huffman, 2001, *1885*
Harold & Eloise Jarboe, 2001, *1900*
John G. Jorns, Barney Jorns, Paul E. Jorns, KC Jorns, 2007, *1885* (Kingman County)
Henry & Minnie Narjes, Raymond & Lillian (Narjes) Krusemark, Don & Franzie Krusemark, 2007, *1907*
Lois B. Haas Larson & Suzanne L. Crane, 2003, *1903* (Decatur County)
Philip & Christine Lunt, 2001, *1885*
Marion F. Mardis, Elwin R. Mardis, Robert K. Mardis, Shawn M. Mardis, 2005, *1890*
Bob L. Moore, 2013, *1877*
Stephen & Gloria Mueller, 2005, *1890*
Neumann-Wheatley Farms, Inc., 2010, *1899*
Matt Novotny, 2002, *1890*
Charles W. Novotny, 2003, *1893*
Ronald R. & Brenda Jo Piester, 2005, *1905*
Arden Reiman, 2001, *1901*
Roger H. Russell & Rick C. Russell & Ryan K. Russell, 2006, *1885* (Stafford County)
Jo Ann Schepmann, 2002, *1902*
Warren & Lori Schepmann, 2005, *1905*
Jo Ann Schepmann, Warren F. Schepmann, 2011, *1910*
Donald & Dorothy Trinkle, 2001, *1893*
Morgan & Deb Trinkle, 2015, *1909*
White Lunt Farms, 2002, *1899*

Rawlins County

R.H. Bud Bearley, 2001, *1890*
Bernita Burk Beesley & Mike Sramek, 2005, *1904* (Gove County)
Eulaine Marie Erikson Benda, 2005, *1905* (Decatur County)
John S. Burk, Arthur Burk, John R. Burk and Aaron M. Sis, 2014, *1909*
Dr. Richard D. & Thyra K. Davis, 2000, 1893 (Shawnee County)
Ruth A. & Randel A. Drake, 2005, *1890* (Pratt County)
Lynn D. Hagler, 2000, *1893*
Doris Hayward, 2000, *1899*
Francis & Violet Holle, 2001, *1892*
Joe & Charlene Kramer Family, 2014, *1909*
Lavern Morgan, 2001, *1884*
Gerald Olson, Evelyn Olson, Janice Baer, 2006, *1889* (Decatur County)
Carolyn M. Peterson, 2000, *1887*
Samson Farm, 2002, *1888*
Dorothy, Douglas, Richard, Beverly, Mark & John Wolters, 2000, *1897*

Reno County

Edward J. & Elner M. Beck, 2000, *1896*
The Clark Family, 2009, *1909*
Helen Cobb, 2001, *1896*
Helen M. Cobb, 2006, *1893*
John L. Conkling Ancestral Farm, 2010, *1884*
Velma Davidson, Gale & Diane Fowler, 2001, *1885*
Velma (Baughman) Davidson, 2001, *1885*
Gordon & Carol Ehling, 2002, *1886*
Robert & Lavonne Farney, Jack and Kay Farney, Logan Farney, 2016, *1883*
Wallace A. & Armilda (Pat) Harden, 2001, *1884*
Miles & Joyce Hartman, 2013, *1892* (Rice County)
G.T & Rose Holcomb, 2014, *1890*
Keith J. Howard, 2017, *1885* (Rice County)
Ed & Marge Jones, 2014, *1878* (Rice County)
Edna Ruth Schwartz Kemp, 2009, *1884*
Jeff & Gayla Moeckel, 2001, *1893*
Merlyn D. & Melva Joy Moeckel, Douglas & Debra Moeckel, 2007, *1899*
Nuest Family Farm, 2008, *1873*
Nuest Family Farm, 2008, *1874*
Kathleen Virginia Olmstead Goble, 2000, *1873* (Sedgwick County)
Cameron & Clinton Peirce, 2001, *1884*
Charles W. & Louise Roth, 2009, *1909*
Norman & Cindy Roth, 2007, *1899*
Charles W. & Louise Roth, 2010, *1897*
The Schlickau Family, 2006, *1904*
Wilbur G. Kinnamon, Bob & Jo Stratton, 2016, *1889*
Joe & Emma Clara Stucky, Stucky Red Bluff Inc., 2007, *1900*
Swanson Brothers/Harold & Donna Swanson, 2011, *1868*
Mark & Virginia Tonn, Justin Tonn, Ashley Tonn, Erhart Tonn, 2007, *1885*
Delbert & Sherry Truesdell, 2013, *1911*
Larry and Cheryl White, 2017, *1899* (Stafford County)

Republic County

Dean & Dave Anderson, 2000, *1878*
Ward Aurand, Glen Aurand & Clay Aurand, 2005, *1882*
Lorene Baxa, 2003, *1891*
John C. & Julia A. Blackburn, 2005, *1903* (Jewell County)
Anver O. and Dorothy Bowman, 2003, *1890*
Hubert Brokesh, 2000, *1893*
Bill & Mary Bruchie, 2000, *1887*
Norma Jean Brunsilius, Larry C. Brunsilius, Morris K. Brunsilius, 2005, *1879*
Norma Carlgren, 2002, *1877*
Ivan & Aileen Carlgren, 2002, *1877*
Aileen & Ivan Carlgren, 2002, *1880*
Robert & Rebecca Craig Jr., 2007, *1892*
Lawrence Dooley, 2004, *1904*
Frank Dumback, James Deane Dunback, Steven J. Dunback, 2015, *1907*
Larson Homestead - Olaf & Christina Larson, Johan Gustaf & Caroline Larson, Ralph W. & Joyce C. Larson, Marilyn (Larson) & Ellis Dunstan, 2007, *1871*
Dwight Edberg, 2003, *1862* (Dickinson County)
Dwight Edberg, 2003, *1885* (Dickinson County)
John Eickman, 2000, *1885*
Jeff Filinger, 2000, *1884*
Richard Fogelberg, Dianne Fogelberg, 2009, *1889* (Cloud County)
Fredheim Farm, 2004, *1902*
Karl D. Fry, 2015, *1903*
Henrietta Grauerholz, Deanna Wittstruck, Dana Grauerholz, Sandra Braden, 2003, *1903*
Herbert & Erma Hedstrom, 2001, *1879*
Oleta & Calvin Hobson, 2000, *1880*
Gerald & Patricia Holmberg, 2005, *1905*
Bonnie (Brokesh) Holmsten, 2000, *1882*
Paul C. Hood and Ken Hood, 2003, *1877*
Franklin J. Hostinsky, 2001, *1897*
Ronald & Janice Janasek, 2012, *1875* (Riley County)
Dana C. & Mary A. Johnson, 2000, *1900*
Ann Johnson Sparke, 2000, *1898*
Willie B. and Dorothy (Konovalski) Kennedy, 2007, *1879*
Darrel & Marcia Klima, 2014, *1895*
Ray (Sonny) & Judy Kuhlman, 2004, *1902*
Richard C. Lachman, Roger D. Lachman, and Margaret A. Lachman, 2010, *1910*
Davey Lahodny & Beverly Waguespark, 2005, *1902*
Charles Lundblade, 2000, *1898*
Vernon & Sylvia Mai, 2003, *1900* (Washington County)
Robert J. Mainquist, 2000, *1897*
RJ & Kathleen Marsicek, 2005, *1893*
Barbara Palecek, Janice McCaulley, Linda Shelly, Pamela Windhorst, Victoria Nylund, 2006, *1906* (Ellsworth County)
Ernest J. Mikesell, 2000, *1870*
Madeline Miller, John O Miller Trust, 2005, *1870*
Jacob Myers, 2002, *Homestead*
Glenn L. Pelesky, 2000, *1899* (Washington County)
Rick & Melinda Pierson, Lynn Blecha, 2015, *1907*
Jacqueline D. Ramsey, Michael T. Dewey, D. Keith Dewey, 2006, *1902* (Cloud County)
Beverly Blazek Reynolds & Allen R. Reynolds, 2004, *1882*
John Rhine & Dorothy Rhine, 2002, *1883*
Donald & Ruth Rizek, 2001, *1879*
Russell Farm Partnership, 2002, *1880*
Norm & Helen Sankey, 2004, *1890*
Gary & Lori Scofield, 2000, *1880*
Scott Homestead, Inc, 2004, *1871*
Kenneth M. & Phyllis L. Sherwood, 2000, *1882*
Marvin & Donna Shoemaker, 2001, *1890*
Marvin & Donna Shoemaker, 2001, *1893*
Marvin & Donna Shoemaker, 2001, *1898*
Elmer D. Sothers, 2000, *1870*
Ann Johnson Sparke, 2005, *1905*
Donald & Katherine Wilber, 2000, *1886*
Alice Marie Rhine Wineinger, 2009, *1893* (Washington County)

Rice County

Robert & Carolyn Behnke, 2006, *1903*
Mark & Angie Beneke, 2008, *1908*
John Caywood, 2002, *1888*
Delmar and Noema Conner, 2007, *1876*
Rex & Julie Engelland, 2002, *1880*
Kermit D. Engelland, 2002, *1884*
Carolyn S. Heter & Eunice M. Haslam, 2000, *1878*
Charles E. Fair, Shirley P. Fair, 2007, *1884*
John Fuller Farm, 2005, *1887*
Geneva Esaw Gene, 2000, *1898*
Wayne & Karen Habiger & Family, 2006, *1900*
Ralph J. Hathaway, 2000, *1878*
Ralph J. Hathaway, 2000, *1878*
Carolyn S. Heter, 2000, *1883*
Carolyn S. Heter & Eunice M. Haslam, 2000, *1874*
Hunsinger Enterprise - Elma M. Hunsinger, 2009, *1892*
Ed & Marge Jones, 2014, *1880*
Don & Rosemary Jordan, 2000, *1879*
Walter B. and Rosalie R. LeClerc, 2002, *1899*
James L. & Judy K. Loesch, 2010, *1905*
Janet Ellis Maskil, 2017, *1914*
Leslie Mathews, 2011, *1876*
John & Mary Oden, Arthur & Dora Oden, Wilma & Anita Oden, Arthur & Jan Oden, Jon & Heather Oden, 2015, *1904*
Felix & Barbara Orth, 2008, *1885*
Marion & Janice (Foote) Peterson, 2000, *1881*
Carol Bayer Reynolds, Esther Delimont, Floyd Bayer, Liz Bomba, Shirley Schendzielos, Louise Bayer Carlson, 2007, *1886* (Rush County)
Lindsay R. & Rhonda R. Schmidt, 2008, *1908*
Lowell Smith, 2009, *1909*
Wayne & Ann Smith, 2010, *1880* (McPherson County)
Marceline Parks & Patricia Williamson, 2007, *1880*

Riley County

Inez Arents, 2001, *1892*
Harold & Bonnie Bailey, 2001, *1883*
Evelyn Belin, Kelvin Belin & Camilla Shaughnessy, 2000, *1880*
Harlan, Arlo & Duane Boda, 2000, *1874*
Ernest Diskau, 2001, *1880*
William G. & Zoe Ann Gugenhan, 2008, *1863* (Clay County)
Terry and Tammie Henry, 2007, *1861*
Homer Hoch, 2001, *1881*
Larry & Diane Hoobler, 2001, *1872*
Larry & Diane Hoobler, Tonya Hoobler, 2017, *1871* (Wabaunsee & Johnson County)
Thomas E. Kimball Family Trust, Judy R. Kimball Trustee, Sharon & Steve Eckhoff, 2013, *1855*
Roy & Debbie Larson, 2007, *1890*
Carl & Jane Larson, 2008, *1894*
Mark & Jayne Lund, 2001, *1877*
Phil D. Moore, 2004, *1870*
Rossell Farm, LLC - Rhonda (Rosell) Baer and Ann (Rosell) Ingham, 2011, *1861*
Ralph Rothisberger, 2001, *1870*
Harwood & Kathy Shawver, 2011, *1909*
M Charlotte Shawver, 2011, *1909*
Tony A. & Sharon L. Sump, 2014, *1868*
David & Janet Vinduska, 2016, *1880*
Donna M. & Harry E. Whitney, 2015, *1901*
Dale & Margie (Peter) Wohler, 2001, *1859*

Rooks County

Verna Arpin & Loretta Kruse, 2005, *1905*
Tim & Brenda Benoit, 2011, *1905*
Edward & Bethy Bigge, 2003, *1900*
Elizabeth Brobst, 2003, *1900*
Bernard Carsten and Family, 2008, *1904*
Kenton Watts Casad, 2003, *1896*
Jay Casey Family, 2002, *1883*
George W Colburn, 2004, *1886*
Trudy & Virgil Deges, 2003, *1903*
Sharon Fink Denton, Janet Elizabeth Converse, 2017, *1915* (Riley County)
George Duncan, 2002, *1900*
LeRoy & Velma Dunn, 2001, *1898*
Gregg Gartrell, 2003, *1902*
Junia L. Hale, 2014, *1901*
Elton & Lola Hunter, 2001, *1898*
Robert R. Jones, 2005, *1892*
Harold & Lyla Lowry, 2016, *1909*
Harold & Lyla Lowry, 2016, *1909*
Harold & Lyla Lowry, 2016, *1908*
Orren Marcotte, 2007, *1907*
McClellan Farms Partnership, 2004, *1898*
Dwight & Marilyn Miller, 2015, *1892*
Dwight & Marilyn Miller, Kenton & Dixie Miller, 2016, *1887*
LLM Farm - Mongeau, 2003, *1862*
L. Lynn Miller, 2003, *1885*
Royce Muir, 2003, *1878*
James & Ann Ochampaugh, 2000, *1892*
Carl Orr, 2003, *1886*
Alvin & Venda Poore, 2005, *1905*
Leonard Rathbun, 2003, *1873*
David Reed and Denise Murchie, 2005, *1903*
Steven Kent Riffel and Randal Eugene Riffel, 2012, *1912*
Mildred Robertson, 2003, *1892*
Mitch & Cindy Ross, 2007, *1905* (Phillips County)
PH Roy, Cecil and Patti Roy, 2017, *1903*
Darrell & Avonne Rubottom, 2015, *1910*
Grant I. Stewart and Grace N. Satran, 2010, *1906*
Donald G. Schneider, 2009, *1891*
Alvin & Betty Schrandt, 2001, *1900*
Harold Simons, 2003, *1890*
Stamper Farms, 2009, *1891*
Duane & Carol Steeples, 2003, *1896*
Darrell Sutor, 2000, *1885*

Rush County

Euriel Albers, 2002, *1882* (Ness County)
Billy T. and Alyce M. Allen, 2002, *1886*
Evelyn Butherus Avey, 2000, *1888*
Mark Bannister, Ted Bannister, Bill Bannister, 2007, *1906*
William E. Baus Trust, 2004, *1901*
Rosalee Bieber, Arlyn Bieber, Margaret Oborny & Richard Bieber, 2000, *1883*
Clifford C. Bizek and Charlotte I. Biays, 2007, *1880*
Wilbur & Marie Brack, 2000, *1887*
Dale & Nancy Brady and Calvin & Irene Brady, 2008, *1892*
Harlan D. and Shirley L. Burgardt, 2014, *1890*
Eldora Conard, 2004, *1904* (Pawnee County)
Loris Erbes Irrevocable Trust & LaVina Erbes Revocable Trust, 2003, *1901*

LaVina Erbes Trust-Loris E. Erbes Trust, 2004, *1904*
Luanne Lebsack French & Arlene Lebsack Rolfs, 2012, *1907* (Barton County)
Jay & Maureen Georg, 2015, *1892*
Loren & Viola Heiser, 2006, *1898*
The Jones Family Farm, 2003, *1903*
Bruce & Ellen Kershner, 2004, *1886*
John A. Koriel and Mary M. Horton, 2003, *1901*
Frank & Elaine Littler and Scott & Amber Littler, 2012, *1878*
Edwin H. & Sharon K. Mauler, 2011, *1892* (Barton County)
Fr. Charles, Eugene, Dennis & Marion Mazouch, 2008, *1892*
Virgel & Kathryn Miller, 2001, *1897*
Duane and Linda Moeder, 2008, *1908*
Edward & Verna Mohr, 2001, *1883* (Barton County)
Edward & Verna Mohr, 2001, *1886* (Barton County)
Edward & Verna Mohr, 2001, *1900* (Barton County)
Edmund L. & Carol A. Oborny and Brian L. & Jennifer M. Oborny, 2009, *1906*
Lee Oelkers, 2002, *1886*
Lawrence Oelkers, 2002, *1892*
Matthew Pechanec, Gregg Pechanec, 2017, *1917*
Marjorie Raynell Weiser Penka, Leroy D. Penka and Bradley R. Penka, 2003, *1889*
James R. Pivonka, 2000, *1888*
Glenn K. & Lavina Scheuerman, 2005, *1897*
Glenn K. & Lavina Scheuerman & Elnora Steitz, 2005, *1901*
Robert Calfant Jefferis, Walter Jefferis, Lenore Baker, Enola Kirby, Gene & Cheryl Schlegel, 2015, *1888*
Lee Roy Schraeder, 2000, *1898*
Shirley Seltmann, 2000, *1888*
George J. Seuser Jr., 2000, *Homestead*
Richard & Donna Smolik, 2004, *1900*
Arthur Stegman, 2000, *1888*
Peter & Anna Enslinger, Joseph & Perpetua Enslinger Suppes, and V. Charles & Raeann Suppes, 2014, *1908*
Terry Urban, 2006, *1904*
C.B. (Clarence) Vesecky, 2000, *1887*
James & Leann Webs, 2002, *1890*
Eleanor Wells, 2000, *1873*
Kirk Wells & Sharon Karst, 2002, *1873* (Ness County)
Samuel L. Witte, Donald E. Witte & Wreatha A. Witte, 2005, *1901* (Barton County)

Russell County

Helen A. Anschutz, 2011, *1901*
Bar S Ranch, 2008, *1900*
Randyl & Alice Becker, Wesley & Irene Becker, 2001, *1901*
Marlene Becker-Steinle, 2000, *1884*
Everett W. & Daisy M. Berry, 2000, *1884*
Esther & Herbert Bolte, 2001, *1885* (Saline County)
Gene & Alice Borell, 2001, *1879*
Gene & Alice Borell, 2001, *1899*
Gene & Alice Borell, 2001, *1899*
Ray E. Bosch & Louise Greenberg, 2001, *1890*
Bouska Family Farm, 2002, *1860* (Ellsworth County)
B&H Partnership - Wilmer Boxberger, 2001, *1898*
Mr. & Mrs. Frederick Boxberger, 2001, *1899*
Gerald and Radonna Brandenburg, 2010, *1893*
Clarence Crawford, 2001, *1890*
Dumler Homestead Trust, 2001, *1878*
Victor Dumler, 2001, *1897*
David L. Ebel, 2009, *1906*
Rolland & Mildred Ehrlich Family Trust, 2015, *1913*
George Eschbaugh, 2000, *1876*
John M. & Judith A. Finkenbinder, 2014, *Homestead*
Mildred M. Finnesy and Joseph P. Finnesy, 2007, *1906*
William & Jean Fritschen, 2001, *1864*
Peter & Katharina Fritschen, Leonard T. & Isabelle M. Fritschen and Lawrence K. & L. Kathleen Fritschen, 2010, *1898*
Lavina Gray, 2001, *1891*
Mamie B. Hall, 2000, *1900*
Kevin L. Hampl and Eldon D. Hampl, 2012, *1908*
Merton H. and Mary Jane Hickman, 2012, *1912*
Cathy J. & Steven W. Jacobs and Linda M. & William L. Jacobs, 2000, *1899*
Luerman - Kilian Farm, 2002, *1885*
Lorraine Krug, 2001, *1887*
The Krug Farm, 2009, *1876*
The Krug Farm, 2009, *1898*
Ruth E. Lanterman, Rita A. Lingg, 2017, *1877*
Etta M. Machin, Gerald L. Machin, Marjorie L. Chick, 2003, *1901*
Meharg Meadows Rita Harzman Lingg, 2003, *1873* (Reno County)
Alvin & DeeAnn Matheson, 2000, *1898*
Stella F. Nesbitt & Helen A. Anschutz, 2000, *1893*
Galen & Rudene Niedenthal, Linda & Carl Johnson, 2004, *1902*
Vernon & Lillian Nowak, 2001, *1881*
Estella M. Crabtree Nuss, 2011, *1899*
Verlan Olson, 2001, *1884*
Steven R. & Sharon B. Opdycke, 2000, *1877*
Ronald & Joyce Pertl, 2002, *1893*
Caroline L. Peterson, Norma M. Strecker, 2017, *1915* (Barton County)
Richard J. Polcyn, 2000, *1900*
Marie Ptacek, Frank C. and Anna Ptacek, Frank C. and Anna Ptacek Heirs, Ernest and Florence Ptacek, Ron and Virginia Ptacek, 2017, *1916*
L.L. and Jayne Reilly, 2002, *1890*
Dr. Lionel L. Reilly & L.L. Reilly, 2002, *1900*
Stephen & Raeleen Reinhardt, 2013, *1900*
Marion & Donna Schmidt, 2001, *1899* (Ellis County)
Margaret E. Soukup and Eugene E. Zamrzla, 2002, *1860* (Ellsworth County)
Margaret F. Soukup & Eugene Edward Zamrzla, 2005, *1860* (Ellsworth County)
Edward Stritesky, 2001, *1900*
Marlene Becker-Steinle and Ulanda K. Nichols, 2000, *1888*
Wallace N. Weber, 2009, *1893* (Ellis County)
Richard & Rosa Lee Whitmer, 2007, *1903*

Saline County

Rodney L. & Sharon K. Anderson and John R. & Mary Anderson, 2000, *1895*
Joseph Woodward Armstrong, Robert Lawrence Armstrong, Joseph Allen Armstrong, Joseph Woodward Armstrong, and Ross Curtis Armstrong, 2014, *1904*
Richard Buhler, 2013, *1876*
Scott Carlson and Kim Carlson, 2013, *1867*
Leon & Norma Jean Hahn, 2000, *1875*
James W. Hanson, 2006, *1903*
Darrell N. Holmquist, 2000, *1868*
Audy & Virginia Hopper, 2001, *1900* (Thomas County)
Esther Hyrup, 2001, *1899*
Charles P. Martin, 2011, *1874*
John & Kimberly Mick, 2001, *1888*
Scott Nichols, 2000, *1897*
Chester Peterson Jr., 2000, *1880*
Richard Rundquist, 2005, *1876*
Swisher Homestead, 2003, *1883*
David and Denise Stein, Kyle and Erica Moyer, 2017, *1890*
Elsie Thiel & Marthano Thiel, 2003, *1887*
Carl, Donald & Norman Tillberg, 2000, *1886*
Rex & Ramona Bengtson Tjaden, Sarah & Britta Tjaden, 2007, *1900*
Maurine Weisgerber Trust - Mary Kay and Brett May, Jim, Sonja, John and Paul Weisgerber, 2010, *1876* (Shawnee County)
Keith Weller, 2001, *1875*
Edna Weller, 2001, *1899*
Weller Family, 2017, *1875*

Scott County

Arlene Beaton-John & Colleen Beaton, 2004, *1903*
Marvin & Ruth Daniels, 2002, *1885*
Bar-X, Inc., Larry & Mildred Dearden; Rick & Joyce Dearden; Alan & Glenita Dearden; Thad & Jenna Dearden; Alene Dearden, 2006, *1906*
Alta Krebs, 2001, *1900*
Elizabeth Miller, 2001, *1901*
Olive Newcomer, 2002, *1900*
William & Robert Novak, 2001, *1886*
Rudolph Family Farm, 2009, *1909*
Ellender Strickert, 2001, *1893*

Sedgwick County

Joe & Evelyn Armour, 2000, *1875*
Richard Basore, 2017, *1901*
Leon & Lavon Black, 2004, *1896* (Harvey County)
George Bowman, 2002, *1902*
Roger & Patricia Bunck, 2004, *1901*
Larry Carter, 2003, *1875*
Mrs. Robert J. Clark, Family Farm, 2014, *1909*
Robert & Ruth Curless, 2010, *1910*
Philip Dosien, 2002, *1882*
Clair & Eleanor Evans, 2003, *1900*
The Gerlach Three: Robert C., DeWayne, Frank, 2015, *1898* (Sumner County)
John Gorges, 2007, *1892*
John & Barbara Hahn, 2003, *1901*
Norman & Lottie Hand, 2004, *1995* (Harvey County)
Maurine & Raymond Holt, Dan Holt, Ranine Bedwell and Marianne Slagle, 2011, *1911*
Calvin E. Kissick, 2004, *1882*
Martin & Marie Koster, 2002, *1898*
W. Dean & Brenda J. Lorenz, 2000, *1891*
Patrick & Jane Matson, 2000, *1894*
Wanda L. Miller, 2000, *1893*
Mildred Stollei Mitchell, 2003, *1903*
Versa M. Nigg, 2000, *Homestead*
Richard and Patti Pelz, 2017, *1915*
Richard Ramsdale, 2001, *1890* (McPherson County)
Henry W Robben, 2011, *1889*
Raymond Rosenhagen, 2001, *1899*
Duane & Isabella Sanders, 2001, *1895*
Leon & Denise Seiwert, 2010, *1900*
Louis J. Sittler and Alice J. Sittler, 2010, *1909*
John W. Struthers, 2000, *1893*
Max Tjaden Family, 2010, *1907*
Herman Tjaden Jr., 2000, *1882*
Norma & John Ulbrich, 2003, *1900*
Hugh E. Williamson, 2016, *1874*
Melvin R. Woolf, Allen R. Woolf, 2017, *1886*
Mrs. Eldor (Kay) Wulf, 2000, *1896*
Young and Sons Farms, 2012, *1910*
Betty (Korber) Ziegler, 2000, *1899*

Seward County

Bloom Farm, 2007, *1907*
Duane P. and Lynda Bozarth, 2017, *1914*
Robert Davies Jr. & William T. Davies, 2000, *1890*
Jamie Eakes, 2014, *1904*
Mike Fitzgerald and Mark Fitzgerald, 2008, *1908*

Eugene Kneeland Keating, 2000, *1887*
Robert and Sharen Keating, 2017, *1917*

Shawnee County
Clinton Acheson, 2003, *1857*
Ralph E. Atwood, 2000, *1860*
Harold J. Bernhardt, 2016, *1915*
William Carls, 2002, *1900*
Mildred Ruth Edwards Conger, 2000, *1857*
Jennie M. Decker, 2010, *1866*
Elizabeth George Edmund, 2000, *1868*
Betty Jean Bolz Fenton, 2004, *1865*
Emil & Marita Lutz, 2010, *1898* (Douglas County)
William Ray Lynn, 2009, *1858*
Larry D. Marken, 2000, *1879*
Wilbur E. McRoberts, 2000, *1869*
Ralph & Irene Miller, LeRoy G. & Patricia A. Miller Family, 2016, *1910*
Gary R. Priddy, Beth E. Priddy, Paul A. Priddy, Diana L. Hatcher, Elizabeth J. Hatcher, 2009, *1868*
Roy & Mary Ann Reser, 2006, *1898*
Gene Slusser, 2000, *1860*
Hayden K. Smith & Alberta Smith, 2000, *1882*
Arthur & Cleota Todd, 2000, *1876*
Dean & Elaine Wilson, 2000, *1893*
Louis Winter, John Winter, M.D. Winter and Marie Winter-Zima, 2014, *1914*

Sheridan County
Albers Enterprises Inc., Donald & Ellen Albers, 2012, *1911*
Paul L. & Vicki L. Babcock, 2011, *1902*
Wes & Casey Bainter, Gayle & Ruth Bainter, 2010, *1894*
L.J. Wright Ranch, Kenneth & Lorrine Clark, Kay Clark & Roberta Clark-Meckel, 2006, *1892* (Graham County)
Sammie & Glenna Coburn, 2001, *1889* (Gove County)
Craig & Suzanne Cooper, 2001, *1892*
David L. Cooper, 2011, *1885*
Larry & Irene Cressler, 2006, *1906*
Foster Family Trust - Edith E. Foster & Andrew J. Foster, 2009, *1880* (Finney County)
Florence Gilchrist, Vernon & Karen Flanagin, 2004, *1900* (Thomas County)
Mary K. Summerson Haffner, 2010, *1892*
Vivian A. Herrmann, Herrmann Family Trust, 2009, *1887*
Howard & Mary Lou Heskett, 2011, *1879*
Lee & Janelle Cooper, Vic & Ardith Karnes, 2011, *1894*
G. Steven & Kathy Launchbaugh, 2009, *1887*
Gerald D. Lewis and Dan D. Lewis, 2011, *1897*
Margaret Richardson Mehlinger, 2015, *1915*
Willard F. Morgan, 2011, *1897*
Sheep Head Ranch L.L.C.- Richard H. Mosier, Manager, 2008, *1887*
Thomas A. Conard, Jack L. Conard and Fred L. Moss, 2011, *1906*
Flossie Mowry, 2001, *1885*
William L. Patmon, Sr., Maxine Patmon, William L. Patmon, Jr., Michelle Patmon, Lane A. Patmon and Amy Patmon, 2011, *1888*
Harriet Richardson, 2016, *1915*
Herman J. Schlageck, Jr., 2012, *1887*
Charles B. & Joan (Farber) Taylor, 2013, *1899*
Charles B. & Joan (Farber) Taylor, 2013, *1902*
Charles B. & Joan (Farber) Taylor, 2013, *1905*
Charles B. & Joan Taylor, 2013, *1911*
Loyd E. and Judith R. Tilton, 2007, *1906* (Logan County)
Velda Transue, 2001, *1891*
Edwin J. Trueblood, 2011, *1902*
James Walden, 2015, *1892*
Nortbert Weis, 2017, *1915*
Germaine M. Weis, 2017, *1915*
Steven C. and Lana Sue Robben Wilson, 2012, *1901* (Thomas County)

Sherman County
Marie Edwards, 2006, *1905*
Bill & Geraldine Gannon, 2005, *1885*
Lowell Guyer, 2000, *1890*
Lowell Guyer, 2008, *1908*
Minnie Belle Imhof Jones, 2005, *1886*
Frances Jones, 2000, *1886*
Kirby Farms, 2014, *1907*
Genevieve Matthews, 2005, *1903*
James McClelland Estate, Marie Edwards & Rhonda Skrdlant, Trustee, 2006, *1893*
Robert and Ann Middleton, 2007, *1891*
Carroll A. Mogge, 2002, *1888*
Carl H. Nelson, 2014, *1911*
Elmer & Joyce Purvis, 2000, *1894*
Colleen (Brooks) Rall, 2014, *1895*
Colleen (Brooks) Rall, 2014, *1901*
Jack SeCoy, 2006, *1905*
Townsend Family: Richard, Leo, Sonya, 2009, *1901*

Smith County
Robert & Roy Atwood, 2002, *1893*
Bell Family, 2002, *1894*
Phyllis Bell, 2006, *1871*
Kenneth & Margaret Burian, 2002, *1892*
Harry & Virginia Dean, 2001, *1872*
Cleon Dettmer, 2000, *1893*
Francis & Ada Gaines, Brian & Linda Dettmer, 2006, *1872*
Dick L. Detwiler, 2000, *1872*
Gilbert Frieling, 2006, *1906*
James Absaln Hagman, Raymond Dow Hagman, Jerry Raymond Hagman, 2016, *1916*
Duane Hawkins, 2002, *1885*
Maydeen (Libby) Hawkins, 2002, *1899*
Darin & Jessica Hawkins, 2011, *1886*
Gene & Avis Hayes, 2000, *1882*
August Henning, 2004, *1893*
Elzene Kirchhoff, 2002, *1879*
Arthur & Melba Kuhlmann, 2003, *1903*
Joe, Dale & Dwayne Lambert, 2000, *1899*
Treva Pauline Lambert, 2002, *1894*
Larry Long, 2002, *1892*
Terry & Katrina Long, 2002, *1893*
Ava Lee Barcus Maydew, 2006, *1879*
Ava Lee Maydew & Larry Westerman, 2008, *1890*
E. Glenn McCormick, 2001, *Homestead*
Gerald & Winifred Minnick, 2016, *1890*
Lyle & Linda Morgan, 2002, *1882*
Lyle & Linda Morgan, 2002, 1888
Emery Poore, 2000, *1882* (Osborne County)
James W. Reneberg, 2014, *Homestead*
Richard Sasse - Sasse Farms, 2001, *1885*
David & Sherrill Sasse, 2003, *1893*
Marianne Schlatter, 2001, *1896*
Robert Arthur Schoen, 2003, *1887*
Terry Schoeni, 2002, *1902*
Marlin D. Smith, 2000, *1872*
John, Jeannine and Mary Sweat, 2013, *1883* (Osborne County)
Williams Farms Inc., 2002, *1886*
Ron & Marlene Wolf, 2004, *1877*
Bonnivie Youngquist, 2001, *1872*
John and Laura Zabe , 2007, *1906*

Stafford County
Judy (Williams) Algrim, Janet (Williams McCoy and James Williams Jr., 2013, *1906*
Wayne & Lula Mae Alpers, 2003, *1892*
Charles Andress, 2005. *1884* (Barton County)
Brian Anshutz, 2002, *1890*
Vernon & Olive Bartlett, 2013, *1886*
Oliver R. Batman & Marilyn Batman, 2004, *1882*
Lita Harrison & Michael E. Christie, 2016, *1899*
Sam and Sharon Crissman, Jr., 2003, *1901*
Beverly K. Cross, 2004, *1883* (Barton County)
Lee & Mary Curtis, 2001, *1885*
Herman L. Debes, William E. & Joelene Debes, and Caroline F. & Aaron J. Moeder, 2008, *1908*
John L. Drew, 2001, *1900* (Barton County)
Carol Rychlec Esfeld, 2000, *1883* (Barton County)
Eleanor & Jim Fertig, 2002, *1897*
Lee Figger, 2001, *1893*
Don & Jeanne Fisher, 2002, *1870*
Randy & Kim Fritzemeier, 2004, *1900*
Thomas Gates, 2001, *1877*
Barbara Grimmett, 2002, *1877*
Clayton E. and Barbara J. Buckle Grimmett, 2002, *1897*
I. Leo Groseclose, George Groseclose, Clarence Groseclose, Maggie Groseclose Volker, 2007, *1901*
Reginald D. Harrison, JoNita Harrison, Reginald Wade Harrison, Deana Harrison Delp, 2016, *1898*
Merlyn & Carol Hartsell, 2001, *1900*
Geraldine (Wendelburg) Hildebrand, 2005, *1891*
Jeff & Gerry Hildebrand, Jody, Joshua & Jo Michal, John Hildebrand, and Jordan Hildebrand, 2013, *1909*
Cecil Hornbaker, Harold Hornbaker, 2001, *1897*
John J. Keenan, 2017, *1862*
Edward L. Kirkpatrick, 2002, *1897*
Lamb and Sons, Inc., 2001, *1878*
Carol Curtis Long, 2001, *1878*
Jon & Rhonda McClure, Willard McClure, 2017, *1906*
Carol & Shirley McNickle, Frieda McNickle Shuler, Tim Alan McNickle, Julie A. McNickle, Denise K. McNickle Bullock, 2007, *1888*
Carol & Shirley McNickle, 2007, *1907*
Marion & Anita Meschberger, 2001, *1889*
Richard Meyer, 2007, *1894*
Eldwyn Minks, 2002, *1880*
Morris Family Trust - Morris Farm, 2016, *1905* (Ness County)
Robert B. Morris - Mathews Farm, 2016, *1905* (Ness County)
Darrel & Christine Murphy, 2005, *1898*
Darrel & Christine Murphy, 2005, *1902*
Darrel & Christine Murphy, 2005, *1905*
Donald & Debra Newell, 2001, *1889*
Carrol & Doris Newell, 2004, *1884*
Curtis & Ruby O'Dell, 2002, *1889*
Loyd Edmund Ratts, 2007, *1885*
David Charles Rewerts and Julie Eileen Rewerts, 2013, *1883*
Irma F. Taylor & Rick C. Russell, 2006, *1901*
Bette Charles Scholey, 2005, *1862*
Betty Jean DeBusk Seibert, 2004, *1904*
Charles Smith, Jr., 2002, *1883*
Merle D. & Marilyn E. Teichmann, 2002, *1897*
Donald G. Waters and Ryan Waters, 2002, *1901*
Doris Witt, 2001, *1887*

Stanton County
Roger and Mary Canny and Canny Farms, 2005, *1885*
Jonell & Jerald Schenk, 2009, *1909* (Mitchell County)
Jonell & Jerald Schenk, 2009, *1909* (Mitchell County)

Stevens County
Mahlon C. Horner, Vallie Horner Burrows, Ernest Burrows, Everett Burrows and Janet Burrows Biernacki, 2014, *1913*

Steven R. & Glenda L. Davis, 2013, *1913*
Marlin & Erma Heger, 2006, *1906*
Vance Lahey, 2002, *1899*
McClure Farms, 2011, *1906*
Stephen R. Morris & Family, Gregory W. Morris & Family, Jana M. Loibl (Morris) & Family, and Elinor R. Findley (Morris) & Family, 2008, *1907*
Dorothy Reynolds & Art Reynolds Family, 2006, *1894*

Sumner County

Keith Albright, 2003, *1899* (Sedgwick County)
Zelda L. Basinger, 2005, *1879*
Juanita Gross, 2005, *1871*
Harold & Mildred Hainsworth, 2000, *Homestead*
Jennie Karesh Jenista, 2000, *1881* (Harper County)
Ed Larson, 2008, *1906*
Joseph & Emma Nixon, Joseph Merlin & Neva Nixon, Valerie Nixon Lawrence and Marilyn Lane, 2009, *1907*
Melvin & Mary Mason, 2010, *1889*
Dennis & Dan Metz, 2000, *1870*
Jason L. Metz, 2000, *1894*
Dennis & Marilyn Metz, 2002, *1901*
Dorinda and Albert Neises, 2017, *1915*
Edith C. Peck, 2000, *1874*
Edith C. Peck, 2005, *1874*
Harold L. Reece and Esther L. Reece, 2009, *1882*
Harold L. Reece and Esther L. Reece, 2009, *1882*
Howard W. & Martha R. Strickland, 2015, *1911*
Robert J. Tompkins, 2005, *1883*
Dale Weishaar, 2005, *1879*
Lloyd & Lee Weishaar, 2005, *1880*
Ray & Merylene Wentworth, 2015, *1874*
Gary R. and Toni Wolff, Lary D. and Kay Wolff, Patsy F. and Mark McCoy, 2017, *1917*

Thomas County

Velma F. Barnett, 2013, *1879*
Henry Bremenkamp Jr., 2009, *1904*
Eileen Dible and Ronald & Janette Dible, 2011, *1902*
William J. & Donna L. Dible, 2015, *1914*
Cleo Duffey, Ray Duffey and John J. Duffey, 2009, *1892* (Rooks County)
Richard Epard & Neva Epard, 2000, *1896*
Leon E. Hoyt, 2000, *Homestead* (Sherman County)
Ernest S. & Anita E. Kistler, 2005, *1905*
Lynn H. and Randy A. Koons, 2017, *1916*
Ava Lee Maydew, 2015, *1915* (Smith County)
Elon Miller, 2000, *1898*
Willard Miller, 2001, *1901*
Sada Munkres, 2000, *1887*
Rundel Family Trust, 2012, *1912*
Doyle Earle Saddler, 2000, *1885*
John & Betty Schroeder, 2001, *1878*
Jess & Anita Thiel, 2016, *1894*
Doug & Joan Todd, 2015, *1915*
Triple T Farms, Inc., 2014, *Homestead*

Trego County

Gene Bellairs, 2002, *1894*
Eileen Bellairs, Steven Bellairs and LaVerna Bellairs DeWitt, 2013, *1912*
Douglas & Deborah Clore, 2002, *1898*
Carroll D. Cook, 2000, *1893*
Merlin & Barb Deines, 2016, *1916*
Loren & Phyllis Gugler, 2002, *1901* (Ellis County)
Robert & Eleanor Halbleib, Marty & Jan Stevens, Chuck & Monica Halbleib, Jim & Bobbie Williams, Dan & Nikki Halbleib, 2009, *1904*
Mary Lou (Blakely) Harvey, 2009, *1886*
William S. Harvey, 2009, *1879*
Darlene F. Hille, 2013, *1906*
Hille Farms Inc- Charles G. Hille and David Hille, 2013, *1901*
Larry and Jean Hixson, 2008, *1908*
Richard Hladek Family - Richard & Mildred Hladek, Cindy Nussbaum, Mark Hladek, Christy Schmitz, 2015, *1899*
Clara Kroeger, 2001, *1887*
Eric & Amanda Littlechild, 2011, *1908*
Frank E. Mussemann, Nathalie E. Musseman, 2017, *1916*
Marvin and Frances Nimz, 2017, *1915*
Patrick Walter O'Toole, 2000, *1899*
Maureen Hillman Pastine, Gordon Brian Hillman, Donna (Hillman) Smith, David Lee Hillman and Ella Beth (Hillman) Ohlemeier, 2011, *1905*
Leonard & Irene Purinton, 2007, *1878*
Fernetta Rhoden, Bradley T. Rhoden and Terra Glendening, 2007, *1907*
Joseph D. Satran, James G. Satran, 2015, *1915*
Brad & Margaret Schmitt, 2016, *1915*
Allen & Lesly Ziegler, 2002, *1897*

Wabaunsee County

B E Murphy Farms Trust - Andrea M Andrews, Trustee, 2011, *1903*
Robert & Patricia Andres, 2003, *1884* (Morris County)
Leona Bandel, 2000, *1900*
Victor & Arnold Bandel, 2001, *1894*
Merrill Beach, 2001, *1854*
Carl & Amanda Brase, 2000, *1890*
Boyd Ivan Davis, 2010, *1885*
Donald L. Dieball, 2000, *1872*
Fred & Viola Gnadt; Gandt Bros. (Ervan & Alan), 2001, *1887*
Alan & Arlene Gnadt, Ervin & Marleen Gnadt, 2016, *1913*
Helen E. Grensing, 2000, *1878*
Donald Hafenstine, 2001, *1887*
Oliver J. Hess, 2000, *1858*
Duane & Diane Hund, 2007, *1902*
Jerry E. Johnson, 2002, *1892*
Johnston Family, 2014, *1902*
William & Paulette Laging, 2009, *1893*
Frances Irene Lassman, 2000, *1881*
Don & Gretchen Lehman, 2005, *1899*
Edna Heder Lietz & Elizabeth A. Heder, 2000, *1881*
Mayer Ranch, 2010, *1861*
Kip & Brenda Meseke, 2009, *1880*
Clarence & Lourine Miller Family, 2006, *1878*
Lourine Miller Trust: Doris Birmingham, Lois Hartman and Joy Miller, 2009, *1898*
Dan & Dorothy Moege, 2002, *1885*
Earl & Adolph Moege, Jr., 2002, *1889*
John R. & Benita L. Peoples, 2000, *1879*
John R. & Benita L. Peoples, 2000, *1883*
Donna Reuter, 2014, *1914*
Jonathan & Christina Ringel, 2013, *1902*
Erma C. Hoch Rollf, 2002, *1901*
Rowley Farm, 2014, *1902*
Robert Louis, Denise Rose Sage, Robert Leroy Sage, 2002, *1885*
Arnold & Iris Schmitz, 2000, *1874*
Cleo & Julie Schultz, 2003, *1898*
Keith E. and Carol R. Schultz, 2004, *1873*
Irene Thoes, 2008, *1855*
Irene Soelter Thoes, 2013, *1909*
John & Shannon Wertzberger, 2009, *1902*
Evelyn F. Zeckser, 2009, *1908*

Wallace County

Rick Cline, 2001, *1899*
Iley & Evelyn Sexson & David & Bonnie Sexson, 2006, *1903*

Washington County

Sonja Anderson Meier, Richard & Margie Anderson, 2007, *1893*
Clarence Burk Family Trust, 2001, *1880*
George H. & Lena K. Buss and Delbert E. & Geraldine O. Buss, 2008, *1908*
Dale & LaVonne Cook, 2001, *1874*
Rodney Doebele, Charles Doebele, Wayne Doebele, Jerry Doebele, Connie Doebele and Gary Behrends, 2010, *1880* (Shawnee County)
Dunavan Farm, 2003, *1883* (Brown County)
Larry Durst, 2001, *1885*
Michael E. & Nora L. Earhart, 2002, *1883*
Earl Eichman Family, 2001, *1899*
Shirley & Clifford Ericson, 2001, *1879*
Gauby Farms, 2001, *1880*
Edwin Hiesterman, 2000, *1883*
John C. Hogue, 2000, *1871*
Marcella Holbrook, 2005, *1884*
Kenneth William & Orcena Jurey, 2010, *1904*
Lorene & Robert Kenaston, 2000, *1898*
Delores Koss, Mike & Mary Koss, 2002, *1898*
Gustav Metz, 2013, *1912*
L. Joe Meier & Michael A. Meier, 2000, *1895*
Morris & Pauline Nelson, 2003, *1889* (Republic County)
Vernon & Eva Helms, Brad and Callie Ohlde, 2007, *1895*
Kenneth Olson, 2000, *1862*
Leland, Louise, & Karl Ostlund, 2004, *1860*
The Pannbacker Family, 2014, *1879*
Dale & Evelyn Peterson, 2001, *1869*
Donald & Barbara Peterson, 2002, *1880*
Donald & Barbara Peterson, 2002, *1902*
Ralph & Delma Roper, 2001, *1899*
Dewayne Rose, 2004, *1879*
Ralph Rothlisberger, 2000, * (Riley County)
Nina Rusco, 2000, *1876*
Albert C. Ryser Family, 2013, *1900*
June (Stone) Sheppard Family of the late John Stone, 2001, *1876*
Marvin A. & Donna J. Shoemaker, Duane L. & Rosalie J. Line, 2002, *1902* (Republic County)
John M. Solbach III, M. Patricia Solbach, 2009, *1881* (Douglas County)
Rodney A. and Adelbert E. Stewart, 2002, *1902*
Lois Thompson, Janis Milam, Kathy Lower, James Rahe, 2017, *1898*
Marvin L. VanWinkle, Gary & Rita VanWinkle, 2008, *1894*
Wilbur McClaughry, Marvin Van Winkle and Bob Van Winkle, 2012, *1911* (Nemaha County)
Shirley Walter, 2001, *1877*
Melvin & Frances Wienck, 2009, *1900*
Wendell & JoAnn Wilgers, 2005, *1902*
Patrick Zach, 2001, *1891*

Wichita County

Kenneth R. Bishop, 2010, *1902*
Helen Smades, Noreen Walk, Jack D. Smades, 2002, *1900* (Pawnee County)

Wilson County

Bambick Family LLC, 2017, *1917* (Johnson County)
Russ & Judy Clough Branden, 2010, *1902* (Haskell County)
Campbell Family Farm (John F., Nettie and son Paul), 2006, *1897*
Glenn & MaryLou Canty Payne, 2000, *1870*
Bill Crawshaw, Bob Crawshaw, 2010, *1878*

David & Sue Ella Creager, 2008, *1894*
David & Donna Erbe, 2003, *1872* (Coffey County)
Marjorie Anne Frankenbery, Rebecca A. Timmons, Wm. Marlen Frankenbery, 2017, *1909*
Darrell & Joyce Garner, 2004, *1879*
James A. Doll, Andrew Huffman, George Huffman, Leo Huffman, Gerald F. Huffman, Gerald L. Huffman, Carol Ott and Kathleen Scalise, 2008, *1881*
George W. Katzer, 2015, *1914*
George W. Katzer, 2016, *1915*
Lloyd W. & Carol D. Kebert; Jack W. & Shelly R. Kebert, 2006, *1889*
Jean & David Luther, 2011, *1900* (Neosho County)
Maxwell Family, 2001, *1888*
Nelson Family, 2002, *1893*
Orrin & Loretta Odell, 2013, *1882*
Billy Max Pryor, Marilyn Pryor, Clifford Pryor, 2007, *1905*
Arthur & Sidna Small, Randall & Nicole Small, 2015, *1915*
William K. Stevens & Mary V. Stevens, 2006, *1877*
Robert & Helen Travis, 2001, *1890*

Woodson County

The Brock Farm, 2002, *1879* (Neosho County)
Calvin Brodman, 2001, *1887*
Patrick D. Collins, 2005, *1855*
Chris & Lurita Culver, 2000, *1886*
David L. & Amy K. Ellison, 2000, *1862*
Charles & Edna Fulhage, 2001, *1881*
Vivian Eagle Gaughan & H. Charles Eagle, 2000, *1900*
Douglas & Karen Saferite Gleue, 2013, *1911* (Coffey County)
Haas Estate Farm, 2002, *1888*
Noel & Julie Jackson, 2000, *1882*
Harold Johnston, 2001, *1901*
Kathleen Krueger and Richard Krueger, 2010, *1907* (Johnson County)
O.L. & Lila Lauber, 2000, *1883*
John W. Lorance and Mary M. (Lorance) Davidson, 2017, *1892*
Randall & Brenda Louderbaugh, 2000, *1866*
Massoth Farm, 2001, *1878*
Daniel Mulsow, 2000, *1900*
Ronald L. Neifert, Linda A. Neifert and C.S. Pendlay, 2013, *1912*
Carol & Don Peterson, 2013, *1909* (Stafford County)
Lauren & Ruthann Pringle, 2002, *1897*
Perry Rhea, 2002, *1902*
Herbert E. Schaede, 2000, *1873*
Rory Smith, 2001, *1878*
Thomas Smith, Jr., 2001, *1873*
Gerald & Elizabeth Stockebrand, 2001, *1857*
Lauren & Barbara Stockebrand, 2001, *1893*
Terry S. Wells, 2000, *1899*
Tom & Linda Wille, 2000, *1882*

Wyandotte County

Dale and Joyce Drescher, 2002, *1887*
Retha Lallier, 2000, *Homestead*

A farmer and his wife pulling beets from their garden in Sheridan County, 1939. Russell Lee photo courtesy of the Library of Congress.

Index

I

J

K

L

U

V

W

Y

Z